Decarbonizing Urban Residential Buildings

Building Energy Research Center
of Tsinghua University

Decarbonizing Urban Residential Buildings

China Building Energy and Emission Yearbook 2025

Springer

Building Energy Research Center of Tsinghua University
Building Energy Research Center
Tsinghua University
Beijing, China

ISBN 978-981-95-8199-3 ISBN 978-981-95-8200-6 (eBook)
https://doi.org/10.1007/978-981-95-8200-6

This Springer imprint is published by the registered company Springer Nature Singapore Pte Ltd.
The registered company address is: 152 Beach Road, #21-01/04 Gateway East, Singapore 189721, Singapore

Preface

Residential buildings are among the key battlegrounds for the energy transition. The rooftops and surrounding open spaces of urban residential complexes represent important resources for developing photovoltaic (PV) power generation. The rapidly increasing number of private electric vehicles in community parking lots constitutes a vital storage resource for the emerging power system. Meanwhile, the growing variety of household appliances in residential buildings can serve as flexible demand-side resources to participate in peak shaving. The central question is how to fully mobilize and integrate these resources so that buildings, in addition to providing living space, can also fulfill new functions of energy production, storage, and regulation. When combined with energy conservation, this "four-in-one" integration of production, storage, consumption, and regulation represents a new mission entrusted to buildings by the zero-carbon energy system.

At the same time, residential buildings still consume large amounts of natural gas for cooking, domestic hot water preparation, and decentralized heating through wall-hung boilers. Natural gas, as a fossil fuel, emits carbon dioxide upon combustion; furthermore, it is itself a potent greenhouse gas, with leakage impacts on climate change even greater than combustion. A complete shift toward electrification—eliminating natural gas supply in residential buildings and realizing "fuel-to-electricity" substitution—remains controversial today, but in the long run it will inevitably become another essential task for the residential sector.

However, achieving "production-storage-consumption-regulation" integration and "fuel-to-electricity" substitution is difficult at the level of an individual household, and even for a single building. Measures such as rooftop PV installation, "one parking space–one charger," and comprehensive electrification should instead be implemented at the scale of the residential community. After decades of development, China's urban residential sector has largely adopted the community as its basic organizational unit. Moreover, the quality of cultural amenities, infrastructure provision, and management practices in these communities plays a decisive role in shaping residents' living experience. Once material needs are largely satisfied, factors such as the residential environment, cultural life, and convenience of facilities will become central to determining residents' sense of happiness.

For this reason, the Ministry of Housing and Urban-Rural Development has advocated for building "good houses," "good communities," and "good neighborhoods" to meet the people's aspirations for a better life. Here, "building" should no longer be understood merely as improving construction standards and physical quality, but rather as innovating management models and operational mechanisms. The challenge is to operate and retrofit the residential communities already established over the past two decades in ways that align with the needs of the energy transition, while also providing comfortable and convenient living environments, a positive cultural atmosphere, and age- and child-friendly services. Achieving this requires institutional innovation, enabling communities to become the fundamental organizational unit for urban residents and restoring a sense of belonging and mutual reliance.

Over the past three decades, Chinese urban management has shifted from being workplace-based to community-based. At present, community infrastructure systems—covering electricity, heating, natural gas, water supply, drainage, parking, waste, landscaping, cleaning, and security—are fragmented across property management firms and specialized service companies, often leading to coordination problems and a lack of unified leadership. This fragmented governance is ill-suited to emerging needs such as PV installation, revenue-sharing, and the deployment and maintenance of charging infrastructure. As a result, the organizational and managerial framework has become the primary obstacle to expanding new energy facilities in residential communities. Meanwhile, many commercial housing projects built in the early years of housing reform are now due for major repairs and functional upgrades, further exposing deficiencies in current management mechanisms. The long-standing challenges surrounding heating metering and billing reform in northern China also illustrate these systemic issues.

By adopting the residential community as the basic unit, it becomes possible to secure heating supply from multiple sources—including district heating networks, various heat pumps, and even gas boilers—thus breaking monopolies and introducing market-based competition. Different communities could then adopt diverse metering and billing schemes according to residents' preferences. Fiscal subsidies for heating could also be better targeted toward low-income communities, replacing the current "across-the-board" approach. This shift could finally resolve the disputes over heating billing reform that have persisted for more than two decades. To achieve this, new organizational and managerial frameworks must be designed that address systemic problems comprehensively through institutional innovation, rather than piecemeal solutions.

Urban residents spend over 70% of their time within their communities. Therefore, through innovative mechanisms, it is crucial to holistically address issues such as cultural and recreational services, eldercare and childcare, building repair and retrofitting, infrastructure construction and operation, and associated cost accounting and billing mechanisms. Only then can residents truly live in fulfilling environments—an essential foundation for meeting the people's pursuit of a better life.

Over the past two decades, rapid urbanization and large-scale urban construction have defined China's socio-economic development. With large-scale housing and infrastructure now largely complete, the focus of urban development must shift from new construction to the repair, retrofitting, operation, and management of existing buildings. From the perspective of operational management, challenges in residential buildings have expanded from single buildings to entire communities. Whether in terms of governance models, PV deployment, or charging infrastructure, the residential community must become the fundamental unit for analysis, planning, and implementation.

Against this background, this book presents our understanding of these issues and offers preliminary ideas for potential solutions. The focus of inquiry shifts from new construction to the maintenance, retrofitting, and management of existing buildings, and from the single building to the residential community as the basic unit. The topics discussed here—ranging from new perspectives on community infrastructure under the energy revolution and carbon neutrality strategy to insights gained from years of field surveys and practice in community management—have been rarely addressed in previous volumes of the Annual Report on China Building Energy Efficiency. Admittedly, our views remain tentative and require further exploration and practice. The primary contribution of this book is to raise these questions, not to provide definitive answers.

We sincerely hope that broader society will pay closer attention to these issues, contributing collective wisdom to the search for community governance models suited to China's realities. Achieving sustainable development in communities and neighborhoods is essential for providing the fundamental conditions of a good life. The energy revolution and the carbon neutrality strategy pose higher demands and new expectations for building energy efficiency in China. Meeting them will require joint efforts to open a new chapter in low-carbon and energy-efficient building development.

Beijing, China Yi Jiang

Acknowledgment

This publication was prepared by the Building Energy Research Center (BERC) of Tsinghua University. The editor of this book is Dr. Shan Hu.

The lead authors were Professor Yi Jiang, Prof. Da Yan, Dr. Shan Hu.

Other authors also contributed important sections for this report, and they were:

Building Energy Research Center of Tsinghua University

Yi Jiang, Shan Hu (Chaps. 1 and 9)
Ziyi Yang, Zhenghua Wang, Yang Zhang (Chap. 1)
Shan Hu, Ziyi Yang (Chap. 2, Sect. 3.1)
Da Yan (Sects. 3.2, 3.3, 4.1, 5.4, 7.1, 7.2, and 8.1)
Xuyuan Kang (Sects. 3.2, 4.1, and 7.2)
Xiaocheng Liu (Sect. 4.1)
Xiaohua Liu (Sects. 4.2, 4.3, 5.1, 5.2, and 5.3)
Ji Zhang (Sects. 4.2 and 4.3)
Shuo Liu, Hao Li (Sects. 5.1, 5.2, and 5.3)
Xiaocheng Liu, Zhihan Su (Chap. 6)
Dongmei Zheng (Sects. 7.1 and 7.2)
Baolong Wang, Wenxing Shi (Sect. 8.4)
Ying Xu, Yujie Fan, Xiaopeng Ren (Sect. 8.5)
Xiaofeng Li (Sects. 8.6, 8.7, and 8.8)

Invited Authors

Zhou Hui, China Building Technology Center (Sects. 3.1, 3.2, and 3.3)

Mingyang Qian, Donghua University (Sects. 3.3, 5.4, and 9.5)

Qinqin Deng, Lixin Sun, Hanyu Yang, China Academy of Building Research (Sects. 3.4 and 8.3)

Hua Liu, Jianguo Xiong, Fan Yuan, Gree Electric Appliances, Inc. of Zhuhai (Sect. 7.3)

Hua Liu, Jianguo Xiong, Baiyang Zhao, Gree Electric Appliances, Inc. of Zhuhai (Sect. 7.4)

Jingjing An, Beijing University of Civil Engineering and Architecture (Sect. 8.1, Chap. 9)

Chi Feng, Zhi Yuan, Shan Gao, Chongqing University (Sect. 8.2)

Yan Li, Beijing Tsinghua Tongheng Urban Planning & Design Institute Co., Ltd. (Sects. 8.6, 8.7, and 8.8)

Bin Hao, Shenzhen Institute of Building Research Co., Ltd. (Chap. 9)

Huihuang He, China Gezhouba Group 1st Engineering Co., Ltd. (Sect. 9.3)

Yujian Xia, Pengcheng Li, Haoxin Lin, China National Institute of Standardization (Chap. 10)

The report was finally compiled by Shan Hu.

Executive Summary

Building Construction Related Embodied Energy Use and Emissions in China have been Decreasing Since 2016

In 2023, China reached a population of 1410 million, with the urbanization rate rising from 37.7% in 2001 to 66.2% in 2023. Rapid urbanization drives continuous development of the building construction. In 2023, the total building stock in China reached 71.6 billion m^2, including 33.1 billion m^2 of urban residential buildings, 22.4 billion m^2 of rural residential buildings, 16.1 billion m^2 of public and commercial (P&C) buildings. The floor area for northern urban heating is 17.3 billion m^2.

The embodied energy use and embodied CO_2 emissions of China's civil buildings have been decreasing since 2016. In 2023, the embodied energy use of civil buildings in China is 0.5 Gtce in 2023, and embodied carbon emissions of civil building construction is 1.4 $GtCO_2$. Adequate planning and speed of construction of the entire building stock, a shift from new construction to maintenance and renovation of existing buildings, low-carbon building materials and construction methods are important issues for the low-carbon transition of China's building sector.

Energy Use and Carbon Emission Intensity of Building Operation in China Continues to Grow Steadily, but Remains Lower Compared to Developed Countries

In 2023, the total commercial energy consumption of building operation in China was 1.17 Gtce, and the total carbon emissions during building operation were 2.2 $GtCO_2$. The electricity consumption of building operation exceeded 2.5 PWh in 2023, leading to a corresponding increase in the indirect carbon emission from electricity use to 1.35 $GtCO_2$. On the other hand, the indirect carbon emission from heating during building operation was 450 million tons of CO_2, which showed a gradual but steady growth trend.

China's building operation energy consumption per capita and per square meter remains significantly lower than USA, Canada, Europe, Japan, and South Korea. Specifically, China's equivalent electricity consumption per capita of building operation was approximately one-fifth of that in USA and Canada, and half of that in Japan and South Korea. Similarly, the equivalent electricity consumption per square meter of building operation in China was one-third of that in Canada and half of that in USA, Europe, Japan, and South Korea. As a result, China's building carbon emissions per capita and per square meter are lower than those in most developed countries.

Shifting China's Urban Residential Sector Toward Electrification and Retrofit

Traditional low-rise housing has been replaced by mid- and high-rise apartments, with large, high-density compounds now the mainstream form. Housing supply has largely met demand, and the focus is shifting to upgrading and managing the existing building stock.

Electricity has become the dominant household energy source. Air conditioning, heating, domestic hot water, and diverse appliances have driven electrification. Cooking and water heating are gradually shifting from gas to electricity, supported by the spread of induction cookers, rice cookers, and electric water heaters. While frugal traditions moderate overall consumption, the growth of high-consumption households highlights the need to manage both technology and behavior to curb demand.

Natural gas remains important, especially for cooking, heating, and hot water, but the trend is toward fully electrified living. The spread of solar and heat pump technologies is diversifying domestic hot water supply and reflecting improved living standards.

Accelerating retrofits is now recognized as a critical pathway to achieving carbon peaking and neutrality goals in the building sector. These trends define the priorities for low-carbon development and sustainable household energy demand.

Advancing from Consumers to Prosumers: Electrification, Efficiency, and Flexibility

Urban residential buildings will shift toward full electrification, active grid participation, and carbon reduction, driven by PV-Energy Storage-Direct current-Flexibility (PEDF) systems, electric vehicles (EVs) with smart charging infrastructure, and dynamic Cr-based regulation.

Urban homes can achieve significant decarbonization and efficiency gains without sacrificing comfort by prioritizing high-efficiency electrification, phasing out fossil appliances, and upgrading electrical capacity. Electrification is advancing in cooking, domestic hot water, air conditioning, and heating. High-efficiency heat pumps, energy storage multi-split air conditioners, smart electric water heaters, and air-source heat pump underfloor heating are core technologies. These improve efficiency and comfort while enabling thermal storage and load shifting to support grid flexibility. Intelligent envelopes and retrofitting enhance energy performance, optimize ventilation and indoor air quality, and mitigate heat island and cold island effects. The broader trend is replacing fossil fuels with high-efficiency electrification and intelligent control.

The passive role of "energy consumers" is giving way to "produce–store–consume–regulate" models, allowing residential buildings to integrate production, storage, flexible use, and grid interaction. Rooftop PV and distributed renewables enable on-site generation; batteries, hot water tanks, and electric vehicles provide storage; efficient appliances support demand-side optimization; and flexible regulation through smart controls and thermal inertia stabilizes the grid. A "self-consumption-first" rooftop PV pathway—supported by storage and demand response—offers resilient economics and grid benefits. Thermal storage via heat pump/electric water heaters, modest comfort set-backs on inverter HVAC, and time-shifting of smart appliances create low-cost, high-availability flexibility. This structure enhances renewable absorption, alleviates peak loads, and balances supply and demand.

The shift from "generation follows load" to "source-grid-load-storage" interaction requires PEDF systems and dynamic electricity pricing to encourage flexibility and reduce curtailment. Market mechanisms and carbon pricing incentivize household participation. EVs play a dual role as transport and distributed storage.

A dynamic carbon emission responsibility factor (Cr) provides real-time signals for demand response, linking household consumption to grid carbon intensity. Smart meters and appliances can autonomously adjust operation, reducing carbon responsibility without sacrificing comfort. Pilots show up to a 15% reduction in carbon responsibility per kWh.

Integrating Residential Compound Charging Systems and Energy Management

By 2050, China may host 300 million EVs, offering vast capacity for peak shaving through smart, bidirectional charging. As EV ownership grows, residential compounds have become the main charging sites, making infrastructure planning critical. Current challenges include insufficient facilities, unauthorized installations, and uncoordinated fast charging, which strain distribution networks and reduce efficiency. EVs offer substantial storage potential, as large battery capacities exceed daily travel needs. Coordinated and bidirectional charging can shift loads to off-peak hours, relieve grid stress, and enhance renewable integration—cutting peak load fluctuations by up to 75%. A "one charging point per parking space" system, supported by unified DC bus architecture and adaptive control, enables smart regulation of charging power. Coupled with time-of-use pricing and carbon responsibility signals, this ensures grid stability and lowers emissions. Integrating rational deployment, intelligent control, and V2G applications transforms residential charging into a dual solution—meeting mobility needs while unlocking EVs as distributed energy storage to support low-carbon urban transition.

Advancing Community Governance for Low-Carbon and Livable Urban Residential Compounds

After more than two decades of housing reform, China's urban living pattern has shifted from integrated work-living compounds to separated residential communities. Changes in physical form and commercialization of housing property rights have reshaped urban communities while creating new governance challenges. The core issues are the lack of robust governance structures and unclear property, management, and revenue rights of public facilities, which limit stakeholder participation and hinder reinvestment.

Future governance should establish compound-level entities representing residents' and owners' collective interests. Guided by party-building and supported by neighborhood and owners' committees, such organizations would unify facility operation, income distribution, and policy implementation. They should coordinate energy, environmental, public service, and new energy infrastructure while serving as the main interface with government, enterprises, and social organizations. Revenues from shared spaces and facilities should be transparently reinvested into maintenance, retrofits, and clean energy projects such as PV and charging systems. Government provides policy frameworks and financial guarantees, while enterprises deliver technical services. This model ensures orderly asset management, aligns with national carbon goals, and enhances residents' quality of life.

Enhancing Standards and Labels for Accelerating Energy Efficiency Transition

China's energy efficiency standards and labeling system has been a cornerstone of demand-side management since the 1990s. Around 60 standards have been issued, covering household appliances, commercial and office equipment, lighting, and industrial devices. Major appliances such as refrigerators, air conditioners, and washing machines have undergone multiple rounds of efficiency upgrades, supported by coordinated implementation of standards, labeling, and incentive programs. The share of Grade 1 and 2 products has grown substantially, reflecting clear market optimization.

Looking ahead, the "15th Five-Year Plan" calls for continuous updating and expansion of standards, closed-loop management of implementation and supervision, and strengthened testing and regulatory capacity. Energy labels will be enhanced to disclose more comprehensive green and low-carbon product information, providing consumers with transparent choices while supporting carbon peaking and neutrality goals.

Contents

List of Figures

List of Tables

Chapter 1
China's Building Energy Use and GHG Emissions

1.1 Basic Situation of China's Building Sector

1.1.1 Urban and Rural Demographic

In recent years, urbanization has grown rapidly in China. In 2023, China had an urban population of 933 million and a rural population of 477 million, with the urbanization rate rising from 37.7% in 2001 to 66.2%, as shown in Fig. 1.1.

China's urbanization rate has been increasing year by year but the rate of growth has gradually slowed down, and the direction of population mobility has shifted from urban-rural migration in the immediate vicinity to inter-urban migration on a larger scale. According to relevant studies, the overall pattern of population movement in China's provincial areas continues to show a flow from the west to the east and from the north to the south, but in recent years there has been a phenomenon of population return to the central and western provinces.[1] Changes in population flows and economic growth in central and western provinces and their capital cities in recent years reflect China's regional economies moving toward greater coordination.

The development pattern of cities and towns has also gradually shifted from incremental expansion to quality enhancement along with population movement and urban development. According to a study conducted by the Institute of Spatial Planning & Regional Economy, National Development and Reform Commission P.R.C, for a typical country, after the urbanization rate reaches 60%, the overall framework of the city is basically established, the main infrastructure is basically completed, and the mode of urban development shifts to a mode in which

[1] Zhikai Gao. New trends in China's population mobility and the coordinated development of the regional economy.

Building Energy Research Center of Tsinghua University, *Decarbonizing Urban Residential Buildings*, https://doi.org/10.1007/978-981-95-8200-6_1

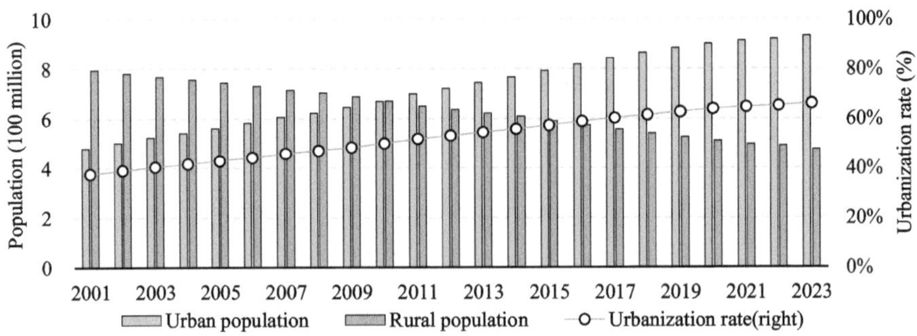

Fig. 1.1 Population growth in China by year (2001–2023)

"incremental expansion and renewal of stock are given equal importance".[2] In August 2021, the Circular of the Ministry of Housing and Urban-Rural Development of the People's Republic of China on Preventing the Problem of Large-scale Demolition and Construction in the Implementation of Urban Renewal Actions (Jianke 2021 No. 63) emphasized that urban renewal should change the mode of urban development and construction, adhere to the concomitant measures of "retaining, changing and demolishing", and focus on retaining and utilizing the enhancement of the main, to strengthen repairs and renovation, to make up for the shortcomings of the city, and focus on the enhancement of functionality, so as to strengthen the vitality of the city. At the same time, the phenomenon of large-scale demolition and construction is strictly controlled; in principle, the demolition of building area in an urban renewal unit (area) or project should not be greater than 20% of the total building area of the status quo; in principle, the ratio of demolition and construction in an urban renewal unit (area) or project should not be greater than 2. From the point of view of the sustainable development of urbanization, the question of how cities can achieve their own renewal while avoiding major demolition and construction should be given sufficient attention.

1.1.2 Building Stock

Rapid urbanization drives continuous development of the construction sector, and the scale of China's construction sector has been expanding. From 2007 to 2023, thanks to the rapid growth of building construction in China, the floor space greatly expanded in urban and rural areas. Specifically, from 2007 to 2014, China's newly built building stock of civil buildings grew rapidly, while it has been declining slowly each year from 2014 to the present. The newly built building stock of urban residential building and P&C building slowly declined from around 3.6 billion m^2 in 2014 to 3.1 billion m^2 in 2023. (Fig. 1.2). Driven by the change of the number of

[2] Jiyuan Wang. Characteristics of real estate transformation trends in typical countries after reaching 60% urbanization rate.

construction projects, the demolished building stock of urban residential areas and P&C buildings increased rapidly from 700 million m^2 in 2007 to approximately 1.6 billion m^{2}in 2018, and then declined slowly to 1.2 billion m^2 in 2023.

In 2023, residential buildings and non-residential buildings made up about 78 and 22% of the newly built building stock of civil buildings in China respectively. According to the difference in building functions, public and commercial (P&C) buildings can be categorized into offices, hotels, malls, hospitals, schools, and others. From 2001 to 2023, the main types completed each year were dominated by offices, malls, and schools. In 2023, the total newly built building stock of these three types accounted for about 66% of that of public and commercial (P&C) buildings, with the proportions of malls, office buildings, and schools at 30, 17, and 19% respectively. For the remaining types, hospitals and hotels had a smaller proportion of 6 and 3% respectively (Fig. 1.3).

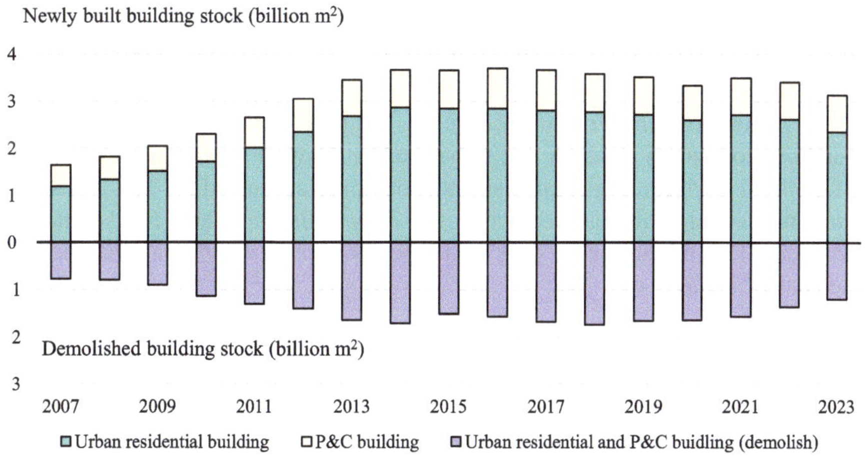

Fig. 1.2 Newly built and demolished building stock of civil buildings in China (2007–2023)

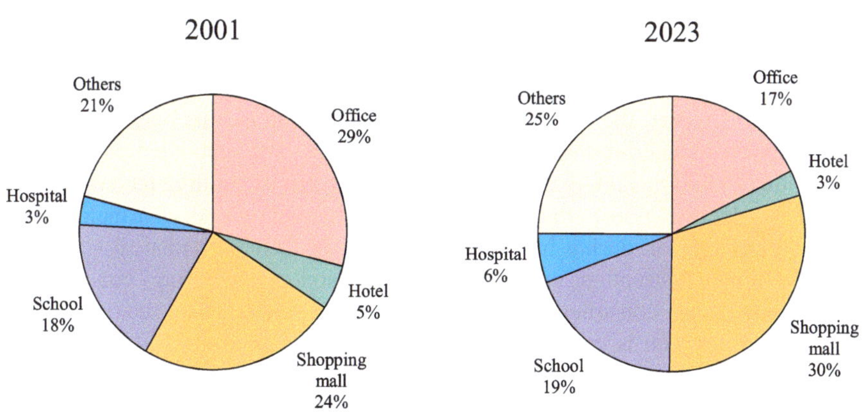

Fig. 1.3 Newly built building stock of P&C buildings according to different functions (2001, 2023)

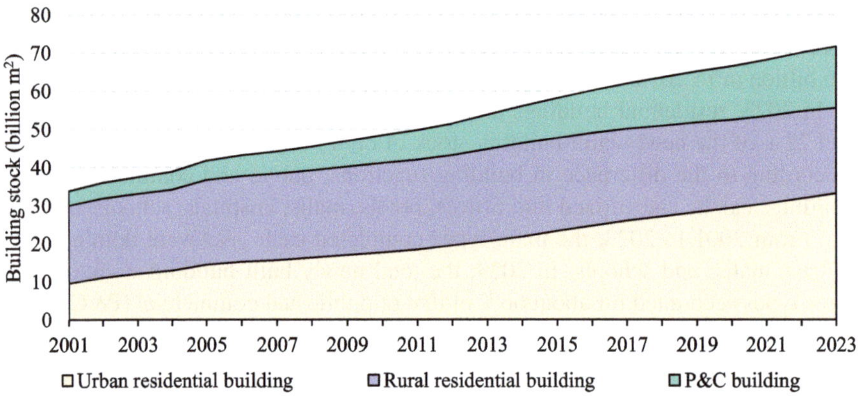

Fig. 1.4 China's existing building stock (2001–2023). Note: The newly built building stock entered in the model is the data under the statistical standards for construction enterprises as specified in the China Statistical Yearbook on Construction. (Source: Estimation results from CBEEM of the Building Energy Research Center (BERC), Tsinghua University)

Large-scale building construction activity has led to the rapid growth of China's building stock every year. In 2023, the total building stock in China was about 71.6 billion m^2, including urban residential buildings accounted for 33.1 billion m^2, rural residential buildings accounted for 22.4 billion m^2, and P&C buildings accounted for 16.1 billion m^2 (Fig. 1.4). The floor area for northern urban heating stood at 17.3 billion m^2.

1.2 Demarcation of China's Energy Consumption and GHG Emissions in the Building Sector

1.2.1 Calculation Method of Energy Consumption in the Building Sector

Energy consumption for the building sector covers different phases of building life cycle. In this report, the embodied energy consumption and the operation energy consumption of buildings are analyzed respectively. The embodied energy consumption of buildings refers to the energy consumption of building material exploration, production, transportation and on-site construction and also includes the energy consumption during building demolition. In China's statistical standards, civil building construction, production building (non-civil building) construction, and infrastructure construction are included in the construction sector, so their energy consumption is collectively known as the embodied energy consumption related to the construction sector. Based on the China Building Energy and Emission Model of BERC, Tsinghua University, this book provides the analysis data of the standard for embodied energy consumption of China's construction sector and the standard for embodied energy consumption of China's civil buildings (see Sect. 1.3.2 for details). The operation energy consumption of buildings, which is the focus of this book, refers to the energy consumption from the operation of civil buildings, including the energy consumed by the provision of heating, ventilation, air conditioning, lighting, cooking, and domestic hot water (DHW) to occupants or users in residential buildings, office buildings, schools, malls, hotels, transportation hubs, recreational and sports facilities, and other non-industrial buildings, and energy consumed by service functions of such buildings. It is very difficult to distinguish the operational energy consumption of buildings that fully serve industrial production processes from the industrial production energy consumption, because the energy consumption in ventilation, air conditioning, and purification of factory buildings including metallurgical factory buildings and integrated circuit or pharmaceutical production factory buildings accounts for a very large proportion in the production energy consumption. However, it is very difficult to include such energy consumption in building energy consumption. Hence, this report does not discuss the buildings serving production processes and only studies civil buildings.

Based on our long-term research on the energy consumption of civil building operation in China and given the difference in heating methods in winter between northern and southern China, the difference in architectural forms and lifestyles between urban and rural areas, and the difference in personnel activities and energy use equipment between residential and P&C buildings, this report divides the building energy use in China into four categories, i.e. northern urban heating (NUH) energy use, urban residential building energy use (excluding NUH energy use), public and commercial building energy use (excluding NUH energy use), and rural residential building energy use, which are defined in detail as follows.

1. **NUH energy use**

   This refers to the energy consumption of heating in winter, including all forms of centralized heating and decentralized heating, in provinces, autonomous regions, and municipalities that adopt centralized heating, including all urban areas of Beijing, Tianjin, Hebei, Shanxi, Inner Mongolia, Liaoning, Jilin, Heilongjiang, Shandong, Henan, Shaanxi, Gansu, Qinghai, Ningxia, and Xinjiang as well as part of Sichuan. Heating is also needed in winter in Tibet, western Sichuan, and part of Guizhou, but should be considered separately as the local energy situations, issues, and characteristics are completely different from those in northern China. The reason for the separate calculation of NUH energy use is that centralized heating has been the main heating method in northern urban areas, including a large number of city-level heating networks and community-level heating networks. Different from other categories of building energy use, in which the calculation is based on the consumption of a single building or a single household, NUH energy use is largely related to the structural form and operation mode of a heating system, and the actual value of energy use is counted and calculated in a unified manner based on the heating system, so NUH energy use is taken as a separate category and treated differently from other categories of building energy use. Based on the form and scale of heat source systems, the current heating systems can be classified into centralized heating systems, which adopt such methods as large- and medium-scale coal-fired combined heat and power (CHP), large- and medium-scale gas-fired CHP, small-scale coal-fired CHP, small-scale gas-fired CHP, large coal-fired boilers, large gas-fired boilers, district coal-fired boilers, district gas-fired boilers, heat pump for centralized heating, residual heat from nuclear power and industrial residual heat, and household heating systems, which adopt such methods as household gas furnaces, household coal furnaces, air conditioners and heat pumps for decentralized heating, and direct electric heating. The main types of energy sources used include coal, gas, and electricity. This report studies the primary energy consumption, including the primary energy or electricity consumption at heat sources and the electricity consumption of equipment (fans and water pumps) serving heating systems. Such energy consumption can also be divided into the conversion loss from heat sources and heating stations, the heat loss and energy consumption of the distribution pipe networks, and the final heat gains of buildings.

2. **Urban residential building energy use (excluding NUH energy use)**

   This refers to the energy consumption of urban residential buildings, except for the heating energy consumption in northern China. In terms of energy end-users, it includes energy consumption of household appliances, air conditioners, lighting, cooking, and domestic hot water, as well as energy consumption of heating in winter in provinces, autonomous regions, and municipalities in the hot-summer and cold-winter (HSCW) zone. The main types of commercial energy sources used in urban residential buildings include electricity, coal, natural gas, liquefied petroleum gas (LPG), and city gas. Decentralized heating is mostly adopted in winter in the HSCW zone, and the energy consumption of the following heating methods all falls into this category: building space heating methods such as air source heat pumps and direct electric heating, and local heating methods such as fire pans, electric blankets, and electrical hand warmers.

3. **Public and commercial building energy use (excluding NUH energy use)**

   The public and commercial (P&C) buildings here refer to buildings where people carry out various public activities including office buildings, commercial buildings, tourism buildings, scientific research, educational, cultural, and medical buildings, communication buildings, and transportation buildings in urban and rural areas. Except for NUH energy consumption, the energy consumption of activities in buildings includes the energy consumption of air conditioning, lighting, sockets, elevators, cooking, and service facilities, as well as the energy consumption of heating of urban P&C buildings in winter in the HSCW zone. The types of commercial energy sources used in P&C buildings include electricity, gas, fuel oil, and coal.

4. **Rural residential building energy use**

   This refers to the energy consumption of rural households, including cooking, heating, cooling, lighting, hot water, household appliances, etc. The main types of energy sources used in rural residential buildings include electricity, coal, LPG, gas, and biomass energy (straw and firewood). The consumption of biomass energy is not included in the national energy macro-statistics. However, as an important part of rural residential building energy use, it will be listed separately in this report.

In this report, actual consumption of electricity other types of energy is counted and calculated separately whenever possible. If they have to be combined, all energy sources will be converted into primary energy sources for addition, namely, the electricity consumption will be converted into primary energy consumption calculated in standard coal based on the annual average coal consumption for power supply in China. As to the CHP method of centralized heating source for building operation, the input fuels are allocated based on the exergy values of output electricity and heat according to relevant provisions of the *Standard for Energy Consumption of Building* (GB/T 51,161-2016). In this report, the conversion coefficient for the exergy of heat is calculated based on an ambient temperature of 0 °C and a supply/return water temperature of 110 °C/50 °C, and the conversion coefficient for heat is 0.22.

1.2.2 Calculation Method of Carbon Emissions in the Building Sector

The embodied carbon emissions of buildings include the carbon emissions from the building material production, transportation, on-site construction, and demolition of civil buildings. In China's statistical standards, civil building construction, production building (non-civil building) construction, and infrastructure construction are included in the construction sector, so their carbon emissions are collectively known as the embodied carbon emissions related to the construction sector. Based on the China Building Energy and Emission Model of BERC, Tsinghua University, this report provides the analysis data of the standard for embodied carbon emissions of China's construction sector and the standard for embodied carbon emissions of China's civil buildings (see Sect. 1.4.2 for details).

Carbon emission during building operation mainly includes the carbon emission from the direct burning of fossil fuels and the indirect use of non-fossil energy during the operation of buildings, which consists of three main types:

1. **Direct carbon emission**: It refers to the direct emission of carbon dioxide in buildings by burning fossil fuels including coal, fuel oil, and gas. The carbon emission can be calculated based on the types of fuels and their different carbon emission factors.
2. **Indirect carbon emission from electricity use**: It refers to the carbon emission during the generation of electricity transmitted into buildings from the outside. The carbon emission can be calculated through the multiplication of the total external electricity used for buildings by the average carbon emission factor of electricity in the power grid, and the PV power generation and electricity consumption in buildings themselves are not counted.

3. **Indirect carbon emission from heating**: It refers to the indirect carbon emission resulting from centralized heating in northern urban areas. The centralized heating systems in northern urban areas adopt combined heat and power generation or centralized coal- and gas-fired boilers for the supply of heat. In this regard, carbon dioxide emitted by coal- and gas-fired boilers fall into the indirect carbon emission from building heating, while carbon emissions of combined heat and power generation plants are allocated according to exergy values of output electricity and heat. In this report, the conversion coefficient for the exergy of heat is calculated based on an ambient temperature of 0 °C and a supply/return water temperature of 110 °C/50 °C, and the conversion coefficient for heat is 0.22. That is to say, 22% of the output heat is treated as the equivalent electricity, which shares the total carbon dioxide emitted from power plants with the output electricity.

1.3 Energy Consumption of China's Building Sector

1.3.1 Building Operation Energy Consumption

The building energy consumption data in this chapter comes from the results of research with the China Building Energy and Emission Model (CBEEM) built by BERC, Tsinghua University, and is used to analyze the development of building energy consumption and carbon emissions in China. In 2023, the total commercial energy consumption of building operation was 1.17 gigatonnes of coal equivalent (Gtce), accounting for about 20% of the total energy consumption in China, and the commercial energy consumption and biomass energy consumption of buildings amounted to 1.22 Gtce (biomass energy consumption: about 0.05 Gtce), with details given in Table 1.1.

From 2010 to 2023, the total energy consumption of buildings and electricity consumption increased dramatically, as shown in Fig. 1.5. The COVID-19 pandemic slowed down various social activities, and the growth in electricity consumption of buildings in 2020 was slower than that in 2019. However, as production and life returned to normal in 2021, the electricity consumption of buildings rose greatly. The electricity consumption of buildings in the whole society exceeded 2.5 PWh in 2023.

In recent years, electricity consumption in China has shown a continuous growth trend. Between 2000 and 2010, electricity consumption grew rapidly, with an average annual growth rate of 12%. After that, the growth rate gradually slowed down, stabilizing at around 7% per year from 2010 onwards. From 2020 to 2023, the growth trend of electricity consumption displayed a "staircase" pattern, marked by two periods of slowdown (2020 and 2022) and two rebounds (2021 and 2023).In 2020, impacted by the COVID-19 pandemic, total electricity consumption grew by only 285.5 billion kWh, with a growth rate of 4%. During this period, the growth of

Table 1.1 Building operation energy consumption in China in 2023

Energy use category	Macro parameter	Electricity (PWh)	Fossil fuel (billion tce)	Commercial energy (billion tce)	Primary energy use intensity
NUH	17.3 billion m^2	0.081	0.198	0.222	13 kgce/m^2
Urban residential building (excluding NUH)	33.1 billion m^2	0.6731	0.95	0.298	785 kgce/household
P&C building (excluding NUH)	16.1 billion m^2	1.3902	0.028	0.448	27.9 kgce/m^2
Rural residential building	22.4 billion m^2	0.3603	0.089	0.197	1083 kgce/household
Total	1.41 billion people 71.6 billion m^2	2.5048	0.41	1.17	

Note: In the table, commercial energy consumption means the energy consumption calculated in standard coal converted from electricity, heat, and fuels, and electricity consumption specifically refers to the electricity consumption in the building energy use

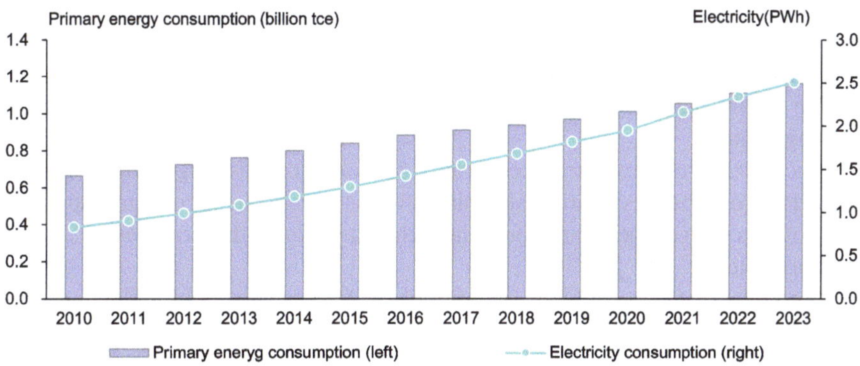

Fig. 1.5 Primary energy consumption and total electricity consumption of building operation in China (2010–2023)

electricity consumption in the secondary industry slowed slightly, electricity use in urban and rural households remained stable, and the tertiary industry's electricity consumption saw the most significant impact, with its annual growth rate falling sharply from around 15% to 2%. In 2021, electricity consumption rebounded significantly, increasing by 808.1 billion kWh, with a growth rate of 11%. This rebound was largely driven by the recovery of the secondary and tertiary industries, both of which exceeded their pre-2020 growth levels, with growth rates of 10% and 18%, respectively. In 2022, electricity consumption grew by 324.4 billion kWh, with a growth rate returning to 4%. During this year, electricity consumption in both the secondary and tertiary industries slowed again, while electricity use in urban and rural households grew rapidly. As a result of extended stay-at-home periods and the rise of remote work and online education, electricity consumption in households surged, with a notable growth rate of 14%. In 2023, electricity consumption achieved a second rebound, reaching 922.38 billion kWh, an increase of 58.66 billion kWh, with a year-on-year growth rate of 6.7%. The tertiary sector made the largest contribution to this growth, with its electricity consumption rising by 183.7 billion kWh, a 17% increase.

Overall, from 2020 to 2023, the increase in total electricity consumption was primarily driven by the secondary industry, contributing about two-thirds of the growth, while the tertiary industry and household electricity use together accounted for about one-third. In the building sector, its growth accounted for about 30% of the total increase, with public buildings contributing roughly 16%, and urban and rural residential electricity consumption each contributing around 7%.

In recent years, the growth of electricity use in public buildings has outpaced that of urban and rural residential buildings, driven by the expansion of commercial activities, data centers, and the increasing demand for digitalization. As the economy resumed and digitalization needs surged, including data centers and 5G base stations, the annual growth rate of electricity use in public buildings reached 10%–15%. At the same time, urban residential electricity consumption showed steady growth, with an annual increase of 5%–10%, supported by rising living standards and the gradual electrification of the residential sector. The fluctuations in urban residential electricity consumption from 2020 to 2023 were also influenced by the record-high summer temperatures between 2021 and 2023, which drove cooling demand.

Since 2020, rural residential electricity consumption has surged and even surpassed that of urban residential use. This growth is largely due to rural grid upgrades and policies such as the promotion of household appliances in rural areas and clean heating systems.

The scale, intensity, and total quantity of the four categories of building energy use are represented in the four blocks in Fig. 1.6 respectively, in which the horizontal axis represents building floor area, and the vertical axis represents energy use intensity per square meter of building. The total area of the four blocks represents the total energy consumption of buildings. In terms of building floor area, urban residential buildings and rural residential buildings have the largest floor area, while the floor areas of NUH and P&C buildings account for about one-fourth and only one-fifth of the total, respectively. However, from the perspective of energy use intensity, the energy use intensity of P&C buildings and NUH is higher than that of the other two categories. Therefore, in terms of total energy use, each of the four categories accounts for about one-fourth of the total building energy consumption. In recent years, the growth in the scale and average energy use intensity of P&C buildings have made their energy consumption the largest proportion of the building energy consumption in China.

Figure 1.7 shows the changes in the total quantity and intensity of the four categories of energy use from 2003 to 2023. The total quantity and intensity in the four categories mainly exhibit the following characteristics:

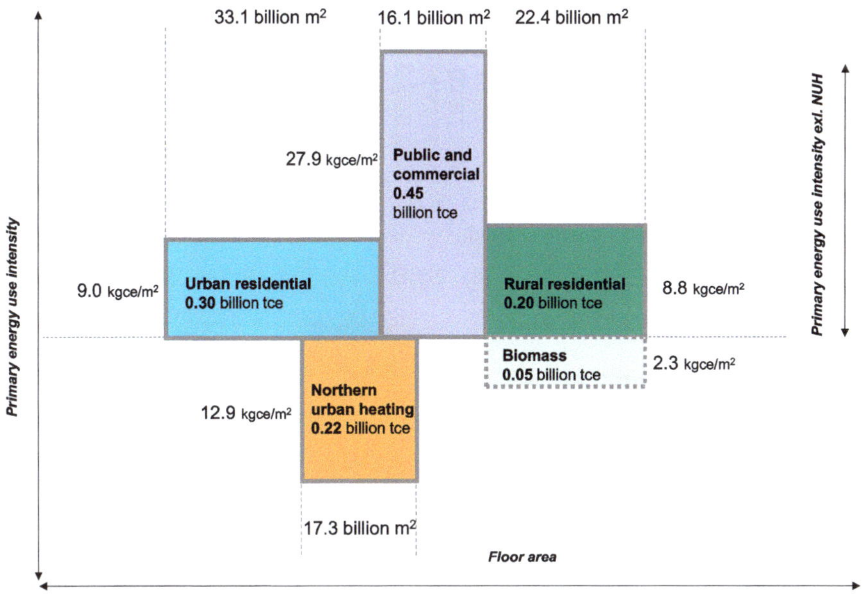

Fig. 1.6 Building operation energy consumption in China (2023). Note: Electricity, heat, and fuels are uniformly converted into primary energy, which is measured in standard coal, and the electricity consumption is converted into primary energy consumption calculated in standard coal based on the annual average coal consumption for power supply in China. The conversion coefficient in 2023 was 302 gce/kWh

- The energy use intensity in NUH was relatively large but has been decreasing with the improvement of the new energy-saving standards and the heat source efficiency in recent years, and the total energy consumption remained stable without any further increase.
- Energy use intensity per unit area of P&C buildings continued to increase. The increasing terminal energy demand of P&C buildings (air conditioners, devices, lighting, etc.) was the major cause for the increase in building energy use intensity. In particular, some large new buildings with large-scale centralized systems have been constructed in many cities in recent years, with energy use intensity much higher than that of similar buildings. As the size of P&C buildings grows, their total energy consumption is still increasing.
- The energy use intensity per household of urban residential buildings increased because there was an increasing demand for domestic hot water, air conditioners, and household appliances. The issue of heating in winter in the HSCW zone also aroused extensive discussions. There was not too much increase in the energy consumption of lighting in residential buildings because of the adoption of energy-efficient illumination devices. The energy use intensity of cooking also remained unchanged. With the further promotion of urbanization and the growth in the size of urban residential buildings, their total energy consumption is still increasing. The commercial energy use intensity per household of rural residen-

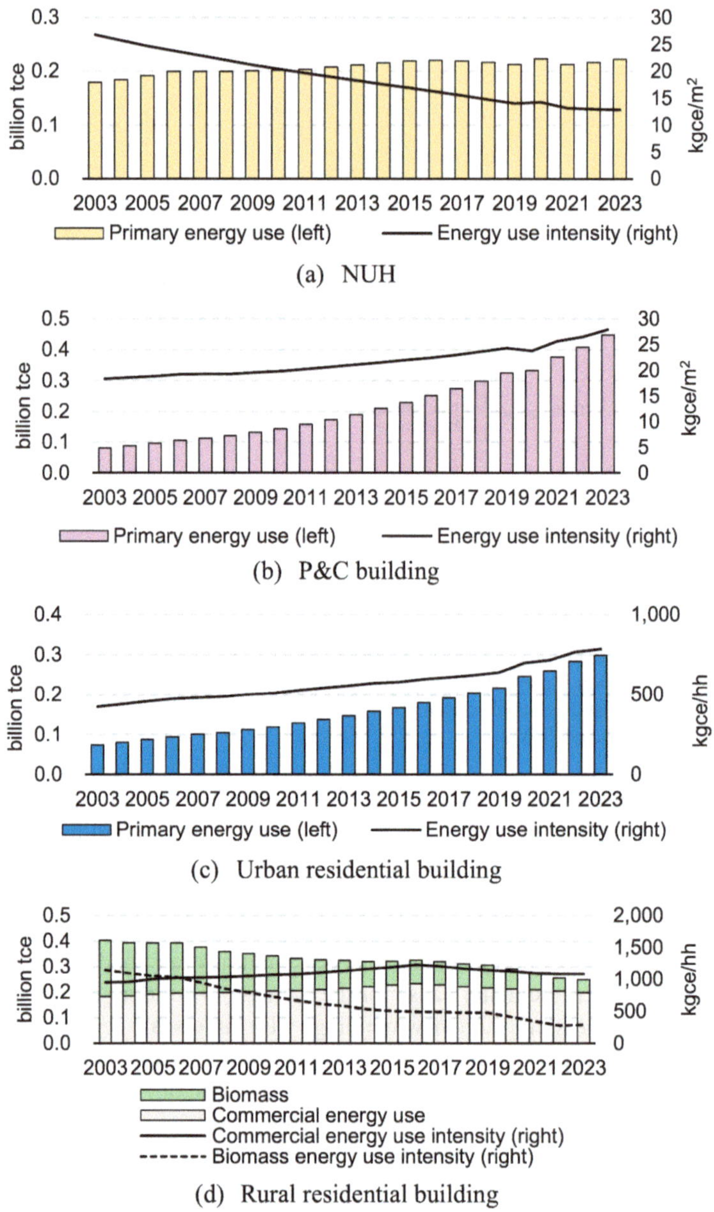

Fig. 1.7 Changes in total building energy use and energy use intensity by year (2003–2023). (**a**) NUH. (**b**) P&C building. (**c**) Urban residential building. (**d**) Rural residential building

tial buildings increased slowly. As the rural population and the number of households slowly decreased, commercial energy consumption in rural areas remained stable. However, as household appliances became more popular in rural areas and the policy of "switching from coal to electricity" was implemented for clean heating in northern China, electricity consumption has increased dramatically in recent years. Meanwhile, biomass energy use has dropped continuously. Hence, the total energy use of rural residential buildings has declined slowly in recent years.

1. **NUH**

In 2023, the energy consumption of NUH was 222 million tce, making up 19% of the total energy consumption of buildings in China. From 2003 to 2023, the NUH area tripled from 6.7 to 17.3 billion m^2, while the total energy consumption increased, but did not double. Obviously, the increase in total energy consumption was less than the increase in building floor area, indicating that remarkable results had been achieved in energy saving. The average energy consumption per unit area of heating was 12.9 kgce/m^2 in 2023, a significant decline from 26.7 kgce/m^2 in 2003. Specifically, the main reasons for the decrease in energy use intensity include the improvement of building insulation, which resulted in the decrease in the heat demand of buildings, as well as the increased share of efficient heat sources, and the improvement of operation management. In recent years, the total energy consumption of NUH has been on a declining trend from the peak around 2017. Due to the COVID-19 pandemic, the heating time was extended to varying degrees in different places in the 2019–2020 heating season. According to the statistical data released in the 2023 Annual Report on Urban Heating Development in China, 86% of cities extended the heating time in the 2021–2022 heating season, so the total energy consumption in NUH went up a little in the 2021–2022 heating season.

Gradual improvement of building envelope performance. In recent years, the Ministry of Housing and Urban–Rural Development of the People's Republic of China has adopted various methods to improve building insulation, including the establishment of building energy efficiency design codes that cover different climate zones and building types, the special examination of energy-saving work that started from 2004, and the renovation of existing residential buildings during the 13th Five-Year Plan period. During the 13th Five-Year Plan period, the energy efficiency design standard for new urban residential buildings in severe cold and cold areas in China was raised to "75% energy-saving standard", approximately 10 million m^2 of ultra-low and near zero energy buildings completed construction, and 514 million m^2 of existing residential buildings and 185 million m^2 of P&C buildings completed the energy-saving retrofit. These methods have greatly enhanced building insulation in China and lowered the actual heating demand of buildings, especially in northern China.

Optimization of heat source structure and significant improvement of heat source efficiency. Recent years have seen a gradual increase in the share of efficient CHP to gradually replace boilers. The proportion of CHP in heat sources for NUH was 42, 48, and 55% respectively in the 3 years. Gas-fired boilers replaced coal-fired boilers. From 2013 to 2020, the proportion of coal-fired boilers dropped from 42 to 13%, while that of gas-fired boilers increased from 12 to 22%. In the meantime, all types of new heat sources kept on growing, with rising proportions of industrial residual heat, residual heat from nuclear power, ground-source heat pumps, and biomass in heating. Heating system efficiency has also been increasing notably in recent years, thus enabling the overall improvement of the efficiency of all types of centralized heating systems.

2. **Urban Residential Buildings (Excluding NUH)**

Urban residential building energy consumption (excluding NUH) in 2023 was 298 million tce, accounting for 26% of the total commercial energy consumption in the building sector. Electricity consumption was 673.1 TWh. With the economic and social development and the improvement of living standards in China, the average annual growth rate of urban residential building energy consumption reached up to 7% from 2003 to 2023, and the terminal electricity consumption in 2023 quintupled that in 2003.

From the view of energy use, cooking, household appliances, and lighting consumed the most energy in urban residential buildings (excluding NUH) in China. Thanks to policies and projects for improving the energy efficiency of cooking, household appliances, and lighting, the terminal energy consumption of these three categories was kept under control, and the total energy consumption has under-gone a slower increase in recent years. Improving energy efficiency and lowering standby energy consumption should become the optimal methods to limit the energy consumption of cooking, household appliances, and lighting. For example, the promotion of energy-saving lamps significantly improved the lighting efficiency of residential buildings. Energy efficiency standards and behaviors need to be upgraded to lower the electricity consumption from long standby time and frequent reheating and restarting of household appliances. The production standards of such appliances as TV set-top boxes, water dispensers, and electric toilet seats need to be improved to lower the waste of energy when they are in standby mode. The methods, for example, include improving the controllability of set-top boxes, enhancing the insulation capacity of water dispensers, and adopting intelligent control of toilet seats. Policy incentives or subsidies shall not cover electric appliances such as clothes dryers, which may change the lifestyle, and we should watch for the energy spikes of these high energy consumption appliances. Even though the energy consumption of winter heating, summer cooling, and domestic hot water accounts for a smaller proportion in the HSCW zone and the energy consumption per household is at a low level, they have been growing rapidly. The annual average growth rate of heating energy consumption in the HSCW zone could be well over 50%. Therefore, saving terminal energy use for those three categories should be our priority in the next stage of energy saving for urban residential buildings. We should avoid the massive adoption of centralized systems, promote decentralized systems in residential buildings, improve the energy efficiency standards of distributed equipment, and prevent drastic energy consumption increases while improving the indoor service level. Refer to the *China Building Energy Use and Carbon Emission Yearbook 2021* for detailed discussions about the energy-saving and emission-reduction pathways for urban residential buildings in China.

3. **P&C Buildings (Excluding NUH)**

In 2023, the total floor space of China's P&C buildings was approximately 16.1 billion m^2, and the total energy consumption of P&C buildings (excluding NUH) was 0.448 Gtce, making up 38% of that of the building sector. Electricity consumption stood at 1.39 PWh. The total area of P&C buildings and the proportion of large P&C buildings were all on the rise, which led to an increase in energy demand. The energy consumption per unit area of P&C buildings grew from 18.3 kgce/m^2 in 2003 to over 27.9 kgce/m^2 in 2023. The energy use intensity increased rapidly, and the total energy consumption surged as well.

In 2020, due to the impact of COVID-19, the operation duration and intensity of public and commercial buildings were subject to pandemic-related control measures, leading to a slight decrease in the average energy use intensity of public and commercial buildings in China. In 2021, the growth rate of energy used in the operation of public and commercial buildings picked up. Since 2001, the newly built building stock of P&C buildings amounted to almost 8 billion m^2, approximately 79% of the current stock. This means that three-fourths of P&C buildings were built after 2001. There are two reasons for this increase. First, a lot of new commercial buildings such as office buildings and commercial complexes have been constructed in recent years. Second, the scale of public service buildings such as schools, hospitals, and sports stadiums has increased due to the necessity for the gradual perfection of relevant infrastructure to promote the building of a well-off society in an all-around way and the improvement of public services. In recent years, the proportion of school and hospital buildings in the new public and commercial buildings has been increasing gradually. In 2021, there were more new hospital and school buildings than new office and hotel buildings.

While the stock of P&C buildings is growing, the number of large-scale P&C buildings is also increasing. In particular, many public and commercial buildings completed in recent years are large top-grade commercial buildings with central air-conditioning. Their electricity consumption per unit area is over 100 kWh/m^2, while the electricity consumption of smaller schools, offices, and stores built in the past is approximately 60 kWh/m^2. The average electricity consumption of public and commercial buildings will continue to increase as the proportion of such new public and commercial buildings with high energy consumption in the total number of public and commercial buildings has been on the rise. Due to the volume and form constraints of such new buildings, the energy use intensity of air conditioning, ventilation, lighting, and elevators in them is much higher than that in general public and commercial buildings. This is also an important cause of the continuous growth in the energy use intensity of public and commercial buildings in China. Refer to the *China Building Energy Use and Carbon Emission Yearbook 2022* for detailed discussions about paths toward energy saving and emission reduction for P&C buildings in China.

4. **Rural Residential Buildings**

In 2023, the commercial energy consumption of rural residential buildings was 0.197 Gtce, accounting for 17% of the total energy consumption of buildings in China. Electricity consumption was 360.3 TWh, and the consumption of rural biomass energy (straw and firewood) was equivalent to about 0.05 Gtce. From 2003 to 2023, urbanization led to a decline of the rural population from 800 to 500 million, and the scale of rural residential buildings was maintained at approximately 23 billion m^2 and has begun to decrease slowly in recent years.

Owing to the higher availability of electricity in rural areas, higher income for rural residents, and more household appliances, the electricity consumption per household in rural areas has increased rapidly. For instance, the number of air conditioners per hundred households in rural areas increased from 16 in 2001 to 106 in 2023, which led to not only the growth of electricity consumption but also a longer peak power load in rural areas during summer. The implementation and promotion of "switching from coal to electricity" in northern China contributed to significant growth in winter heating electricity consumption and peak power load there. Moreover, increasing biomass energy has been replaced by commercial energy, leading to a rapid reduction of the proportion of biomass energy in household energy consumption in rural areas. China released the *Work Plan for Implementation of the PV Poverty Alleviation Project* in 2014, which proposed the development of the photovoltaic (PV) industry in rural areas as an important means of poverty alleviation. A new energy system based on rooftop PV may be built by taking advantage of abundant renewable resources in rural areas. Such a system can realize the net power output to the power grid while meeting the energy demands of rural life, production, and transportation. It can totally cancel the use of fossil and biomass fuels while raising the living standard in rural areas. Thus, it will not only root out the environmental pollution and carbon emission problems caused by the burning of fossil and biomass fuels but also make the production and output of zero-carbon energy another important economic activity in rural areas. Besides, this can make an important contribution to the sustainable development of China's energy system and become an important part of the rural revitalization strategy.

In recent years, with the thorough implementation of haze control measures and clean heating in eastern China, governments at all levels and relevant enterprises have made huge investments to increase the power supply capacity, lay gas pipe networks and change original small household coal-fired furnaces into low-pollution forms in rural areas, which leads to a substantial increase in electricity and gas consump tion. The change in rural energy structure will lead to a fundamental transformation of rural energy use patterns, thus facilitating the modernization of rural areas. This opportunity should be leveraged and scientific planning should be made to revolutionize rural energy supply and consumption and establish a new energy system with renewable energy sources as the mainstay for rural residents, which will play an important role in the current energy revolution of China.

1.3.2 Embodied Energy Consumption of Building Sector

China's continuous advancement of urbanization in the last two decades has also enabled the embodied energy consumption of civil buildings to become an important part of the total energy consumption of the whole society. A large quantity of building materials is needed, the production process of which lead to great energy consumption and carbon emissions. This is one of the key reasons for the continuing growth of energy consumption and carbon emissions in China.

According to the estimation from the BERC, embodied energy use of civil buildings in China amounted to 0.46 gigatonnes of coal equivalent (Gtce) in 2023, accounting for 8% of China's total energy consumption. The embodied energy use of civil buildings in China has peaked at 2016, while due to the slow decrease in the newly built building stock of civil buildings in recent years, its embodied energy use has also dropped gradually since 2016 as shown in Fig. 1.8. In 2023, the embodied energy of urban residential, rural residential, and P&C buildings accounted for 69, 5, and 26%, respectively.

In fact, the construction sector consists of not only civil buildings but also buildings for production purposes and infrastructures such as motorways, railways, and dams. The embodied energy use of the construction sector mainly includes all types of energy use related to the construction of buildings and infrastructures. According to the calculation of BERC, Tsinghua University, the total embodied energy use of China's construction sector in 2023 was 1.45 Gtce, accounting for up to 25% of the primary energy consumption of the whole society. From 2005 to 2023, embodied energy use in China's construction sector grew from approximately 0.4–1.45 Gtce, as shown in Fig. 1.9. Embodied energy from building materials is the mainstay of the total embodied energy use of buildings, in which iron and steel and cement production consumes more than 80%.

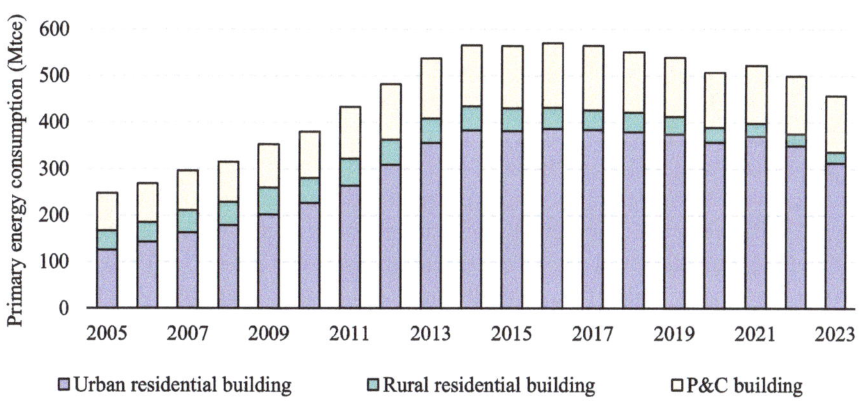

Fig. 1.8 Embodied energy use of China's civil buildings (2005–2023). Source Estimation by BERC, Tsinghua University. This figure only covers civil building construction. The newly built building stock data is based on the data under "the statistical standards for construction enterprises as specified in the *China Statistical Yearbook on Construction*"

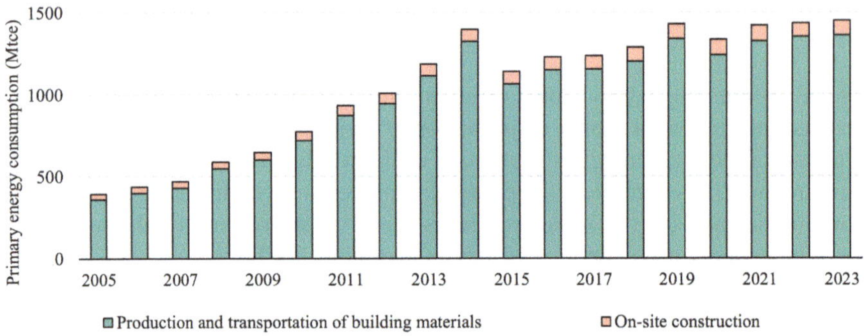

Fig. 1.9 Embodied energy consumption of China's construction sector (2005–2023). Note: The construction sector involves the construction of civil buildings, production buildings, and infrastructures (Data about the consumption of building materials is from the *China Statistical Yearbook on Construction*. (Source: Estimation by BERC, Tsinghua University)

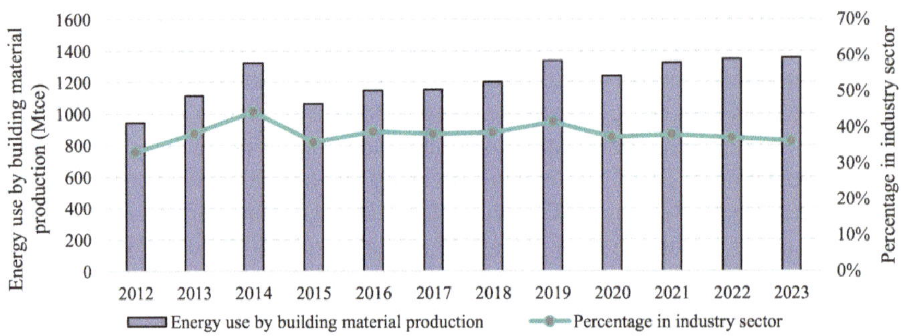

Fig. 1.10 Energy consumption for building material production. Materials for the construction sector here mainly include steel, cement, aluminum, glass and architectural ceramics

The construction demands of rapid urbanization in China have not only driven the growth of energy consumption directly but also determined China's industrial structure which is dominated by traditional heavy and chemical industries including steel and cement. This is also a key reason for the high energy consumption per unit of industrial value added in China.

There was 1.36 Gtce of industrial energy used for producing building materials in 2023. Between 2012 and 2023, building materials accounted for approximately 40% of the total industrial energy consumption, as shown in Fig. 1.10. The fast pace of urbanization in China drove the demand for building materials, which was the key reason why the higher percentage of total energy consumption was represented by iron and steel, building materials, and other traditional heavy industries.

As urbanization and infrastructure development have achieved initial progress in China, the transformation of the construction mode is ongoing. In 2023, the per capita floor area for urban residential buildings was 36 m^2 in China, which was close to that of some developed countries in Asia, such as Japan and South Korea, but still far lower than that of the US. The reason was that during China's urbanization, the main type of building in the urban communities was apartments instead of single houses, such as that of the US. From the perspective of urban form, the utilization ratio of P&C buildings was high in China because of the high-density large city mode of development, so it is unnecessary to follow the per capita P&C building scale in Europe and America. In the future, there will be no more rapid growth of iron and steel, building materials, and other high energy consumption industries, so long as there is less demolition, and the building life cycle can be properly maintained. Therefore, for the next round of urbanization, the demolition of buildings that have not reached the end of life shall be abolished. Technologies to extend the building life cycle shall be invented. Buildings and infrastructures should be properly repaired, and the building life cycle should be extended to facilitate industrial transformation and total energy control.

1.4 GHG Emissions of China's Building Sector

1.4.1 Carbon Dioxide Emissions from Building Operation

The carbon dioxide emissions from building operation are affected by the growth in total energy demand of buildings, the improvement of building energy efficiency, and the adjustment of building fuel types and energy supply structure. Electricity, coal, and gas were the major energy sources for building operation. Electricity accounted for 65% of the total energy use in urban residential buildings and P&C buildings, in which CO_2 was indirectly emitted. The adoption of CHP in NUH could also lead to indirect CO_2 emissions. The percentage of coal and gas consumption was higher than that of electricity consumption for NUH and rural residential buildings. The percentage of coal and gas consumption was about 89% for NUH and the percentage of fossil energy consumption was about 48% for rural residential buildings, which led to massive direct carbon dioxide emissions. In another aspect, as the percentage of zero-carbon electricity has increased in China, the average emission factors have declined tremendously, at 540 gCO_2/kWh in 2023. Besides, the share of electricity consumption in building operation energy consumption gradually increased as well. These two trends have promoted the low-carbon development of building operation energy consumption.

According to the analysis results from CBEEM, in 2023, the total carbon emissions during building operation in China were 2.2 billion tonnes of carbon dioxide (Gt CO_2), equivalent to 1.58 t per capita carbon emissions and 31 kg/m^2 of average carbon emissions per unit area. In the total carbon emissions, the proportions of direct carbon emissions (0.42 Gt CO_2), indirect carbon emissions from electricity use (1.35 Gt CO_2), and indirect carbon emissions from heating (0.45 Gt CO_2) were 19, 61, and 20%, respectively, as shown in Fig. 1.11.

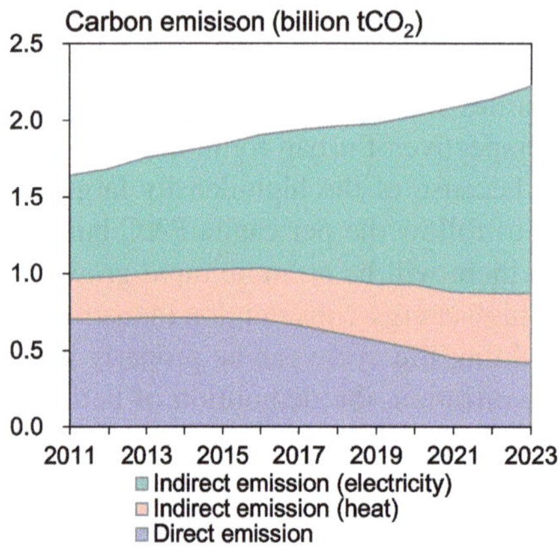

Fig. 1.11 Carbon dioxide emissions from building operation (2023)

1. **Direct Carbon Emission**

   In 2023, the direct carbon emissions of buildings were 420 million tons of CO_2, including about 150 million tons of CO_2 directly emitted from urban and rural cooking, about 190 million tons of CO_2 emitted from household gas- and coal-fired heating boilers,[3] and 80 million tons of CO_2 directly emitted from natural gas consumption for hot water, steam boilers, absorption refrigeration, and other purposes. Emissions from rural areas accounted for more than half of the direct carbon emissions.

   In recent years, with the vigorous promotion of "switching from coal to electricity", "switching from coal to gas" and clean heating in rural areas, the direct carbon emissions in the building sector of China have been declining slowly from the peak around 2015. Direct carbon emissions of the building sector will continue to decrease and no longer reach a new peak as long as electrification is continuously promoted in new buildings.

   To reduce direct carbon emissions of the building sector to zero, the key lies in the point-in-time and intensity of promoting "electrification". It is expected that direct carbon emissions of buildings will be reduced to zero during 2040–2045. Analyses indicate that electrification will not increase the operation cost in 80% of all cases and that the initial investment in equipment will be returned in about 5 years through the reduction of operating cost. Therefore, the major obstacle to promoting electrification in buildings is not economic cost but the change in the concept of energy use and cooking culture. Increasing the publicity of "zero carbon emissions of buildings by electrification" among the public and promoting "switching from gas to electricity" in new and existing buildings

[3] Refers to gas- and coal-fired heating boilers installed in urban and rural residential buildings as well as coal- and gas-fired boilers installed in public and commercial buildings. Such fuels are burned directly in buildings, so the resulting carbon emissions fall into the category of direct carbon emissions from buildings.

are the most important approaches to realizing zero direct carbon emissions from building operation.

2. **Indirect Carbon Emission from Electricity Use**

   In 2023, the electricity consumption by building operation in China was 2.5 PWh, and the indirect carbon emission from electricity use was 1.35 billion tons of CO_2. At present, the per capita electricity consumption of the building sector in China is one-sixth of that in the US and Canada and about one-third of that in France and Japan, while the electricity consumption per unit area of buildings in China is one-third of that in the US and Canada. The difference in lifestyle and building operation is one of the main reasons for the difference in electricity use intensity between China and developed countries.

   In recent years, the carbon emission increase caused by the growth in electricity consumption of buildings has exceeded the carbon emission decrease resulting from the reduction of the carbon emission factor of electricity. The indirect carbon emissions from electricity consumption of buildings will keep increasing before peaking. China should maintain a green and economical lifestyle and building operation mode to avoid any surge in energy use of buildings that once occurred after rapid economic growth in the history of the US, Japan, and other developed countries. When the building floor area in China reaches 75 billion m^2 in 2060, the electricity consumption of buildings should be 3.8 PWh, which will meet the demand of the Chinese people for a good life and energy consumption of buildings. On this basis, the new type of electric power system adopting "photovoltaic, energy storage, direct current and flexibility (PEDF)" technologies should be promoted. When the reduction in indirect carbon emissions from building electricity use caused by the increase of "green electricity" in practical electricity consumption through flexible electricity use is greater than the growth in indirect carbon emissions from building electricity use caused by the growth in the total scale and electricity use intensity of buildings each year, the indirect carbon emissions from the electricity use of buildings in China can peak. Upon overall popularization of the "PEDF" power distribution mode and flexible means of electricity use, the zero-carbon goal for electricity consumption of buildings can be achieved earlier than that for the national power system.

3. **Indirect Carbon Emission from Heating**

   In 2023, the building floor area of NUH in China was 17.3 billion m^2, and the indirect carbon emission from heating during building operation was 450 million tons of CO_2. Recent years have seen continuous growth in the centralized heating area and heating demand in northern China but a continuous decrease in energy consumption and carbon emissions per square meter of heat supply. The total indirect carbon emissions from NUH showed a trend of slow growth. The indirect carbon emissions from heating during building operation can peak by around 2025 through further strengthening the renovation of existing buildings, fully exploiting low-grade residual heat resources, and phasing out scattered coal-fired boilers. Afterward, with the gradual completion of the zero-emission transformation of the remaining thermal power (to be replaced with CCUS and biomass fuel) by the electric power sector, the indirect carbon emissions from building heating can be reduced to zero in sync with the decarbonization process of the power system.

To this end, it is necessary to continue improving the envelope performance of new buildings and the renovation of existing buildings in a strict manner, so that the average heating demand of buildings in northern China can decrease from 0.37 GJ/m^2 at present to below 0.25 GJ/m^2. During 2020–2035, the return water temperature should be lowered through the terminal renovation of centralized heating systems, thus effectively recovering the residual heat of thermal power plants and the industrial low-grade residual heat. The increasing building heating demand will be met by tapping into the potential heating capacity of existing heat sources. CHP transformation will be carried out for the coastal nuclear power in northern China to supply heat to areas within 200 km from the normal of coastal northern China. From 2035, in coordination with the schedule of the thermal power shutdown of the electric power system, seasonal heat storage projects will be implemented simultaneously to solve the problem of reduced heat source power resulting from the shutdown of thermal power plants. Until 2045, the annual residual heat of nuclear power, the annual residual heat of peak shaving thermal power, the residual heat of wind and solar curtailments at the centralized wind and solar PV power bases, as well as the heat emitted annually from the industrial low-grade residual heat will have been collected through seasonal heat storage projects. In this way, the indirect carbon emissions from building heating can be reduced to zero, while zero-carbon emission of the power system is realized.

In view of the four categories, the scale, intensity, and total quantity of carbon emissions from them are represented in the block diagram in Fig. 1.12 respectively, in which the horizontal axis stands for the building floor area and the vertical axis for carbon intensity per square meter for the four categories.

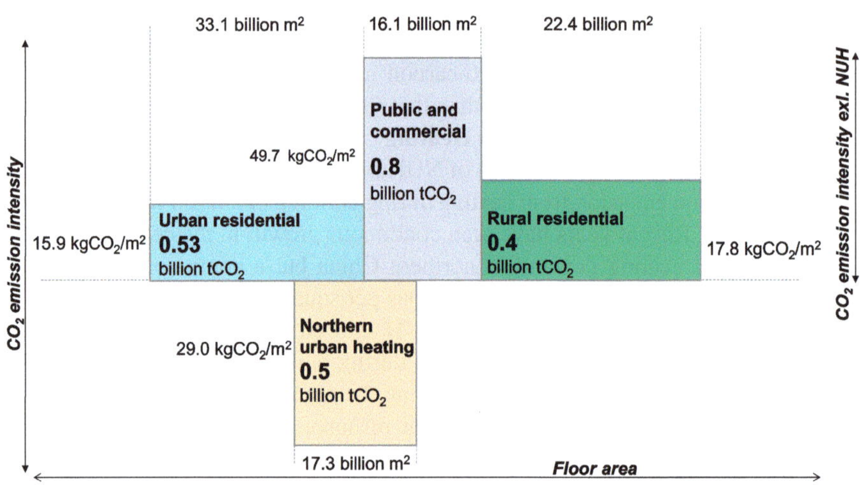

Fig. 1.12 Carbon dioxide emissions from building operation in China (2023)

The total area of the four blocks represents the total carbon emissions. The growth in the total carbon emissions from the four categories is shown in Fig. 1.13. It can be seen that the characteristics of carbon emissions from the four categories are not the same as those of their energy consumption. As public and commercial buildings have the highest energy use intensity, their carbon

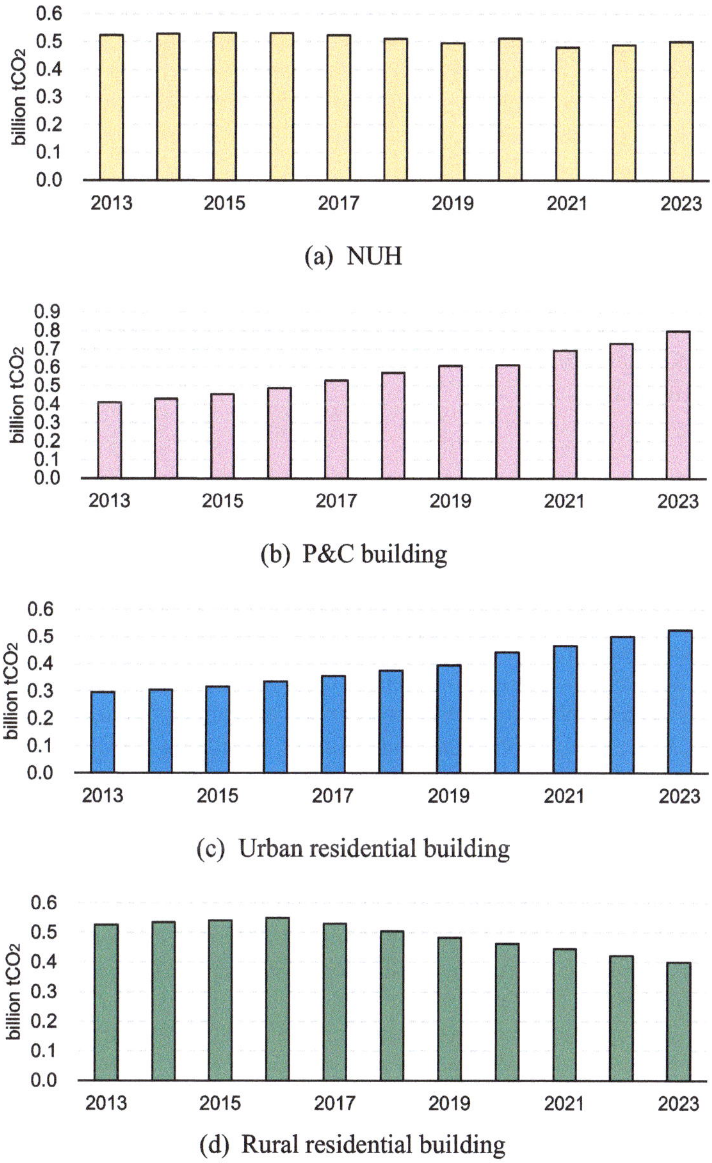

Fig. 1.13 Carbon emissions of building energy use (2013–2023). (**a**) NUH. (**b**) P&C building. (**c**) Urban residential building. (**d**) Rural residential building

intensity per unit area is also at the peak. In 2023, their carbon intensity was 49.7 kg CO2/m2, and their total carbon emissions were still on the rise as their total energy consumption and energy use intensity were growing steadily. For NUH, its carbon intensity was only second to that of public and commercial buildings due to the consumption of a lot of coal. In 2023, the carbon intensity was 29.0 kg CO2/m2, and the carbon emissions peaked and remained stable at approximately 0.5 Gt CO2 due to the increasing heat demand, the improvement of heat supply efficiency, and the consistent speed in the transformation of energy structure. Although there is little difference in primary energy use intensity per square meter between rural and urban residential buildings, the carbon intensity per square meter of rural residential buildings is higher than that of urban residential buildings due to a low level of electrification and a high proportion of coal fuel. The carbon intensity per unit area of rural residential buildings was 17.8 kg CO2/m2. Due to the implementation of "switching from coal to electricity" and "switching from coal to gas" in rural areas, the total carbon emissions of rural residential buildings have peaked and then declined annually in recent years. The carbon intensity per unit area of urban residential buildings was 15.9 kg CO2/m2 and was increasing slowly with the growth in electricity consumption.

1.4.2 Embodied CO_2 Emission of Building Sector

As urbanization continues, the embodied energy use of civil buildings in China also increases rapidly. The construction of buildings and infrastructures not only consumes a colossal amount of energy but also leads to a lot of carbon dioxide emissions. In addition to carbon dioxide emissions resulting from energy consumption, emissions in the cement production process[4] are also an important part.

In 2023, the total carbon emissions from civil building construction in China were about 1.4 Gt CO_2, mainly including carbon emissions from energy use for the manufacturing and transportation of building materials (76%), the industrial process emissions of cement (21%) and energy use during construction (3%), as shown in Fig. 1.14. Although such carbon emissions are included in the industry and transportation sectors, they are driven by the demand of the building sector. Hence, the building sector shall also be responsible for such carbon emissions and reduce its demand to contribute to emission reduction. With the end of the large-scale construction period in China, the scale of newly built building stock has been decreasing each year. The carbon emissions from civil building construction peaked in 2016 and have been declining slowly year by year in recent years.

[4] Refers to carbon emissions from chemical reactions (excluding combustion) for cement produc tion.

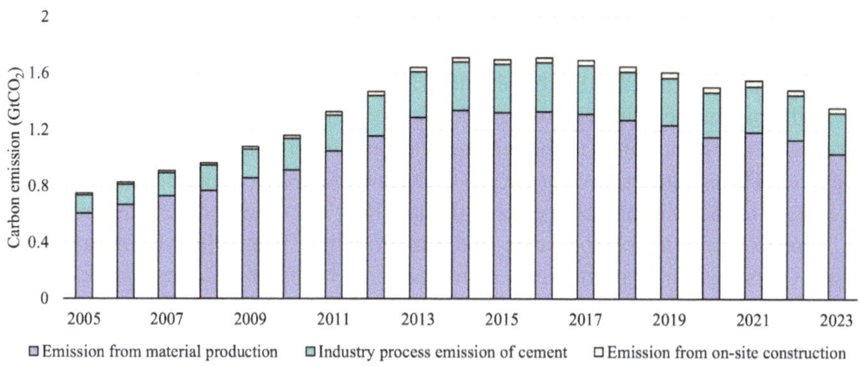

Fig. 1.14 Embodied carbon emissions of civil buildings in China (2005–2023). Note: Civil building construction is included only. (Source: Estimation by BERC, Tsinghua University)

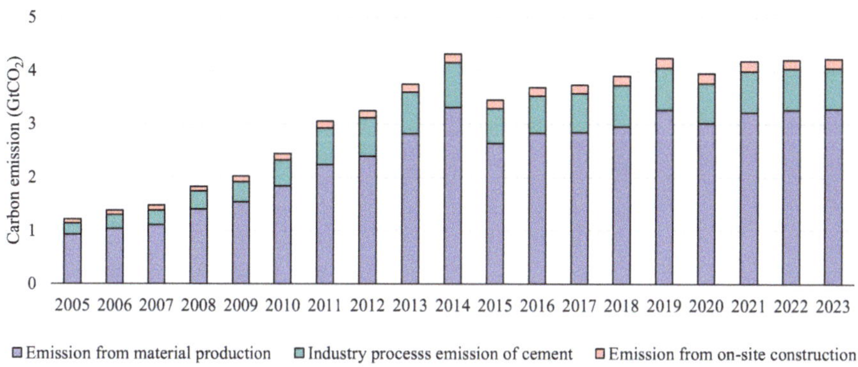

Fig. 1.15 Embodied carbon emissions of the construction sector in China (2005–2023). Note: The construction sector involves the construction of civil buildings, production buildings, and infrastructures. (Source: Estimation by BERC, Tsinghua University)

In fact, as China is still in the urbanization stage, various infrastructures need to be constructed in addition to civil buildings. In 2023, the total carbon emissions from construction in the construction sector were about 4.2 Gt CO_2, nearly one-third of China's total carbon emissions, as shown in Fig. 1.15. Carbon emissions from civil building construction were about 32% of the total carbon emissions from construction in China's construction sector.

To realize zero carbon emissions from building construction as early as possible, the total number and scale of buildings should be controlled reasonably first to minimize excessive construction and avoid large-scale demolition and construction. At present, the total building stock and per capita floor area in China have met the demands of urban and rural residence, production, and living. By 2060, the production and living demands of the future urban and rural populations in China can be met with 40 m^2 of per capita residential floor area, 15.5 m^2 of per capita P&C

building stock, and 75 billion m^2 of total floor area. To realize zero-carbon emissions from building construction in China, the construction speed, total quantity, and scale of buildings need to be planned rationally.

In the meantime, China's construction sector will shift from the large-scale construction of new buildings to the maintenance and functional improvement of existing ones. The construction of houses in China has shifted from increasing the supply of houses to meet immediate needs to demolishing old ones and building new ones to improve building performance and functionality. The "Mass-Demolition-for-Mass-Construction" pattern has become the main mode of the construction sector. Based on the objective of total number planning for civil buildings in the future and in view of the reasonable construction speed, the gradual transformation from "Mass-Demolition-for-Mass-Construction" to "replacing demolition with fine repair" will enable the stabilization of China's construction sector and the gradual reduction of carbon emissions from civil building construction to 0.2 Gt CO_2. Zero emissions from building construction are expected to be realized in 2050 through further application of new building materials and new structural systems and technologies.

Chapter 2
Comparison of Energy Consumption and Carbon Emissions from Building Operation Between China and Other Countries

2.1 Energy Consumption and GHG Emissions in the Global Building Sector

2.1.1 Energy Consumption by Global Building Operation

According to the International Energy Agency's (IEA) calculation of global energy use and emissions of the building sector (as shown in the figure below), in 2022, the embodied energy consumption during the global building construction stage (including building and infrastructure construction) and the building operation energy consumption accounted for 34% of the total global energy consumption, in which the embodied energy consumption during building and infrastructure construction accounted for 4% and the energy consumed during the operation stage accounted for 30%. In 2022, the total global CO_2 emissions (including energy-related and industrial process emissions) were 36.8 Gt CO_2, in which the embodied CO_2 emissions from construction (including building and infrastructure construction) in the construction sector accounted for 7% and the CO_2 emissions from building operation accounted for 27% (Fig. 2.1).

According to BERC's calculation of China's building energy use and emissions, in 2023, China's embodied energy use and operational energy use in the building sector accounted for 28% of the total social energy use,[1] which was close to the global level. However, China's building embodied energy use was 8% of the total social energy use, higher than 4% of the global level. If coupled with the embodied energy use for production building and infrastructure construction, China's building embodied energy use will be up to 27% of the total social energy use. The building

[1] The primary energy consumption method is adopted for conversion. The heat consumption and electricity consumption of buildings are converted by the coefficient of coal consumption for thermal power supply into primary energy consumption, which is then added together with the consumption of other energy types at terminals.

Building Energy Research Center of Tsinghua University, *Decarbonizing Urban Residential Buildings*, https://doi.org/10.1007/978-981-95-8200-6_2

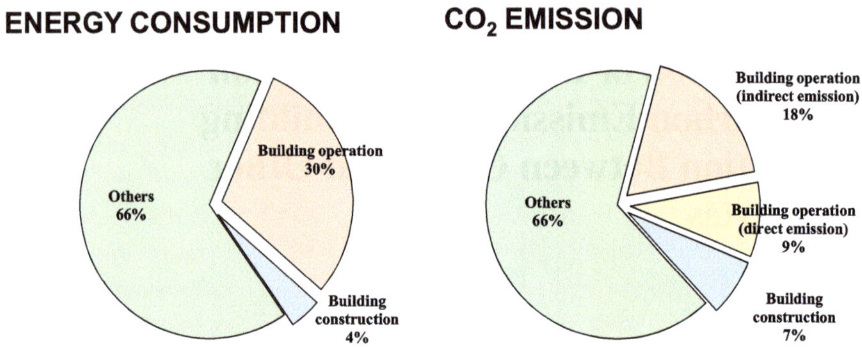

Fig. 2.1 Energy use and CO_2 emissions in the global building sector (2022). Note: The construction sector involves civil building construction, production building construction, and infrastructure construction. This figure uses the terminal energy consumption data provided by IEA, which is obtained through the direct addition of heat consumption for heating, electricity consumption of buildings, and terminal use of various energy types. The electricity consumption is converted to primary energy using a calorific value equivalent method. This conversion method is different from that used in the comparison of building energy consumption among countries in the ensuing part of this research, so it should be treated differently in data comparison. (Source: 2023 Global Status Report for Buildings and Construction, International Energy Agency)

operation energy use accounted for 20% of the total and was still lower than the global average. In the future, the share of China's building sector energy use will continue to increase with economic and social development and the improvement of living standards.

In terms of CO_2 emissions, in 2023, China's total social carbon emissions (including energy-related and industrial process emissions) were about 12.5 Gt CO_2, in which the embodied CO_2 emissions from building construction and the CO_2 emissions from building operation accounted for approximately 29%, including 11% from building construction and 18% from building operation (Fig. 2.2). If the energy-related CO_2 emissions were considered only, the CO_2 emissions from energy consumption of the whole society in China could be about 11 Gt CO_2 in 2023, including about 20% from building operation.

As China's urbanization is still ongoing, the major share of energy consumption and emission has come from building and infrastructure construction. The share of embodied energy consumption from building construction in China is higher than the global average and also higher than that of member countries of the Organization for Economic Co-operation and Development (OECD) that have already been urbanized. However, compared with those in OECD member countries, the building operation energy consumption and carbon emissions in China are at a lower level. As urbanization becomes slower in China, the proportions of building operation energy consumption and related emissions in the social total will further increase. China will gradually shift the focus of its building energy efficiency and low-carbon work from the low-carbon development of new buildings to the low-carbon operation of existing ones.

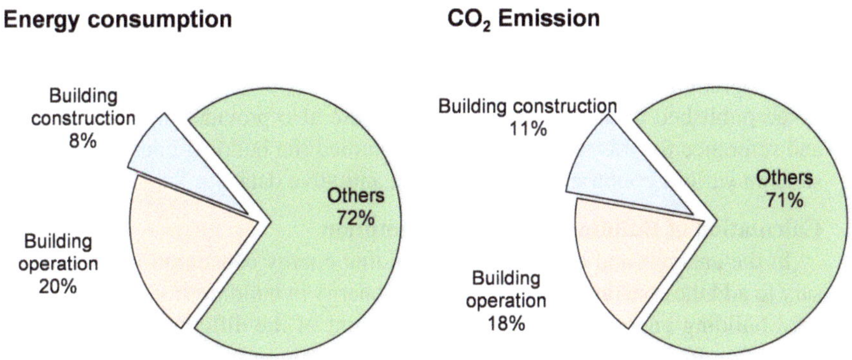

Fig. 2.2 China's building energy use and CO_2 emissions (2022). Note: The construction sector involves the construction of civil buildings, production buildings, and infrastructures. The diagram on the right illustrates the structure of China's total social carbon emissions (including energy-related and industrial process emissions). (Source: Estimation with CBEEM of BERC, Tsinghua University)

2.1.2 Boundaries and Comparative Study Methods of Building Energy Consumption and Emissions

Comparing building energy consumption among countries is an important means to know the building energy consumption in China, analyze its future development trend, and design the paths of building energy saving. In this section, the building operation energy consumption and carbon emission data of countries in the world are collected and analyzed through comparison. To ensure data comparability and better reflect actual energy use, the energy consumption data in this study only included commodity energy and did not include biomass energy that was not circulating.

Two types of data are collected to compare building energy consumption among countries. The first type includes population, the number of households, and the floor area of buildings. The other is building energy consumption data, which mainly includes the total consumption of electricity, heat, coal, natural gas, and other fuels during building operation. The building energy data of countries around the globe collected in this study mainly comes from two sources:

1. Databases of international organizations and agencies: mainly including IEA, Odyssee, World Bank, and Eurostat databases;
2. Official statistics of countries: For example, Japan's data mainly comes from the *Statistical Handbook of Japan* and the *Japan Statistical Yearbook* published by the Statistics Bureau of Japan. The data on the US is mainly sourced from the periodic surveys conducted on representative buildings of the country and the statistics released per year by the Energy Information Administration (EIA). Canada's data mainly comes from Natural Resources Canada. The data on South Korea mainly comes from the building information statistics of the Ministry of

Land, Infrastructure, and Transport and the KOSIS data. The data of India is mainly from the National Statistical Office (NSO) and the Ministry of Statistics and Programme Implementation (MoSPI).

3. Some published research reports and literature also provide important support and reference for this study as they have studied the building energy and emissions in various countries and provided quantitative data.

1. **Calculation of Building Energy Consumption**

   In the analysis and comparison of building energy consumption, it is necessary to add the consumption of all types of energy in buildings together to get the total building energy consumption, on account of the different percentages of electricity, fuels, and heat used for building operation in various countries. At present, the end-use energy consumption method and the primary energy consumption method are commonly used for calculation. In the context of the low-carbon energy transition, the development trend of building energy use in various countries is to achieve full electrification. With the gradual increase in the percentage of electricity in the energy structure, it will be more meaningful to convert all categories of energy use into electricity and then add them together to get the total energy consumption of buildings. Therefore, the total energy consumption of buildings is calculated in this section through the conversion of various types of energy into electricity. To decouple the level of building energy use and the level of the energy conversion system, a unified energy conversion factor benchmark is used for the conversion between each fuel and electricity in this study. Under the principle of the energy conversion factor benchmark, the total global building energy use can be directly allocated to the global primary energy, and the energy conversion systems have positive and negative values respectively to reflect their efficiency (high or low) and energy structure (good or bad), and the sum is zero. The energy conversion factor benchmark theoretically means the global average conversion level, namely the global average of the power generation capacity of each fuel. The energy conversion factor benchmark used in this section are listed in the table below (Table 2.1):

2. **Calculation of Carbon Emissions from Buildings**

   The data on carbon emissions from building operation of different countries in this section is sourced from IEA and the calculation results of the CBEEM model by BERC, Tsinghua University. When calculating the total carbon emissions from building operation, the direct carbon emissions, indirect carbon emissions from electricity use, and indirect carbon emissions from heating in

Table 2.1 Energy conversion factor benchmark in the calculation of total energy consumption of buildings in various countries

Energy	Unit	Conversion factor benchmark
Coal	gce/kWh	300
Oil	goe/kWh	191
Natural gas	Nm^3/kWh	0.2
Heat from boiler	kWh/GJ	133
Heat from CHP	kWh/GJ	70

buildings were considered. When calculating indirect carbon emissions from building electricity use, the total carbon emissions from electricity generation in each country were divided by the total electricity generation to obtain the average carbon emission factor for electricity use in each country. The carbon emission factor was used to calculate indirect carbon emissions from building electricity use. Carbon emissions from building heat use were calculated with the building heat use and the carbon emission factor per unit of heat. In the study of the building operation energy consumption in each country, various types of energy were converted into electricity by taking a unified conversion factor benchmark as the conversion coefficient. In the study of carbon emissions from building operation in each country, the real carbon emission factor instead of a unified carbon emission factor was used to calculate the real carbon emissions because carbon emissions from buildings are closely related to the energy structure and must be discussed together with the energy structure and the energy conversion system.

For carbon emissions during building operation, each country proposed a goal to reduce carbon emissions in the building sector. The technical pathways and priorities to achieve carbon neutrality in buildings differ between countries. To quantify and analyze the various problems confronted by each country in their attempt to achieve carbon neutrality in the building sector, the carbon emission factors for electricity and heat of each country were used in the calculation. Therefore, differences in the energy mix and efficiency among countries will affect the total amount and intensity of carbon emissions during building operation.

2.2 Energy Consumption and Carbon Emissions in the Building Sector of Different Countries

2.2.1 Building Operation Energy Consumption of Different Countries

Three indicators were selected to compare building energy use across countries: total amount, energy use per capita, and energy use per floor area, as illustrated in Fig. 2.3. The building energy use in this figure was calculated by adopting the electricity-equivalent method to calculate the total energy use of building operation. Energy use intensity per capita and energy use intensity per floor area are shown on the horizontal and vertical axes, respectively. The size of the bubbles represents the total energy use for building operation in specific countries. The bubble chart demonstrates that total energy use for building operation in China was similar to that in the US, although the energy use intensity remained at a lower level. The energy consumption per capita and per floor area in China were far lower than those in the US, Canada, Europe, Japan, and South Korea. The equivalent electricity

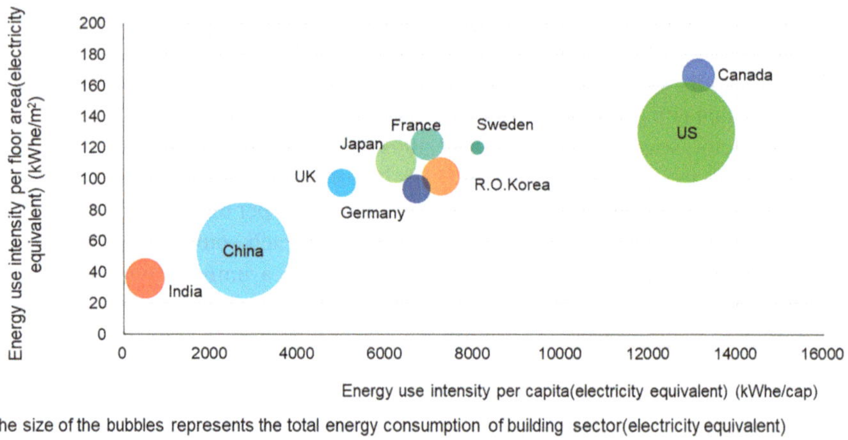

Fig. 2.3 Comparison of building operation energy consumption in different countries (electricity equivalent method). Source: CBEEM of BERC, Tsinghua University; World Energy Balances, Energy Efficiency Indicators database (2023 edition) of IEA; WDI database of the World Bank; Satish Kumar (2019) of India

consumption per capita for building operation was about one-fifth of that in the US and Canada and about half that in Japan and South Korea. The equivalent electricity consumption per floor area for building operation was one-third of that in Canada and half that in the US, Europe, Japan, and South Korea. In the context of tackling climate change and reducing carbon emissions, most countries are carrying out energy transformations, including promoting electrification in the building sector and replacing fossil fuel-based energy with renewable electricity. China needs to develop a different route from developed countries to achieve the target for low carbon emission and energy-saving in the building sector. This would pose a significant challenge to China's low-carbon and sustainable development in the building sector. Meanwhile, many developing countries are experiencing rapid changes in building energy use. China's building energy development pathway will serve as an important reference for many countries' choices, which will further influence global building energy development.

Comparing the proportion of electricity in the final energy consumption of the construction sector across various countries from 2002 to 2022, as shown in Fig. 2.4. Sweden, the United States, Japan, and Canada have consistently maintained a relatively high level of electrification in the construction sector, exceeding 40% since the beginning of this century and still maintaining a steady growth trend. France and South Korea have experienced rapid growth in electrification rates, rapidly increasing from around 30% in 2001 to now surpassing 40%. The electrification rates in the construction sectors of the United Kingdom and Germany have remained relatively stable, with slow growth mainly due to these two countries still retaining a certain

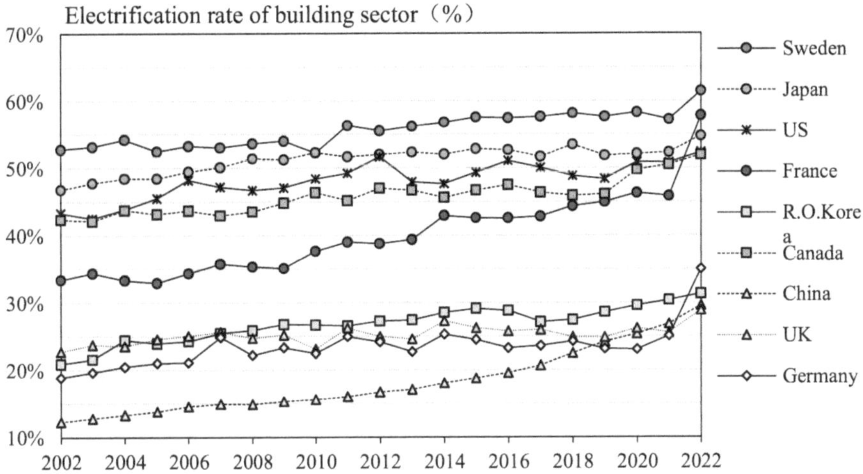

Fig. 2.4 The electrification rate of energy use in the construction sector by country (2001–2022)

proportion of fossil energy for building heating. China's electrification rate in the construction sector has rapidly increased from 18% in 2002 to 43% in 2022, surpassing both the United Kingdom and Germany, and is currently in a phase of rapid growth.

2.2.2 *Carbon Emissions from Building Operation in Different Countries*

The per capita total carbon emissions and the proportion of carbon emissions from the construction sector by country are shown in Fig. 2.5. From the figure, it can be seen that China's current per capita total carbon emissions (including industrial, construction, transportation, and electricity sectors) are slightly higher than the global average but still lower than countries like the United States and Canada. In terms of per capita carbon emissions from building operations, it is also slightly higher than the global average but significantly lower than developed countries. This is mainly because China is still in the process of industrialization and urbanization, and the proportion of carbon emissions from building operations to the total societal carbon emissions is still lower than that of developed countries. In recent years, China's pressure to address climate change has been increasing, and the construction sector also needs to achieve low-carbon development and peak emissions as soon as possible. How to achieve this goal is another major challenge for the development of the construction sector.

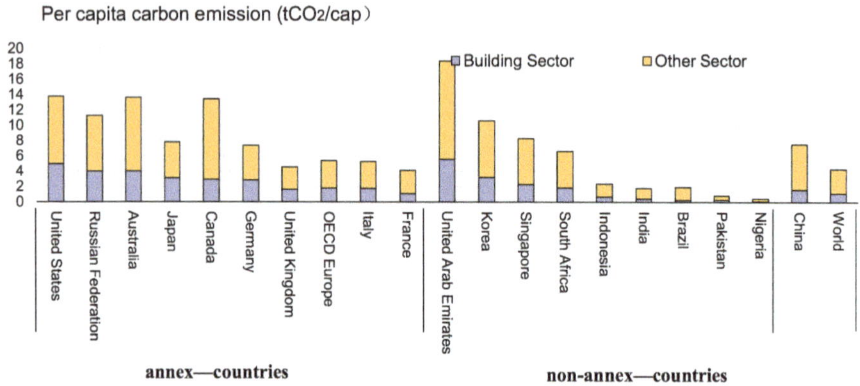

Fig. 2.5 Per capita carbon emissions comparison by country (2022)

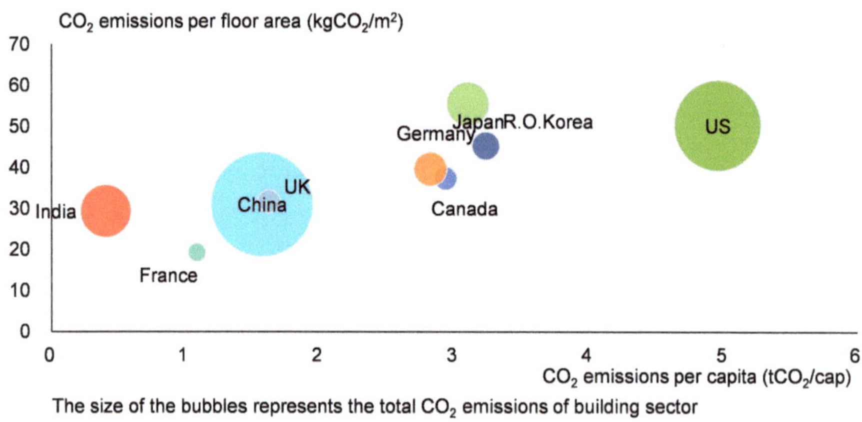

Fig. 2.6 Comparison of per capita carbon emissions of different countries (2022). (Source: Data of countries in 2022 as provided in the CO_2 Emissions from the Fuel Combustion Highlights 2024 database, IEA. Data from China are the results of CBEEM of BERC, Tsinghua University)

Several countries have set their own goals for achieving carbon neutrality and their paths to achieving carbon neutrality in the building sector. Reducing carbon emissions in the building sector is also one of the important fields to realize carbon neutrality in the whole society. Figure 2.6 shows the total carbon emissions from building operation (bubble chart area), per capita carbon emissions (horizontal axis), and carbon emissions per unit building floor area (vertical axis) of different countries, which are converted according to the energy structures of these countries. The bubble chart of carbon emissions demonstrates that carbon emissions in the building sector are affected not only by the total energy consumption but also by the energy structures of these countries. The per capita carbon emissions and carbon emissions per unit floor area from building operation in China are lower than those in most developed countries due to China's low building operation energy consumption. However, the energy structure of France is dominated by low-carbon nuclear power; although it has higher building energy use intensity than China does, its carbon intensity is lower than that of China. This also shows that the low-carbon transition of both energy systems and building energy use structure should be achieved in addition to the improvement of energy saving and energy efficiency of buildings on the path toward carbon neutrality.

In addition to the comparative analysis of carbon emissions from building operations in various countries, analyzing the trends in carbon emissions is also crucial. Figure 2.7 compares the trends in carbon emissions from building operations from 2002 to 2022. The dashed circles represent carbon emissions data from 2002, while the solid circles represent data from 2022. This figure reflects the changes in per capita carbon emissions and per unit area carbon emissions from building operations in various countries over the past 20 years. Based on these trends, these countries can be classified into three categories. (1) The first category includes countries such as the United States, Canada, Germany, the United Kingdom, and France. A common characteristic among these countries is that the total carbon emissions, per capita emissions, and per unit area emissions from building operations have all shown a downward trend. This is partly due to the reduction in per capita and per unit area energy consumption and also because these countries have actively promoted the transformation of their energy structures, vigorously developing zero-carbon electricity. (2) The second category includes countries like South Korea and Japan. Over the past 20 years, their total carbon emissions have increased, as have their per capita emissions, while per unit area emissions have decreased. The analysis suggests that although the total carbon emissions from building operations in

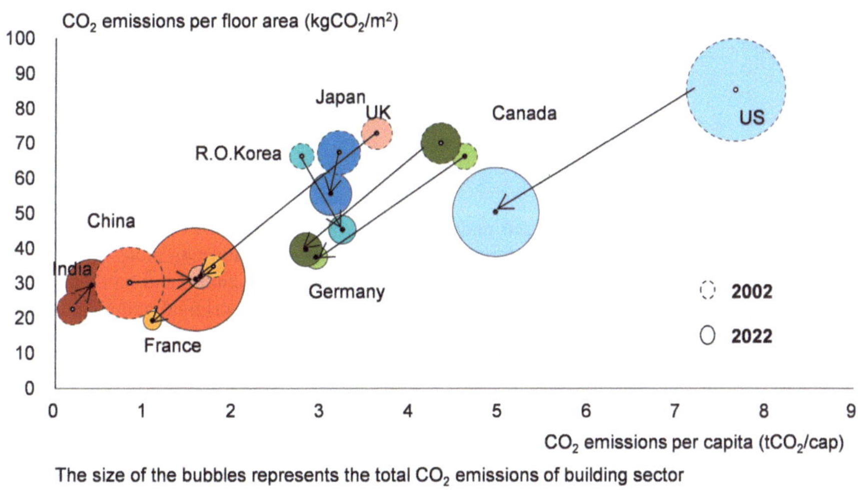

Fig. 2.7 Comparison of the trends in carbon emissions from building operations between China and Foreign Countries (2002 and 2022). (Source: Data of countries in 2021 as provided in the CO_2 Emissions from the Fuel Combustion Highlights 2024 database, IEA. Data from China are the results of CBEEM of BERC, Tsinghua University)

Japan and South Korea have grown slowly in recent years, their population growth rates have been extremely low. Japan's population has even shown negative growth. The growth rate of the population is lower than that of carbon emissions, while the growth rate of building area is higher than that of carbon emissions, leading to inconsistent trends in per capita and per unit area carbon emissions in these two countries. (3) The third category includes countries like China and India. Over the past 20 years, total carbon emissions, per capita emissions, and per unit area emissions have all shown an increasing trend. China and India have both been in a period of rapid development over the past 20 years, with increasing energy intensity. In order to achieve carbon peaking as soon as possible, developing countries such as China and India should simultaneously control the total energy consumption while vigorously promoting the low-carbon transformation of their energy systems.

Chapter 3
Current Energy Use in Urban Residential Buildings

3.1 Current Status and Development of Urban Residential Buildings

3.1.1 Urban Residential Buildings

The urban residential buildings discussed in this book refer to those located in urban and town areas. Urban areas denote districts within municipal jurisdictions or cities without district divisions, including residents' committees and other zones where the urban construction of district or municipal government administrative centers extends. Town areas refer to county government administrative centers outside urban areas, as well as other towns, encompassing residents' committees and other zones where the urban construction of town government administrative centers extends. The energy use of residential buildings refers to the energy consumed by residents from air conditioning, heating (i.e., distributed heating, excluding central heating in northern urban areas), cooking, DHW, lighting, and household appliances to meet the demands of living, studying, and resting. The main energy sources are electricity and gas. With the rapid adoption of electric vehicles (EVs) in recent years, EV charging points in urban residential compounds have been integrated into residential building distribution systems. This trend has led to the inclusion of EV charging point analysis in this book.

China is undergoing rapid urbanization, and the urban population is growing by leaps and bounds. The population has been growing by 16 million in urban areas annually. From 2000 to 2023, the urban population nearly doubled from 459 to 933 million. The traditional family structures in China have also changed amid urbanization and social and economic development. Traditionally, a typical Chinese family consists of a husband and a wife with their offspring, and it is commonplace to see three to four generations, sometimes even five generations of family members living under the same roof. After China's reform and opening up, traditional extended families gradually transformed into smaller nuclear families to better suit

Building Energy Research Center of Tsinghua University, *Decarbonizing Urban Residential Buildings*, https://doi.org/10.1007/978-981-95-8200-6_3

modern lifestyles. A smaller, simpler, and more diverse family structure has become the main feature of a typical Chinese household since then. According to the *China Population Census Yearbook*, the number of people per Chinese household dropped from an average of 3.11 in 2000 to 2.61 in 2020.

With the rapid development of urbanization, many residential buildings have risen up in urban areas to meet the demand for accommodation from new residents in cities. Between 1990 and 2000, most of the residential buildings in China were multistoried with small floor space, around 60–70 m^2, and the highest building was below seven floors. This was one way to meet the huge demand for accommodation during urbanization at that time. The economic development and improvement in people's living standards have also led to an increasing proportion of large-unit-area dwellings in newly constructed residential buildings. Between 2000 and 2010, new residential buildings shifted focus from fulfilling basic housing needs to offering medium-sized apartments in mid- and high-rise buildings. During this period, most of the apartments in the buildings were approximately 80–90 m^2, with 60–70 m^2 of floor space. More high rises can be seen during this period. In recent years, rapid urbanization in China has led to fast urban population growth. On the other hand, as urban land resources have been declining, the cost of comprehensive land development has increased. To improve returns, developers have built numerous high-rise, high-density residential buildings. In recent years, driven by multiple factors, high buildings have dominated urban residential buildings in China. In the meantime, the plot ratio and the height of buildings are increasing as well.

China's housing system has undergone over 20 years of reform, and the urban residential pattern has also shifted fundamentally. In the early stages of reform and opening up, the implementation of the Danwei (work unit, mainly SOEs) welfare-oriented public housing distribution system gave rise to a "work unit compound" model characterized by the integration of working and living spaces. This model, with work units uniformly managing housing and providing public services, fostered a stable acquaintance-based social governance pattern despite limited physical conditions. This pattern was radically changed through the housing commercialization reform in 1998. Since then, commodity housing compounds have gradually become the mainstream settlement pattern. As of 2023, about 70%–85% of urban residents had been living in residential compounds furnished with property services, and the per capita housing area in urban areas surged from 15.5 m^2 in 2000 to 35.5 m^2, reaching the level of developed countries. Current urban habitable spaces exhibit two prominent characteristics. In terms of physical form, verticalization and high-density configurations have emerged as defining features of new residential buildings. Beijing's Tiantongyuan and some other super-large communities can accommodate a population of up to one million. Such gated communities have their spatial boundaries defined by fences or green spaces and contain internally shared facilities such as roads and clubhouses for owners, forming an "externally enclosed, internally communal" space structure. Regarding governance structure, the privatization of housing has transformed the roles of residents from work unit welfare recipients to property owners, leading to the decentralization of governance authority to property companies and other market entities. However, incomplete supporting mechanisms in the institutional transition period have resulted in recurring conflicts, including property rights disputes between developers and owners, controversies over property service quality, and the absence of resident self-governance mechanisms. A case in Beijing, where the average annual elevator maintenance cost in a compound exceeded RMB one million, has reflected the complexity of communal facility management. The new residential pattern has presented multiple governance challenges. For facility management, issues such as elevator maintenance, fire-fighting access occupation, and parking space distribution have come into sharp relief. From the perspective of the relationship between neighbors, high-rise residential buildings have eroded traditional acquaintance-based social ties. In the meantime, national initiatives for building energy efficiency retrofitting require consensus from over 90% of owners, which often leads to a deadlock in decision-making on public interests. In the face of challenges in the latter half of urbanization, China has fully shifted the focus of housing development from new housing construction to stock management and renewal. China's current urban residential building stock, spanning 33.1 billion m^2, represents not only the main arena for urban governance but also an important vehicle for transitioning from "quantity-driven housing supply" to "quality-oriented dwellings." Therefore, attention should be directed to the emerging community governance issues resulting from the compound-based residential pattern. Further discussion on this is provided in Chap. 8 of this Report.

The main types of units for existing residential buildings in China are small- and medium-sized apartments with 60–80 m^2 and 80–100 m^2/household, respectively. According to the national demographic census in 2015, the average space per household in urban areas was 92 m^2, and the median was 80 m^2. The unit space per household for urban areas was smaller than that of town areas and rural areas, which were 118 m^2 and 119 m^2, respectively. The number of floors for China's residential buildings has been increasing, which is a reflection of increasing contradiction among land resources, demographics and ecology, and the environment. With the rapid growth of China's economy, urbanization has been greatly enhanced, which leads to a higher concentration of population in cities. On the other hand, people's livelihood has been improving, fueling an aspiration for better housing conditions. Therefore, demand surges in both quantity and quality. However, most of the cities in China are experiencing the crisis of resource shortages posed by rising populations and diminishing land resources, and ecological challenges. Admittedly, populous cities with high building density will remain common for years to come in China, which is an inevitable reality we must accept. In light of energy and resource constraints, the unit space of residential buildings in China should be properly contained, and the volume of urban residential buildings needs to be controlled.

Amid urbanization, the total floor area of urban residential buildings increased by nearly fourfold from 2001 to 2023. In 2023, the total floor area of urban residential buildings was 33.1 billion m^2. The living standards for urban residents have been greatly enhanced as well. The per capita residential floor area (total floor area of urban residential buildings divided by total urban population) in China increased from less than 20 m^2 in 2001 to 35.5 m^2 in 2023. According to the *China Statistical*

Yearbook and the *China Population Census Yearbook*, the per capita floor area per household for urban residential buildings increased from 22.4 m^2 in 2000 to 38.6 m^2 in 2020. The source of this data is the result obtained from a large-scale sampling survey of urban household families across the country. This figure does not include students, military personnel, and other groups who are not homeowners. This data evidence the improvement of the living standards of urban residents.

The recent development of urbanization in China is gradually meeting the demand of urban residents. According to a 2017 research on Chinese household financial status by Survey and Research Center for China Household Finance, Southwestern University of Finance and Economics, the urban household housing possession rate (the share of households having dwellings among all households) in 2017 was 90.2%, and the urban household homeownership rate (the share of households that own the homes they live in among all households) was 80.8%, which was leading the world. In 2017, the ratio of complete urban residential housing to the total number of households was 1.18, and the ratio of self-owned complete housing to total households was 1.155, meaning that there was at least one house per household. In recent years, vacant housing units in urban areas have become a phenomenon that has attracted much attention. The vacancy rate of second- and third-tier cities is higher than that of first-tier cities. There are two kinds of vacant housing units. The first refers to housing units that have been completed but remain unsold. These housing units are the subject of investigation by the Ministry of Housing and Urban-Rural Development (MHURD), and their vacancy rate is of concern for financial risks and security in the recovery of bank credit funds. The second kind of vacant house refers to the empty housing units among all that have been sold. The vacancy rate of these homes reflects housing stock utilization and is our current primary concern. This phenomenon can be attributed to two main factors: first, individuals who own a single home but have migrated elsewhere for work; and second, those who own multiple housing units but choose not to rent them out. China has yet to provide official data on the definition of this vacancy rate. According to a study in 2017 on Chinese household financial status by the Survey and Research Center for China Household Finance, Southwestern University of Finance and Economics, the vacancy rate of urban residential buildings was 18.4% in 2011, 19.5% in 2013, 20.6% in 2015, and 21.4% in 2017. It can be extrapolated that there were 65 million empty urban residential housing units in 2017. Based on related studies, the natural vacancy rate of China's urban residential buildings was 9.8%. The existing vacancy rate of housing units in China is higher than the natural vacancy rate, and it is even higher in second- and third-tier cities. The large number of vacant housing units has consumed a large amount of housing home loans and dampened the desire for consumption. Meanwhile, it also leads to increasing energy consumption for building material production, construction, decoration, and unnecessary household maintenance (e.g., basic water, electricity, and heating device maintenance). These are the problems we need to tackle to reach carbon peaking and carbon neutrality.

According to the comparison of per capita floor area between China and other countries in Chap. 1, the per capita floor area in China is close to the level of developed countries. This also means that urban residential buildings largely meet the demand of their residents, and the main contradiction in urban living has shifted from an overall shortage to an imbalance in distribution. As urbanization develops, there will be more than one billion Chinese people living in urban areas. Based on 35 m^2/cap of floor area, the total floor area required in the future will not exceed 38 billion m^2. This means that the construction of urban residential buildings has been largely completed in China. Next, the focus of the building sector will be fully shifted from the construction of new buildings to the energy-saving and low-carbon operation of existing buildings and the energy-saving retrofitting and functional upgrading of old buildings.

3.1.2 Energy Use in Urban Residential Buildings

The main energy sources for urban residential buildings in China are electricity and natural gas. With economic and social development and the improvement of living standards in urban areas, electrical appliances are becoming increasingly popular in urban households, and electronic information equipment and electric cookware have also emerged in recent years. The demand for air conditioning in the summer and heating in the winter has greatly increased at the same time, leading to a significant increase in total energy and electricity consumption in urban residential buildings. With the significant rise of electrification in recent years, the share of electricity in primary energy consumption in urban residential buildings in China increased from 63% in 2001 to approximately 70% in 2023. In 2023, the total electricity consumed by urban households nationwide reached 673.1 TWh, accounting for 7.3% of the total electricity consumed. The largest driving force behind the total electricity consumption of urban households in China is the increase in the urbanization rate. The increases in both the average and the per capita electricity consumption of urban households are relatively flat. From 2001 to 2023, the average and per capita electricity consumption of China's urban residential households increased by approximately threefold. However, the average electricity consumption of urban residential households in China was still low compared to the household electricity consumption in developed countries in Europe and the United States, and registered around 1937 kWh/year, as shown in Fig. 3.1. The ownership rate of household appliances such as freezers, electric ovens, and dishwashers, which are commonly used in Europe and the United States, remains relatively low in Chinese households but has been demonstrating accelerated growth in recent years. Therefore, the overall level of electricity consumption in China is still lower than that in Europe and the United States.

Chinese people have been exercising frugality since time immemorial. Small and medium-sized apartment buildings have always been the main type of residential building in China due to land resource limitations. People typically use energy and services only when necessary, turning them off to conserve resources when not in use. Under the current economic environment in China, the Chinese people can afford an American lifestyle with enough disposable income in their pockets.

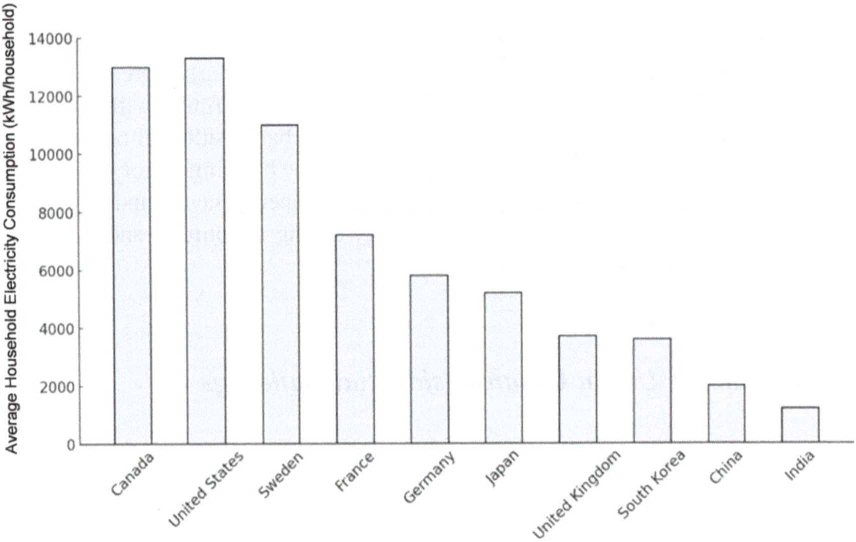

Fig. 3.1 Comparison of household electricity consumption around the world (2022). (Data Source: IEA Database)

However, most Chinese households are maintaining the traditional way of life with low energy consumption, even though they have earned more money. Therefore, it can be concluded that improvement in energy consumption is not limited by the level of their income, but by a traditional way of life and behavior inherited from the past.

In urban residential buildings, there is still room for energy efficiency improvement and energy consumption reduction for all kinds of new household appliances, especially for appliances with a standby mode, such as water dispensers, electric toilet seats, set-top boxes, and other low-power appliances. Together with the amount of electricity consumption during its standby cycle, some of these appliances consume similar amounts of electricity comparable to recognized high-energy-consuming appliances such as refrigerators. It remains an important task to introduce energy-efficient standards targeting these kinds of appliances, as well as new technologies, energy-saving modes, and smart control mechanisms to lower electricity consumption during their standby cycle. On the other hand, subsidizing policies should not cover these high-energy-consuming appliances. Instead, green lifestyle and energy-saving behavior should be promoted among the public to avoid the surge in electricity consumption after the growing popularity of these appliances.

Based on the improving living standard of urban households and the electrification of cooking, heating, and DHW supply, the annual electricity consumption of urban households in the future would be 3600–4200 kWh, which is enough for a happy life in China. According to our study, 3600 kWh/(hh a) of annual electricity consumption is sufficient to support a family of three with high living standards in a 100 m^2 house with fully electrified appliances, as shown in

Table 3.1 Electricity consumption of urban households

Electricity consumption	Family of three members	Measured data	Family information
Annual electricity consumption			
Family without heating	**3600**	**2250**	A family of five members
Family with heating	**4200**		
Category			
空调 **Air conditioner**	**700**	**110**	A Beijing family, three air-cons, part time part space
Electrification of heating in HSCW zones	**800**	**400**	Average 3~5 kWh_e/m^2 in HSCW zones
Electrification of DHW	**600**	**710**	A Beijing family of five members, consumed 710 kWh of power, which was equivalent to 426 kWh of electricity consumption for a family of three
Electrification of cooking	**600**	**114 m^3 natural gas**	A Beijing family
Lighting	**300**	**427**	A Beijing family
Electric appliances	**1400**	**1003**	
Refrigerator	200	130	200 L, level 1 energy efficiency standard
2 TVs	300	263	
PC and entertainment facility	300	300	2 PCs
Electric rice cooker	160	70	Level 1 energy efficiency standard
Kitchen range hood + exhaust fan	40	20	
Microwave oven and other electric cooking devices	100	73	
Washing machine	200	90	Drum type, with level 1 energy efficiency standard
Water dispenser or electric water heater	100	57	

Table 3.1. While the real-world data in the third column in Table 3.1 came from a professor's family at Tsinghua University, the average electricity consumption was merely 2250 kWh/(hh a), lower than the target. The annual electricity consumption would be 4200 kWh if we include an additional 800 kWh/(hh a) of electric heat pump in hot summer and cold winter (HSCW) zones, which was calculated from doubling the 4 kWh/m^2 average in the same year. Therefore, 4000 kWh/(hh a) of average electricity consumption can fully support all Chinese urban households to lead a relatively high standard of life. Given that electrification will be the future of urban household development in China, in which DHW

supply and cooking will be powered by electricity, the energy demand for urban residential buildings will require just 1.7 PWh annually for 420 million households, or one billion residents.

To achieve low-carbon and sustainable development for urban households, considering current energy consumption patterns across sectors, the main tasks for energy conservation in residential areas include:

1. Scope control: Reasonably plan the amount of floor area of residential housing and control the per-unit area of each household. The per capita floor area and the aggregate urban residential building floor area should be contained within 35 m^2 and 38 billion m^2, respectively. At present, China has basically reached the aggregate and per capita levels. Next, the focus should be put on the imbalance in regional and population distribution, as well as the renovation of old residential compounds during urban renewal.
2. Lifestyle: Promote and maintain a green lifestyle and the principle of frugality. Promote "partial-time, partial-space" to avoid "full-time, full-space" due to the limit of building types, energy systems, and service modes.
3. To promote electrification for cooking, DHW supply, and heating in HSCW zones. The goal of zero-carbon emissions can be achieved when there is full electrification in urban households, and renewables are the mainstay for our power grid.
4. To find a comfortable lifestyle based on the building type and renew our effort to promote residential buildings with windows and allow natural ventilation. A large area centralized system for heating, air conditioning, and DHW should be avoided in HSCW zones. The existing decentralized system should be promoted while improving its terminal flexibility, adjustability, and energy efficiency to avoid electricity spikes while improving service levels.
5. For household appliances, the key task is to improve energy efficiency while improving manufacturing standards for appliances with long heating and standby cycles, such as small electric water devices, smart toilet seats, water coolers, and other high-energy appliances. For instance, we could enhance the controllability of the set-top boxes and improve the insulation capacity of water dispensers to avoid wasting energy during standby. Policy incentives or subsidies should not cover electric appliances such as clothes dryers, which may result in changes in lifestyle, and we should watch for the energy spikes of these high-energy consumption appliances.
6. Policy mechanisms and measures: Further improve and implement energy-efficient policy standards and mechanisms, such as the *Standard for Energy Consumption of Building*, multistep electricity price, minimum energy performance standard for all types of household appliances, labeling system for energy-efficient appliances, etc. A market-oriented approach should be adopted to guide the public towards an energy-saving lifestyle, encouraging sustainable behavior and ensuring energy conservation becomes a universal practice.

3.2 Electricity Use in Urban Residential Buildings

1. Total amount and distribution of electricity consumption

   To fully understand the status of electricity consumption in urban residential buildings, a survey was conducted in 2023, covering 173,891 households across 1475 urban residential buildings distributed throughout 16 districts of Beijing. The electricity consumption data was obtained directly from the smart electricity meter marketing data of surveyed households and subsequently analyzed in combination with the basic information of these households, such as floor area and number of family members.

   Figure 3.2 shows the electricity consumption distribution per unit area in the surveyed urban residential buildings. The data reveals a concentrated pattern, with 95.4% of buildings falling within the 14–38 kWh/m^2 range. The average consumption stands at 28.06 kWh/m^2.

   Figure 3.3 displays the distribution of average electricity consumption among surveyed urban residential households. The data reveals a wide dispersion in consumption levels, with 95.4% of households falling between 310 and 7256 kWh annually. The mean consumption is 2895.93 kWh/household.

   Based on data analysis of surveyed samples and the national electricity carbon emission factor, urban residential buildings emit 15.96 kgCO_2/m^2 of electricity-related carbon emissions, with an average of 1647.20 kgCO_2/household.

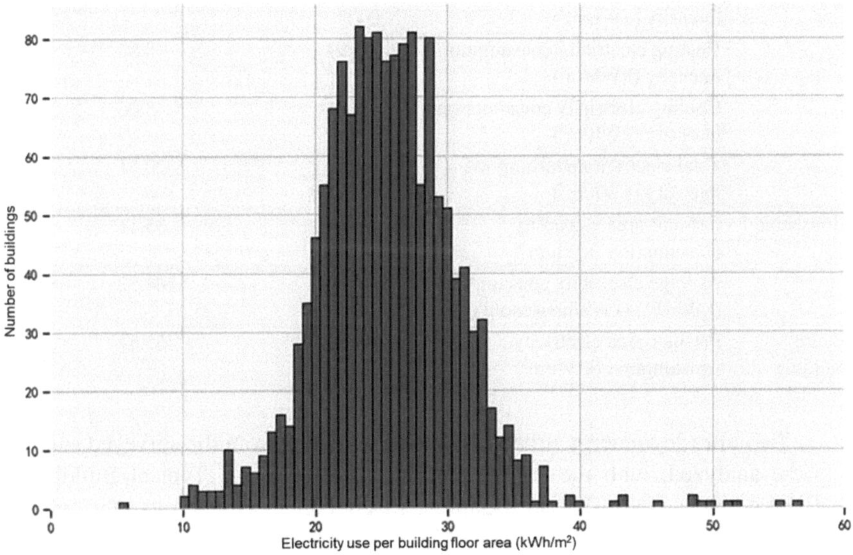

Fig. 3.2 Distribution of electricity consumption per unit area of surveyed urban residential buildings

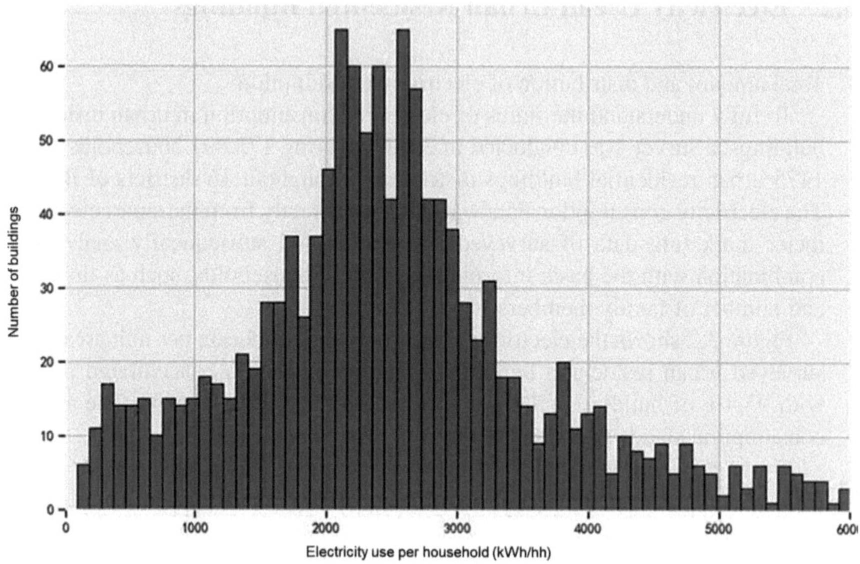

Fig. 3.3 Distribution of average electricity consumption of urban residential households surveyed

Table 3.2 Energy consumption intensity of typical ultra-low-energy buildings surveyed

		Typical ultra-low-energy building 1	Typical ultra-low-energy building 2
Public area	Lighting electricity consumption intensity (kWh/m²)	8.85	1.55
	Heating electricity consumption intensity (kWh/m²)	6.94	8.86
	Cooling electricity consumption intensity (kWh/m²)	6.45	8.06
	Total electricity consumption intensity (kWh/m²)	41.6	20.61
Household	Per-unit-area electricity consumption (kWh/m²)	57.64	85.41
	Average electricity consumption (kWh/户) (kWh/household)	2690.06	4504.52
Entire building	Per-unit-area electricity consumption (kWh/m²)	49.32	55.87

Two ultra-low-energy urban residential buildings from the surveyed samples were analyzed, with the results presented in Table 3.2. Typical Building 1, located in Chaoyang District, Beijing, had a public area electricity consumption of 41.60 kWh/m^2, comprising 8.85 kWh/m^2 for lighting, 6.94 kWh/m^2 for heating (January–April and November–December), and 6.45 kWh/m^2 for cooling (May–October). A survey of 15 households in this building provided data on the indoor floor area and electricity consumption. The average floor area per household was 46.67 m^2, with an average electricity consumption of 2690.06 kWh/household, translating to 57.64 kWh/m^2. Factoring in the number of occupied households and the building's overall conditions, the total electricity consumption per unit area was calculated at 49.32 kWh/m^2, precisely meeting the ultra-low-energy building design threshold ($\leq$50 kWh/m^2).

Typical Building 2 is also located in Chaoyang District, Beijing. Its public area electricity consumption was 20.61 kWh/m^2, including 1.55 kWh/m^2 for lighting, 8.86 kWh/m^2 for heating (January–April and November–December), and 8.06 kWh/m^2 for cooling (May–October). A survey of 15 households in the building revealed an average floor area of 52.74 m^2/household, with average electricity consumption reaching 4504.52 kWh/household, equivalent to 85.41 kWh/m^2. Based on the occupancy rate and overall building conditions, the total electricity consumption per unit area was calculated at 55.87 kWh/m^2, exceeding the ultra-low-energy building design threshold of 50 kWh/m^2. As revealed in the comparison above, the actual per-unit-area electricity consumption intensity of the surveyed typical ultra-low-energy residential buildings still exceeds the specified threshold. From the perspective of public area electricity consumption, property management and other responsible entities still need to improve their operational management to lower the electricity consumption for lighting, cooling, and heating. In terms of household electricity consumption, it is essential to provide residents with more energy-saving guidance and implement incentive programs to promote electricity conservation. This can be achieved through upgrading household appliances to higher energy efficiency levels and encouraging power-saving behaviors among users, thereby further advancing residential electricity conservation objectives.

2. Analysis of electricity consumption compared to international levels

To delve into and compare the differences in electricity consumption distribution between urban residential buildings in China and other countries, the urban residential electricity consumption data of six cities in Jiangsu surveyed in 2014 were used, and residents in the six cities were divided into ten groups by their average annual electricity consumption from lowest to highest. The average electricity consumption for each group was calculated. The data was compared with the national average annual household electricity consumption of Japan, South Korea, the United States, and France. It can be observed in Fig. 3.4 that the overall residential electricity consumption in China is significantly lower than that of Japan, South Korea, France, and the United States. However, for the top 10% of users (Group 10) with the highest electricity consumption in China, their average household usage has reached levels comparable to those of Japan, South Korea, Europe, and the United States. The development of these high-energy-consumption users is noteworthy.

The wide variety of household appliances and differing usage habits among households and users lead to significant imbalances and variations in urban residential electricity consumption. Meanwhile, as household incomes rise, new types of appliances continue to emerge alongside evolving lifestyles. These

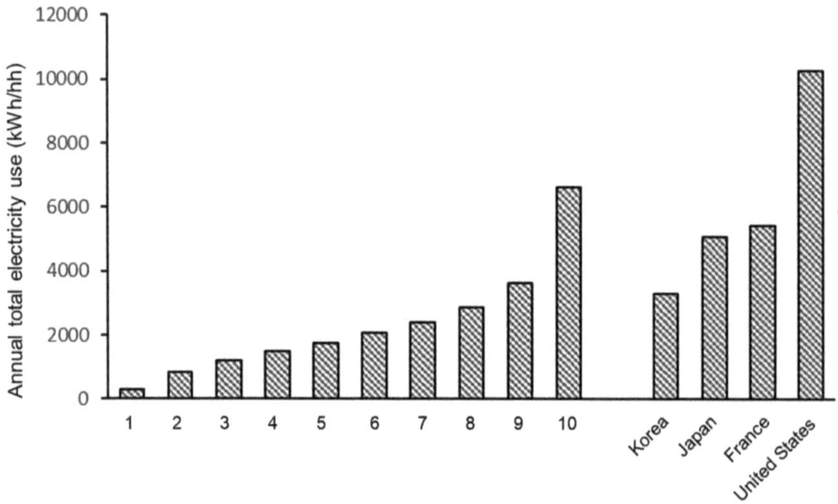

Fig. 3.4 Comparison of average annual household electricity consumption across percentile groups for residential buildings in six Jiangsu cities versus other countries

appliances drive diverse power usage patterns and substantially influence urban residential electricity demand.

3. Difference in electricity consumption mode

The primary factor behind variations in household electricity consumption lies in differing lifestyles. Key contributors to higher usage include diverse appliance operation habits, prolonged standby modes, and varying approaches to air conditioning and heating system utilization.

Household appliances come in diverse types, and homeowners' usage patterns vary accordingly. With rising living standards, urban households are increasingly adopting new appliance categories, driving shifts in lifestyles. Beyond essentials like TVs, refrigerators, air conditioners, and washing machines, energy-intensive appliances such as freezers, dishwashers, dish sterilizers, clothes dryers, water dispensers, and smart toilet seats are becoming more prevalent. These emerging appliances contribute to distinct electricity consumption behaviors and substantially influence urban residential power usage.

To explore the reason for electricity consumption differences among various households, two typical families were selected to represent households with high and low electricity consumption. The electric power of frequently used appliances and the lifestyle of the two households were surveyed. The results show that the family with high electricity consumption, which was located in Changsha, Hunan Province, consumed 6995 kWh of electricity in 2019. The family with low electricity consumption was from Beijing, and they consumed 999 kWh of electricity in 2019. The basic information about the two households is shown in Table 3.3.

Based on the electrical consumption test data and surveys of household lifestyles, the annual total electricity consumption of major appliances for two households is summarized in Fig. 3.5 below.

Table 3.3 Basic information of the selected families

		Family A	Family B
Annual energy consumption		6995 kWh	999 kWh
Family location		Changsha, Hunan Province	Beijing
Building floor area		100 m^2	160 m^2
Family members		2 middle-aged adults +1 student	2 middle-aged adults +1 student
Frequently used Electric appliances	Air conditioner	Split air conditioner*3	Split air conditioner*3
	Heating	Electric heater	District heating
	DHW	Gas water heater + small electric water heating device	Gas water heater + centralized domestic hot water supply
	Other electric appliances	Refrigerator, washing machine, projector, stereo, treadmill, smart toilet seat, sweeping robot, range hood, electric rice cooker, etc.	Refrigerator, washing machine, television, stereo, range hood, electric rice cooker, electric kettle, etc.

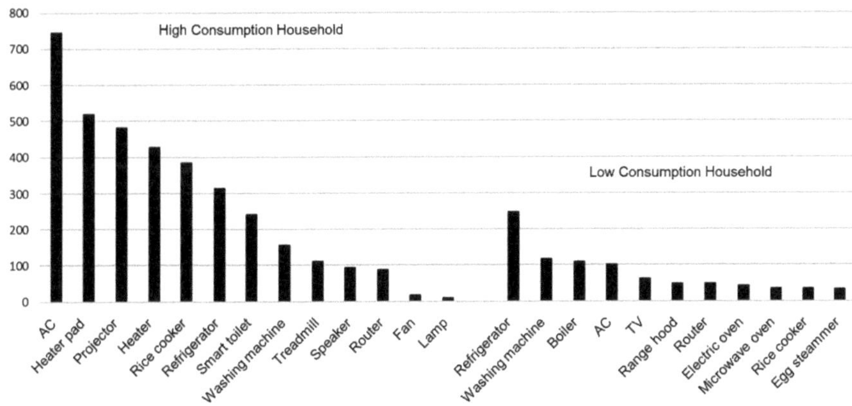

Fig. 3.5 Annual electricity consumption of key appliances in two families (kWh)

By comparing the two families, it can be found that air conditioners consume 745 kWh of electricity per annum in the family with high energy consumption, which is the largest energy consumer among all appliances in this family. In a family with low electricity consumption in Beijing, air conditioners consumed only 100 kWh/year. In addition to the issue of climate differences in different regions, one of the key contributors to power use differences is the use habits of air conditioners in different families. There were two air conditioners in the two bedrooms of family A, and they were used from June to September, while the air conditioner was not used in family B until the hot weather was unbearable for several weeks in July. In addition, family A was in Changsha, Hunan Province, without a district heating system. Therefore, the electric heater used in family A had an annual electricity consumption of 430 kWh.

Aside from air conditioning, there are also several other appliances in family A that consume more electricity than other families, such as small electric water heating devices, projectors, electric rice cookers, and smart toilet seats. The long standby time of these devices all led to a great waste of energy.

Among all the appliances, the small electric water heating device consumed the highest amount of electricity. The DHW in family A was produced by a gas water heater and a 10 L small electric water heating device. The total monthly electricity consumption was 20 kWh in summer and 60 kWh in winter, and its annual electricity consumption for family A was 518 kWh, which was even higher than their annual electricity consumption for heating.

Additionally, Family A owns other high-energy appliances like projectors, electric rice cookers, and smart toilet seats. Each consumes at least 200 kWh annually, comparable to a refrigerator's energy usage. The following sections examine usage patterns for these high-energy devices.

Family A's combined electricity consumption for the projector and stereo system totaled 481 kWh, comparable to a television's usage. The projector operated at 300 W for 3–5 h daily, consuming 1.3 kWh/day. With an average standby power of 1.3 W, its annual standby consumption reached 11.4 kWh. In contrast, Family B's television consumed only 62 kWh annually, operating at 256 W for an average of 40 min daily. The projector's significantly higher electricity consumption primarily stems from its more frequent use and longer operational duration compared to the television.

The electric rice cooker in family A also consumed a large amount of power, 385 kWh/year, which was 11 times higher than family B's electric rice cooker. A 4 L electric rice cooker for a family of three people usually consumes 0.189 kWh of power for one meal, and the annual electricity consumption would be 140 kWh if the electric rice cooker is used twice a day. According to our test, family A not only used an electric rice cooker for normal meals but also preset the cooker for 3 h from 3:00 am to 6:00 am to cook porridge every day; each time, the cooker consumed 0.68 kWh of power (as shown in Fig. 3.6). The standby power of the electric rice cooker is 7.2 W, so the total standby electricity consumption of the electric rice cooker would be 50 kWh/year. Therefore, the difference in electricity consumption of the same household appliance could be as high as ten times.

The smart toilet seat was another high-energy consumption appliance in family A, which consumed 241 kWh/year, equivalent to refrigerators in the average family. As demonstrated in Fig. 3.7, its electricity usage occurs in two scenarios. In the first scenario, electricity is used to heat the water instantly during cleaning, the heating power is approximately 1300~1600 W, and the cleaning cycle is approximately 1 min. In the second scenario, electricity is used for heating to keep the seat warm during winter. The heating power is between 35~40 W during standby mode, and the average daily electricity consumption during this scenario is approximately 0.84~0.96 kWh. On a typical day, there was 1.03 kWh of power consumed by the smart toilet seat, of which 0.864 kWh of power was used to maintain the

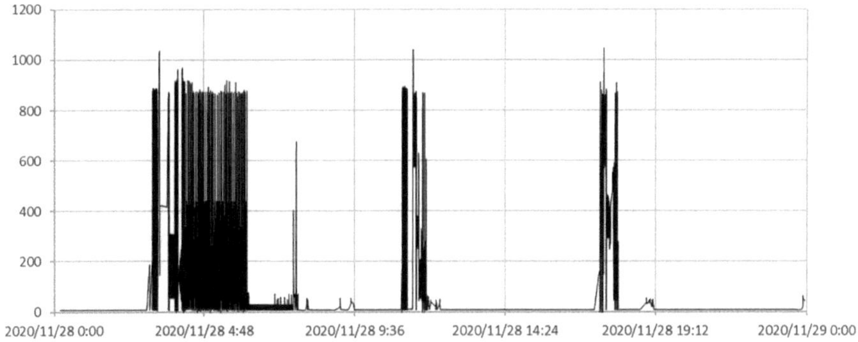

Fig. 3.6 Daily power curve of electric rice cooker in family A (W)

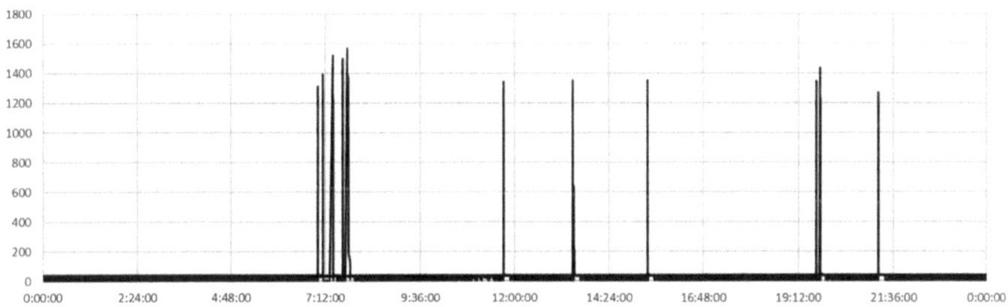

Fig. 3.7 Daily power curve of smart toilet seat (W)

temperature of the seat, accounting for 84% of the total electricity consumption, and 0.166 kWh was used for water heating, accounting for 16% of the total.

Apart from the aforementioned appliances, the annual electricity consumption of the treadmill, stereo, and Wi-Fi routers of family A was also significant, totaling 111 kWh, 94 kWh, and 87 kWh, respectively. The Wi-Fi router operates 24 h/day, and its average operating power was 9.9 W. Both Treadmill and the stereo were used almost every day, and they consumed power even during the standby period. The standby power of the treadmill was 4.2 W, and the annual standby electricity consumption was 34 kWh, which was 30% of the total electricity that the treadmill consumes. The standby power of the stereo was 1.8 W, and its annual standby electricity consumption was 16 kWh, accounting for 17% of the total.

Through an analysis of the electricity consumption composition of this high-energy household, it has been concluded that the main reasons for high electricity usage in home appliances can be categorized into three types, apart from differences in heating and air conditioning usage patterns. The first category includes continuous heating appliances such as electric water heaters, under-sink water dispensers, water dispensers, and smart toilet seats. These devices consume significant amounts of electricity annually due to prolonged standby modes and repeated heating cycles. The second category comprises high-energy-consuming appliances like dryers, dishwashers, sterilizing cabinets, wine coolers, and electric ovens. Although these devices have high energy demands, they are not yet widely adopted in urban Chinese households. Close attention should be paid to their power consumption and corresponding energy efficiency policies. The third category involves standby power consumption from various appliances such as washing machines and televisions. For instance, most appliances in Household A remain on standby when not in use—including projectors, rice cookers, treadmills, and speakers, and their combined standby power adds up to 111 kWh/year. While each appliance's standby consumption may seem negligible, collectively, they become a notable component of household electricity use that cannot be overlooked.

These once uncommon but now increasingly popular household appliances will significantly contribute to rising electricity consumption in urban Chinese homes. Based on their power usage patterns, three types of high-energy appliances deserve attention. Below, we examine each type's current electricity use and potential energy-saving measures:

1. Appliances that require continuous heating: Devices such as electric water heaters, small electric water heating devices, water dispensers, and smart toilet seats operate in constant heating mode to maintain water or surface temperature, leading to significant electricity consumption for heating. The purpose of constant heating is to ensure that the appliance can meet urgent hot water demand in households. According to our research, the "effective" power use for water heating only took a small share of the total amount, and the majority of the power was used to maintain the temperature of the water tank to offset the heat loss. Therefore, it can be regarded as "ineffective" power.

   There are two types of appliances that require continuous heating. The first type of appliance is equipped with a water tank, such as an electric water heater, small electric water heating device, and water cooler. These appliances operate on a constant heating cycle to maintain the temperature of the water tank and stop heating when the set temperature is met. Figure 3.8 shows the power curve of the water cooler during the 2-h standby period. The water tank of the water dispenser was heated every 20 min to sustain the temperature, which consumed 0.015 kWh of power. Based on our calculation, the average ineffective power use of the water dispenser was 46 W, which means that its daily standby electricity consumption was 0.97 kWh, higher than the daily electricity consumption of a refrigerator in an average household. The high electricity consumption of these appliances is mainly caused by poor insulation of the water tank. Therefore, we can conserve more electricity by improving the insulation performance of the tank. Additionally, one of the most popular water heating devices on the market today is the tankless water heater. It would only take approximately less than a minute to heat the water, which is a healthier way to meet hot water demand in households while conserving almost 1 kWh/day of power.

   The second type of appliance is tankless, without insulation. Appliances such as smart toilet seats need to maintain the temperature of the seat at all times, and thermal insulation is unable to offset the heat radiation from the seat. Currently, smart toilet seats on the market are mainly equipped with instantaneous water

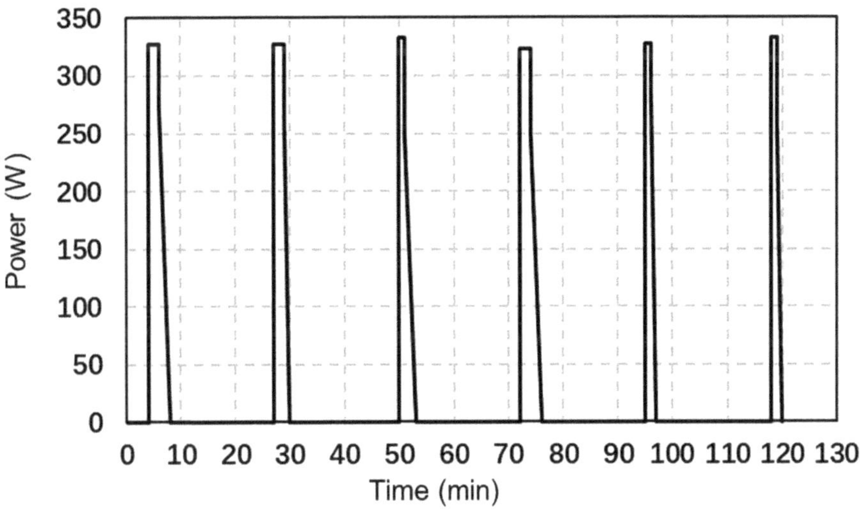

Fig. 3.8 Power curve of the water dispenser for 2 h in a real situation (W)

heating functions due to hygiene, safety, and energy efficiency requirements. The power of the electric water heater is approximately 1600 W, the power of seat heater is in the range of 40~50 W, and the power of the electric dryer is 340 W. Based on the use behavior on the electric toilet seat of family A, the total electricity consumption should be approximately 0.84~0.96 kWh to achieve 24 h of constant heating of the seat, which is 84% of the total power use, and the power consumption is higher than that of the refrigerator. Smart control technology would be useful to reduce the high electricity consumption of these kinds of appliances. Meanwhile, some of the smart toilet seats are equipped with timers, which can be automatically turned on or off according to the user's habits. It can stop heating when no one is home during the day and turn back on when needed. This prevents energy waste from maintaining the toilet seat temperature.

2. High-energy-consuming appliances that are non-essential for daily living: In recent years, as China's economy has grown and people's living standards have risen, urban households are using more home appliances with greater energy consumption each year. In recent years, sales of conventional appliances have stabilized while unconventional ones have grown significantly. For example, sales for kitchen appliances such as microwave ovens, induction cookers, and electric rice cookers are plateauing, while new appliances such as electric ovens and dishwashers have demonstrated fast growth in sales in recent years (Fig. 3.9). Among all household appliances, vacuum cleaners, water purifiers, and air cleaners have maintained steady growth, while sales for smart toilet seats and sweeping robots have surged significantly (Fig. 3.10).

Note: For microwave ovens and dishwashers, sales in 2014 are set at 100; for induction cookers, electric rice cookers, and electric ovens, sales in 2013 are set at 100. For air purifiers and water purifiers, sales in 2014 are set at 100; for

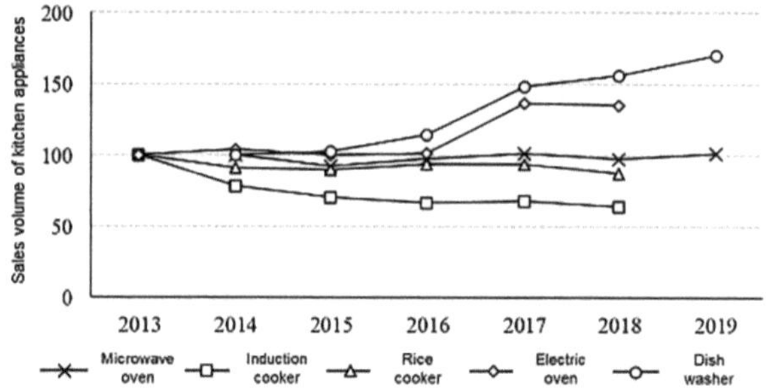

Fig. 3.9 Sales volume of China's kitchen appliances

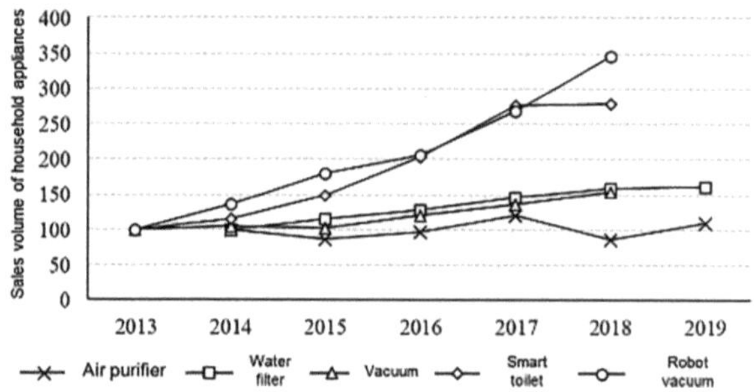

Fig. 3.10 Sales volume of China's household appliances

vacuum cleaners, smart toilets, and sweeping robots, sales in 2013 are set at 100. Data source: chinaiol.com.

Among the newly added appliances, dishwashers, smart toilet seats, wine cabinets, and washing machines with hot water functions are high-energy appliances in households.

In recent years, dishwashers have experienced the fastest sales growth among kitchen appliances. A standard dishwasher operates in three stages: heated washing, disinfection, and drying, with cycle times ranging from 60 to 300 min depending on the selected mode. Currently, most commercial dishwashers rely on high-temperature washing (70–80 °C) for stain removal and sterilization, followed by hot-air drying, which demands substantial energy to generate the required heat. A typical wash cycle consumes 0.64–1.6 kWh of electricity, resulting in an annual energy consumption of 300–500 kWh, classifying dishwashers as high-energy appliances.

A wine cabinet works like a refrigerator, using a compressor to regulate internal temperature and humidity. Typically, these cabinets operate between 4 and 22 °C. Energy consumption depends on their size: standard models holding 30–150 bottles use about 0.3–0.8 kWh/day, while larger units consume roughly as much as refrigerators do.

The electricity consumption of a washing machine with a water heating function is notably higher than that of a normal one. Figure 3.11 demonstrates the power curve of a single washing cycle of a washing machine in a real situation. Its workflow can be divided into three main steps, i.e., heating, washing, and spin-drying. The total electricity consumption for a single washing cycle is

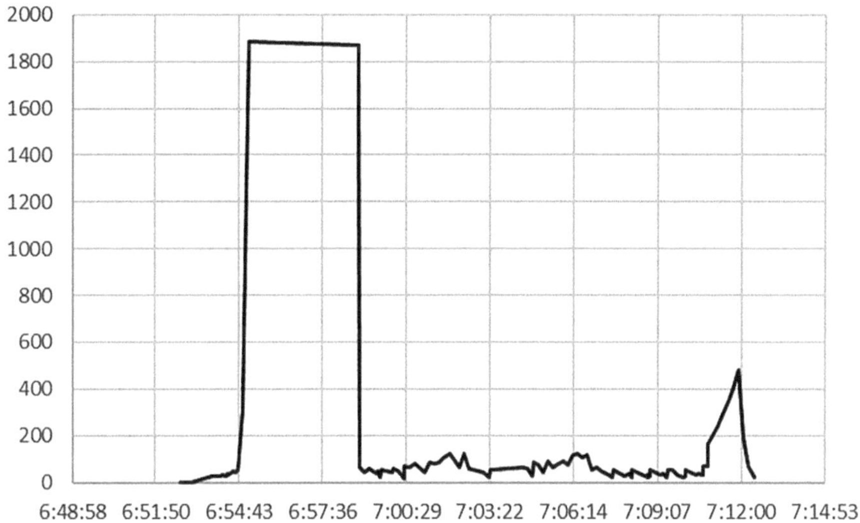

Fig. 3.11 Power curve of a single washing cycle of a washing machine (W)

0.131 kWh. During the 200 s of water heating, the heating power reaches 1900 W, consuming 0.112 kWh of power, which is equivalent to 85% of a full wash cycle's energy usage. As a result, the electricity consumption of a washing machine with a water heating function is five times higher than that of a normal one. Currently, washing machines are equipped with the function to adjust water temperature. Appropriately lowering the hot water temperature or using room-temperature water for laundry can reduce the machine's electricity consumption.

3. Appliances with long standby cycles: For example, the yearly electricity consumption of appliances such as projectors, electric rice cookers, and treadmills during a long standby cycle in family A could be 111 kWh. Apart from refrigerators and Wi-Fi routers, which are switched on at all times in households, there are other appliances with long standby cycles, such as TVs, PC, air conditioners, washing machines, and electric water heaters. The standardized standby power of these kinds of appliances is between 0.5 and 5 W, with 4~40 kWh of electricity consumption per year per single appliance.

From the previous analysis, it can be concluded that there are several main reasons contributing to the high electricity consumption of urban households in China. We need to pay special attention to the electricity consumption of household appliances with a constant heating function. Policies need to be promulgated to enhance the energy efficiency of these appliances. Smart control technologies need to be introduced to avoid unnecessary reheating and heat loss. A green and energy-saving lifestyle needs to be promoted among users to reduce the waste of energy from using these types of appliances. Policy incentives or subsidies should not cover electric appliances such as clothes dryers, which may result in changes in lifestyle, and we should watch for the energy spikes of these high-energy consumption appliances. Energy-efficient technologies of all kinds should be introduced to upgrade appliances with long standby cycles to ensure that the device can be switched to an energy-saving mode during off time to reduce power use. An energy-saving lifestyle should be promoted among the populace to remind people to turn off appliances when unnecessary.

3.3 Natural Gas Use in Urban Residential Buildings

3.3.1 Household Natural Gas Use and Distribution in Urban Residential Buildings

For urban households in China, the demand for natural gas comes mainly from heating, domestic hot water, and cooking. In addition to cooking, the demand for natural gas for heating and domestic hot water varies considerably from household to household and can be divided into four main categories: **gas heating + gas boiled hot water, non-gas heating + gas boiled hot water, gas heating + non-gas boiled hot water, and non-gas heating + non-gas boiled hot water.** Due to the greater energy intensity of heating in northern China, the heating method can have a greater impact on total household gas consumption than domestic hot water and cooking. Therefore, this chapter introduces the research results on the average household natural gas consumption and the distribution in urban areas of Beijing based on different methods of heating. A total of over 100,000 households have been surveyed.

Comparing the gas use of self-heating and non-self-heating households, as shown in Fig. 3.12, it can be seen that the total gas consumption in self-heating

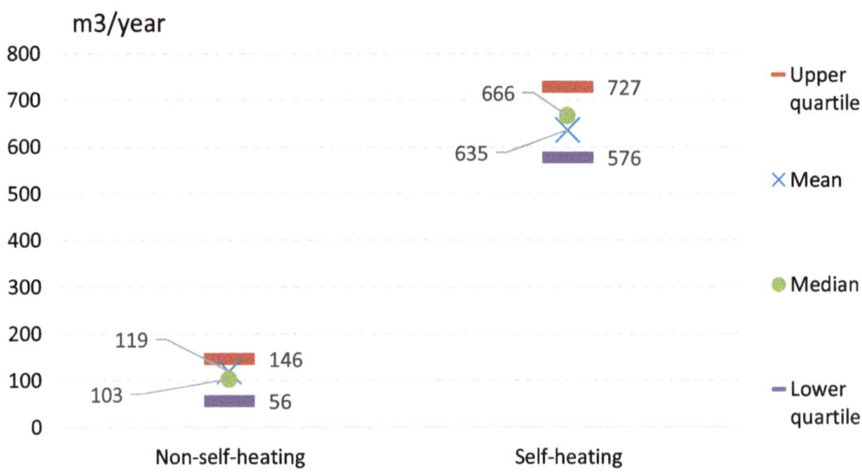

Fig. 3.12 Comparison of natural gas consumption between self-heating and non-self-heating households in Beijing (2018)

households is much higher than that of non-self-heating households. The average annual consumption of natural gas in self-heating households is 666 m^3/year, with a median level of 635 m^3/year. For households with district heating, the average consumption is 119 m^3/year, and the median level is 103 m^3/year. Based on the data above, if the average amount of gas consumed for cooking and DHW in both types of households is similar, the annual heating gas consumption of those self-heating households is approximately 550 m^3/year. Assuming the average household floor area is 90 m^2, the heating intensity is approximately 6 Nm^3/m^2.

Further statistical analysis of the distribution of gas consumption in non-self-heating households is conducted. The sample distribution of gas consumption per unit area in Beijing's residential buildings in 2022 is shown in Fig. 3.13. It can be observed that the natural gas consumption per unit area of residential buildings is also concentrated, which is in the range of 0.35~2.8 Nm^3/m^2 for 95.4% of residential buildings, and the average gas consumption per unit area is 0.81 Nm^3/m^2.

The distribution of natural gas consumption per household of residential building samples is shown in Fig. 3.14. It can be found that the natural gas consumption per household of residential buildings is also concentrated, with 95.4% of residential buildings falling within the range of 9.87~336.71 Nm^3/household. The average natural gas consumption stands at 118.85 Nm^3/household.

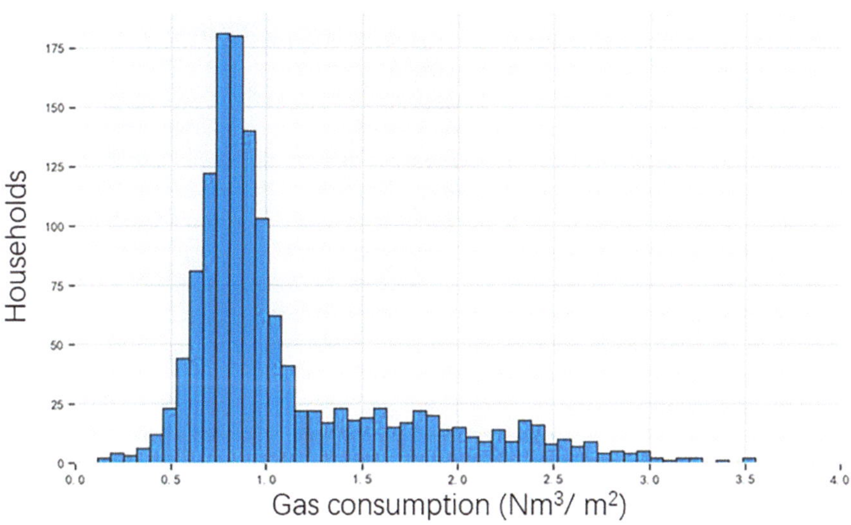

Fig. 3.13 Distribution of natural gas consumption per unit area of residential buildings (2022)

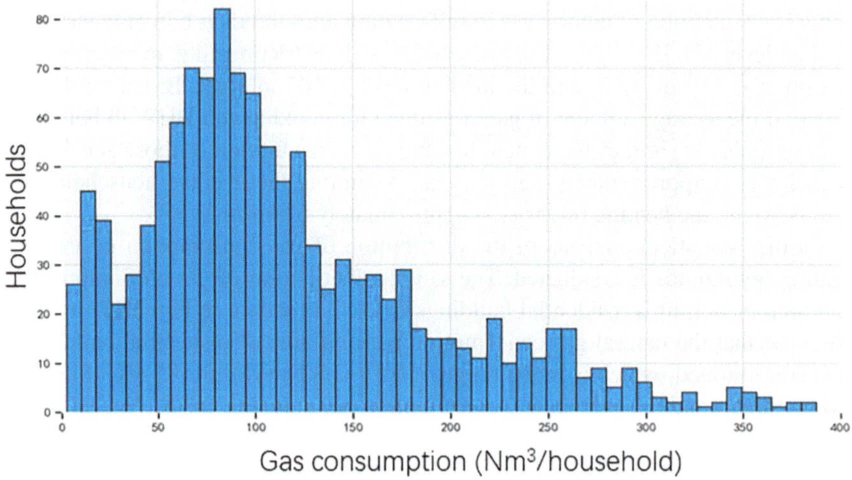

Fig. 3.14 Distribution of natural gas consumption per household of residential buildings (2022)

Fig. 3.15 Household cooking energy usage distribution in Beijing (2018)

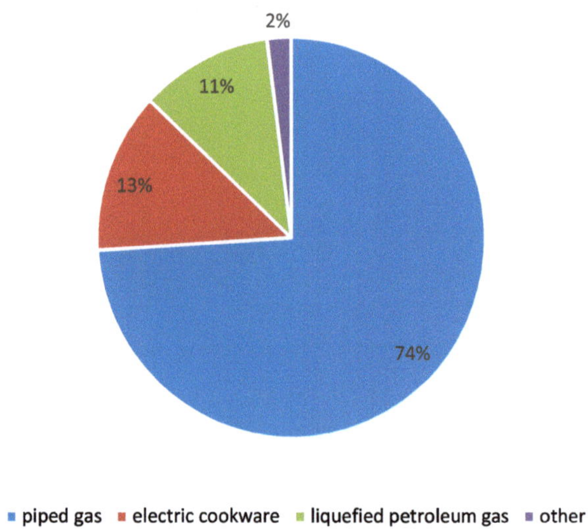

3.3.2 Electrification of Cooking

1. **Types of cooking energy use**

   In terms of cooking energy, as shown in Fig. 3.15, taking Beijing as an example, the majority of households use piped gas as the main cooking energy source, accounting for 74% of the total number of households. In addition, 13% and 11% of households mainly use electric cookware and liquefied petroleum gas (LPG), respectively. These three energy sources together account for 98% of the total, which are the three main cooking energy sources currently used by Chinese households.

From a nationwide perspective, according to the data released in the *China Urban-Rural Construction Statistical Yearbook*, the penetration rate of natural gas in cities, counties, organic towns, and rural areas all showed an upward trend from 2007 to 2022. Natural gas penetration rates in cities and counties rose markedly, reaching 98.06% and 91.38%, respectively, in 2022; while the gas penetration rates in organic towns and villages were still low, at 59.16% and 33.54%, respectively (Fig. 3.16).

2. **Gas consumption intensity in cooking**

To understand the energy consumption intensity of typical households, a questionnaire survey was conducted on the cooking habits of households in Beijing, and the results are shown in Fig. 3.17. It can be found that eating at least one meal (supper) at home every day is the lifestyle of the vast majority of households, accounting for 56%, while 26% of households eat all three meals at home every day, and 16% of households eat at home no more than seven times a week.

To further estimate the energy intensity, a typical household that occasionally eats at home was selected, and their cooking habits were surveyed; the results are shown in Table 3.4 According to their cooking habits, the household's annual consumption of natural gas for cooking is approximately 65 m^3.

3. **Electrification of cooking methods**

Low-carbon energy systems based on renewable energy are the inevitable direction of energy transition in China. The main sources of zero-carbon energy are nuclear power, hydropower, wind power, solar PV, and biomass energy, and the direct form of energy output is shifted from fossil energy to electricity. The shift to low-carbon energy requires full electrification on the demand side, drastically changing how buildings consume energy. Most energy-consuming equipment in buildings is already electrified, but cooking remains one of the few

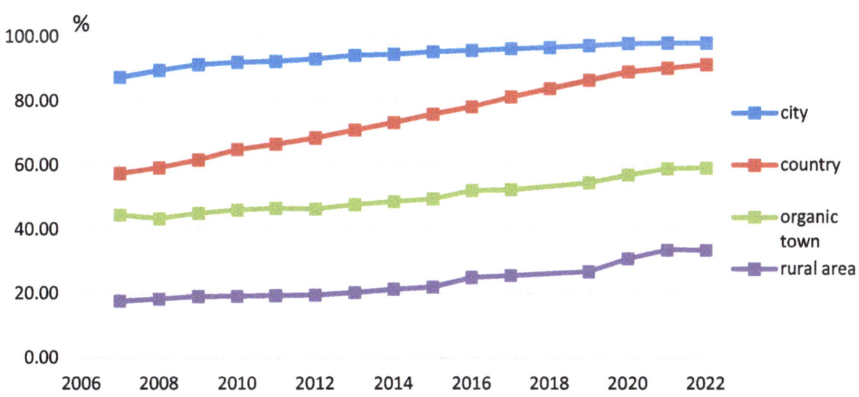

Fig. 3.16 China's gas penetration rate (gas-using population/resident population)

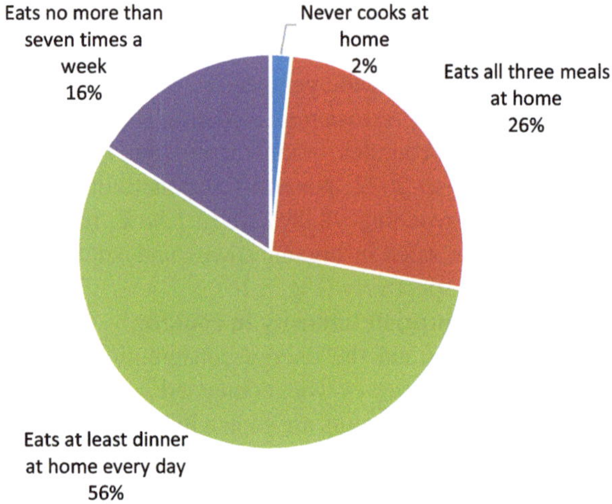

Fig. 3.17 Cooking habits of Beijing households (2018)

Table 3.4 Estimates of gas consumption for cooking in a typical household

Average number of meals at home per week	10
Average time used when cooking a meal (min)	20
The number of gas burners used per meal	1
Thermal power of gas stove (kW)	4
Natural gas consumption per hour (m^3)	0.38
Annual cooking gas consumption (m^3)	65.6

activities still reliant on fossil fuels. To achieve a zero-carbon transition in residential buildings, we must promote household cooking electrification.

Over the past decade, the electrification of cooking has become a defining trend among urban households in China. Annual sales of electric cooking appliances remain high, reflecting strong and sustained adoption. Rice cookers, among the most ubiquitous appliances, epitomize this shift: as early as 2012, ownership surpassed 100 units per 100 urban and rural households. Data from the National Bureau of Statistics shows that microwave oven ownership per 100 urban households rose from 50.6 units in 2013 to 56 units by 2022. This transition is also evident in evolving cooking practices; in Beijing, the proportion of households primarily using electric rice cookers nearly doubled from 7% in 2012 to 13% in 2018, according to comparative surveys (Fig. 3.18).

At present, electric cooking technology is also continuously progressing. Apart from traditional electric rice cookers, induction cookers, microwave ovens, and other electric cookware, new types of cooking utensils that suit the traditional cooking habits of Chinese residents, for example, electric flame stoves that convert electric energy into thermal energy to simulate the flame effect of traditional gas stoves,

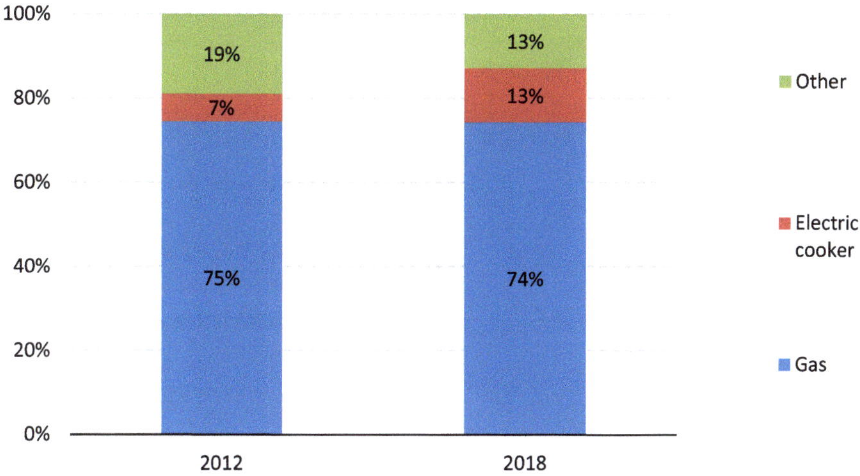

Fig. 3.18 Primary cooking energy sources for households in Beijing

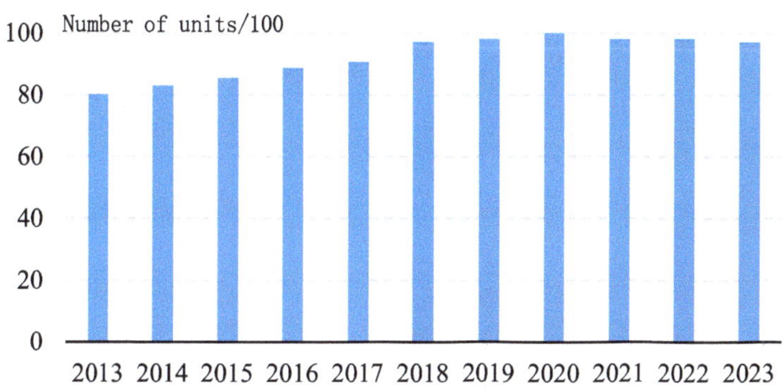

Fig. 3.19 Ownership of domestic hot water equipment in Beijing

are also emerging, which will also play a role in the electrification of cooking methods in China.

3.3.3 Electrification of DHW Heating

1. **DHW Use and Trend**

   With the improvement of people's living standards, the popularity of domestic hot water is growing rapidly, and the number of water heaters owned by every 100 urban households in China grew rapidly from 80.3 units in 2013 to 97.2 units in 2023, marking a near 100% penetration rate in urban households. In 2001, however, the figure was only 52 (Fig. 3.19).

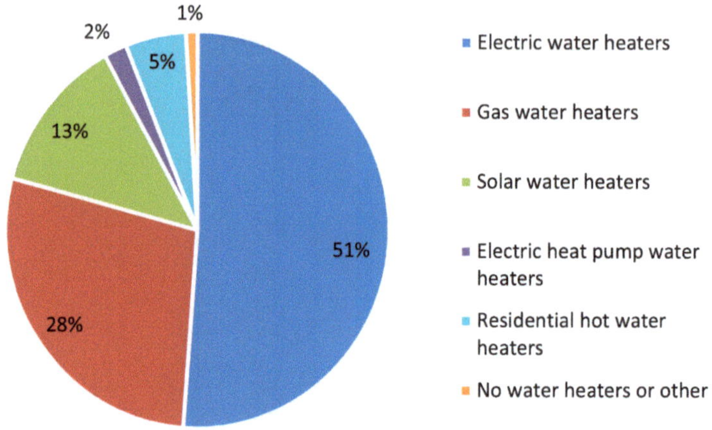

Fig. 3.20 Distribution of domestic hot water equipment in Beijing

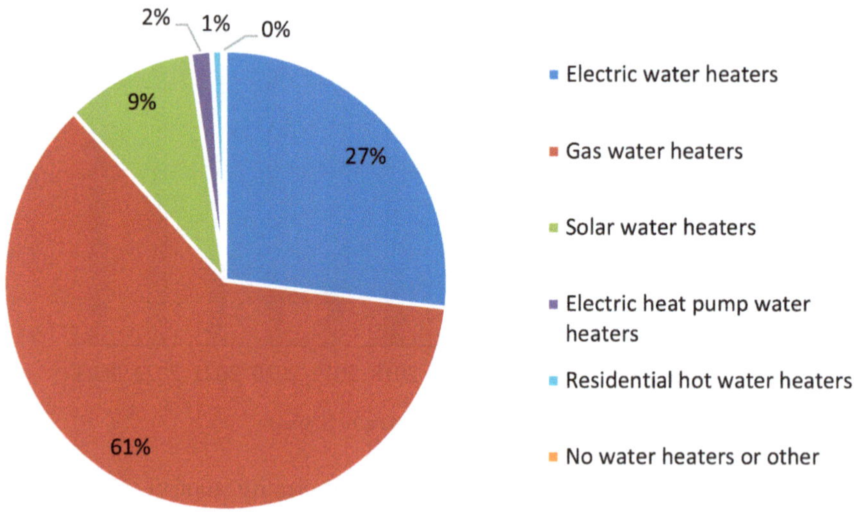

Fig. 3.21 Distribution of domestic hot water equipment in Shanghai

Figures 3.20 and 3.21 show the distribution of hot water equipment for household use in Beijing and Shanghai in 2018. At present, electric water heaters and gas water heaters are the most commonly used domestic hot water equipment in urban households, but there are differences in their distribution in northern and southern China. The main water heating equipment used by Beijing residents is electric water heaters, accounting for 51%, while the main water heating equipment used by Shanghai residents is gas water heaters, accounting for 61%. In addition to the above two types of water heaters, solar water heaters and electric heat pump water heaters are also commonly seen in households. Heat pump water heaters have shown promising development trends in recent years, but they remain in the early market stage with relatively low adoption rates in households.

2. **Energy Intensity of DHW**

   In general, the difference in water consumption is one of the most important factors affecting the energy consumption of hot water in households. Taking Beijing as an example, a survey on the main domestic hot water usage of urban residents in China is shown in Fig. 3.22 and it can be found that the most important domestic hot water usage of urban households is showering, followed by washing face and hands, as well as washing dishes. Only approximately 30% of the total households use domestic hot water for bathing in a tub.

   The abovementioned domestic water use habits are also the main reason why China's household domestic water use is much lower than that of developed countries. Urban residents in China are mainly accustomed to showering, while developed countries such as Japan prefer tub baths. According to studies conducted by BERC, Tsinghua University, the average domestic hot water consumption in China is 50 L/household/day, approximately 25% of Spain's average level, 18.5% of the United States', and 22.2% of Japan's, as shown in Fig. 3.23.

   Based on the abovementioned usage of domestic hot water, we estimated the annual gas consumption of domestic hot water for a typical family using a gas water heater. The basic assumptions and results are shown in Table 3.5, and the annual gas consumption of domestic hot water for a typical family is approximately 80 m3.

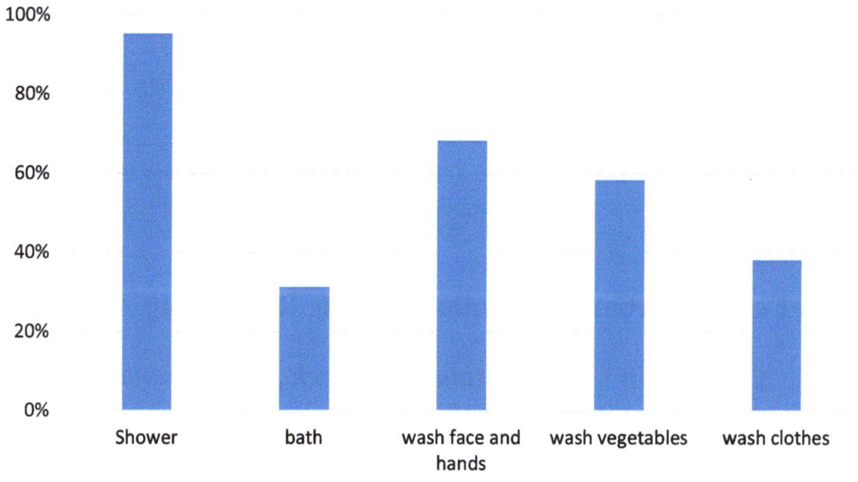

Fig. 3.22 Domestic hot water usage in Beijing in 2018

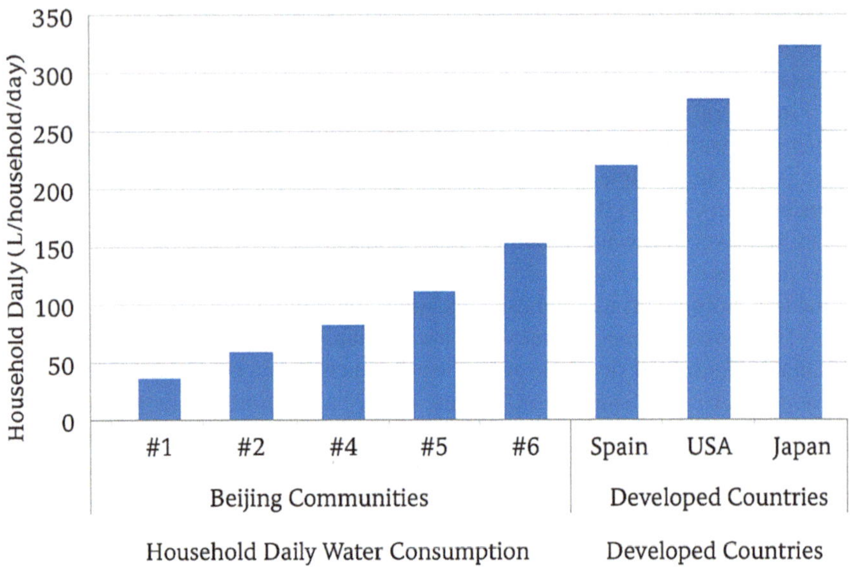

Fig. 3.23 Comparison of water consumption between Chinese and foreign households (L/household/day)

Table 3.5 Estimated gas consumption for domestic hot water in a typical household

Average domestic hot water consumption (L/household/day)	50
Set temperature of gas water heater (°C)	45
Average tap water supply temperature (°C)	10
Gas water heater efficiency	90%
Annual gas consumption for domestic hot water (m^3)	82.8

3.4 Existing Urban Residential Building Retrofitting

3.4.1 Changes in Urban Residential Development Trends

In 2008, China released the *Regulations on Energy Conservation in Civil Buildings*, which defines and puts forward relevant requirements for the energy efficiency retrofitting of existing buildings. During the 11th Five-Year Plan period, China started to steadily implement the energy efficiency retrofitting of existing buildings based on subsidies provided by the government and in combination with projects for people's well-being, such as subsidized housing construction and comprehensive renovation of old urban areas. The retrofitting has been progressively expanded from northern urban heating (NUH) areas to HSCW zones, and from urban areas to rural areas in conjunction with initiatives like clean heating. On October 29, 2020, the Fifth Plenary Session of the 19th Communist Party of China (CPC) Central Committee adopted *The Proposals of the CPC Central Committee for Formulating the 14th Five-Year Plan (2021–2025) for National Economic and Social Development and the Long-range Objectives Through the Year 2035*, which proposes "advancing the urban renewal projects, and strengthening the renovation of old urban residential compounds and the development of communities". In September 2021, the CPC Central Committee and the State Council issued the *Working Guidance for Carbon Dioxide Peaking and Carbon Neutrality in Full and Faithful Implementation of the New Development Philosophy*, which proposes "We need to advance energy conservation retrofits of urban buildings and municipal infrastructure and ensure buildings consume less energy and emit less carbon.". In October 2021, the General Offices of the CPC Central Committee and the State Council issued the *Guideline on Green Development in Urban and Rural Areas*, advocating for the green transformation of existing buildings, alongside urban renewal projects, such as renovating old residential compounds and the retrofitting and seismic reinforcement of dilapidated rural housing. In the same month, the State Council released the *Action Plan for Carbon Dioxide Peaking Before 2030*, which puts forward the task of speeding up renovation of residential and public buildings to achieve energy efficiency and continuing to advance energy-saving and carbon-cutting upgrades of heating pipe networks and other municipal infrastructure. In March 2022, the Ministry of Housing and Urban-Rural Development published *The 14th Five-Year Plan for Building Energy Efficiency and Green Building Development*, which clearly states that strengthening energy efficiency and green retrofitting of existing buildings and improving the energy efficiency standards of existing residential buildings are key tasks during the 14th Five-Year Plan period. The plan specifies the following requirements: In severe cold and cold zones, building-side energy efficiency upgrades will be further advanced in alignment with clean heating initiatives in the north. These efforts include improved insulation and the intelligent regulation of heating pipelines. In HSCW zones, energy-saving retrofits of existing residential buildings will be actively promoted to meet residents' needs for heating, cooling, and ventilation, thereby enhancing both energy performance and indoor comfort. During urban renewal projects in aging residential communities, emphasis will be placed on strengthening energy efficiency measures. A comprehensive model will be fostered: one that integrates enhancements to public spaces, upgrades to age-friendly facilities, infrastructure modernization, and functional improvements to buildings, supporting low-carbon, livable environments. Residents will also be encouraged to choose high-efficiency products when replacing components such as windows, doors, air conditioners, or wall-mounted boilers. Additionally, the plan sets the goal of striving to complete energy-saving renovations for over 100 million square meters of existing residential buildings across China by 2025. In June 2022, the Ministry of Housing and Urban-Rural Development and the National Development and Reform Commission issued the *Implementation Plan for Carbon Peaking in Urban and Rural Construction*, proposing measures to enhance energy efficiency retrofit assessment, develop specialized retrofit plans, upgrading all eligible residential buildings with retrofit potential, and ensuring compliance with current energy efficiency standards. On March 15, 2024, the General Office of the State Council released the Notice on Forwarding the *Action Plan to Speed Up Energy Conservation and Carbon Reduction in the Construction Secto*r from the National Development and Reform Commission and the Ministry of Housing and Urban-Rural Development (GBH [2024] No. 20). As stated in the notice, by 2025, the completed area of existing building energy efficiency retrofitting will increase by over 200 million square meters compared to 2023; by 2027, China will further advance the energy efficiency retrofitting of existing buildings. The part of residential buildings that have undergone energy efficiency retrofitting should achieve an energy efficiency level as specified in the current standards, and the overall energy efficiency of public buildings without energy-saving measures should be improved by more than 20% after retrofitting.

In summary, residential building energy efficiency retrofitting is one of the effective technological paths for China to achieve carbon peaking and carbon neutrality strategies in the building sector. As predicted by industry experts, under the current development pattern, carbon emissions from building operations are expected to peak around 2040, making it impossible to reach the construction sector's goals for carbon peaking and carbon neutrality by 2060. Therefore, a series of technical measures, including residential building energy efficiency retrofitting should be taken to accelerate progress toward these goals in the construction sector.

3.4.2 Changes in Urban Residential Standard Development

China initiated work related to energy efficiency retrofitting standards for existing residential buildings in the 1990s and produced the result, i.e., *Technical Specification for Energy Conservation Renovation of Existing Heating Residential Building* (JGJ 129-2000) in 2000. This standard was superseded by the *Technical Specification for Energy Efficiency Retrofitting of Existing Residential Buildings* (JGJ/T 129-2012) published by the Ministry of Housing and Urban-Rural Development in October 2012. The latter has been used since its implementation on March 1, 2013. After the release of the *Technical Specification for Energy Conservation Renovation of Existing Heating Residential Building* (JGJ 129-2000), the local standards for energy conservation renovation of existing buildings were published successively across the country. For instance, in 2006, Beijing issued the *Technical Specification of Renovation of Energy Efficiency for Existing Residential Building* (DB 11/381-2006) as the first local standard in China. Subsequently, Heilongjiang, Shanghai, Chongqing, and Anhui promulgated local standards concerning energy efficiency retrofitting of existing residential buildings.

In terms of the retrofitting measures specified in existing local standards, Beijing mandates that the heat consumption of buildings after energy efficiency retrofitting should not exceed the threshold specified by the local retrofitting standard and that the performance or effect of the retrofitted part should meet or surpass that set out in the *Design Standard for Energy Efficiency of Residential Buildings in Severe Cold and Cold Zones* (JGJ 26); it also specifies the retrofitting requirements for solar water heating systems, solar PV systems, and public lighting. Shanghai has established threshold values for determining the retrofitting of building envelopes, specifying requirements for the renovation of heating, ventilation, and air conditioning, domestic hot water supply systems, and power and lighting systems. Inner Mongolia adds provisions for solar water heating systems. Ningxia stipulates that retrofitting is required when the heat transfer coefficient of a non-energy-efficient residential building envelope reaches a predefined limit. Clean heating cities in northern China require that the building energy efficiency should be improved by more than 30% after retrofitting.

Since its implementation, the industry standard *Technical Specification for Energy Efficiency Retrofitting of Existing Residential Buildings* (JGJ/T 129-2012) has significantly contributed to advancing energy efficiency retrofitting of existing residential buildings. This standard has effectively guided and regulated key processes, including energy efficiency diagnosis of residential buildings, formulation of energy efficiency retrofitting solutions, and construction quality acceptance. In addition, it has facilitated the smooth advancement of residential building energy efficiency retrofitting during the 12th Five-Year Plan and the 13th Five-Year Plan periods. However, over nearly a decade of implementation, China's vast geography and pronounced climate diversity have resulted in substantial variation in building energy-saving measures across different climate zones. Consequently, regional differences have emerged in building energy consumption benchmarks, criteria for identifying retrofit needs, and the specific goals set for energy efficiency upgrades. Furthermore, ongoing advancements in the thermal performance of building envelopes, improvements in the energy efficiency of equipment and systems, and the introduction of new energy-saving technologies—coupled with the release and enforcement of updated mandatory engineering codes such as the *General Code for Energy Efficiency and Renewable Energy Application in Buildings* (GB 55015) and the *General Code for Maintenance and Renovation of Existing Buildings* (GB 55022)—have significantly reshaped the standards landscape. Under the dual carbon goals, the original industry standard, *Technical Specification for Energy Efficiency Retrofitting of Existing Residential Buildings* (JGJ/T 129-2012), is no longer sufficient to guide current retrofit practices.

3.4.3 Field Research on Energy Efficiency Retrofitting in Clean Heating Pilot Cities

Since the launch of the *Northern Region Winter Clean Heating Pilot Program* (CJ [2017] No. 238) in 2017 by the Ministry of Finance, the Ministry of Housing and Urban-Rural Development, the Ministry of Ecology and Environment, and the National Energy Administration, a total of 88 cities across five batches had received central fiscal support for clean heating by 2022, covering 15 provinces (autonomous regions and municipalities directly under the central government) in northern China. Over the 3-year demonstration period, a total of RMB 107.1 billion in subsidies will be distributed across five batches of pilot cities based on full fund disbursement. These funds primarily support initiatives including "coal-to-electricity," "coal-to-gas," and energy efficiency retrofitting of existing buildings in rural areas. Among them, the energy efficiency retrofitting of residential buildings is the highlight of the user-side retrofitting. The editorial team conducted relevant research on the current situations and solutions of retrofitting in several pilot cities successively, mainly including the building age, current conditions of building envelopes, heating methods, present situation of energy consumption, retrofitting subsidy costs, retrofitting measures, and energy efficiency improvement after retrofitting in some cities. The retrofitting solutions of different clean heating pilot cities in northern China were mainly formulated based on a 30% increase in energy efficiency and subsidy costs. Energy efficiency retrofitting was mostly completed by taking two or more measures. After retrofitting, most of the retrofitted building parts could meet the requirements of the local retrofitting standards in force and achieved a 30% increase in energy efficiency (Fig. 3.24).

Cold City A, for example, was researched regarding the project name, floor area, building type, construction time, building time, building structure, building story, exterior wall structure, shape factor, and south-facing window-to-wall ratio based on the statistics of each project. The research results reveal that most retrofit projects involved four- to seven-story residential buildings constructed between 2000 and 2005. The exterior walls are largely 240 mm hollow brick walls without thermal insulation measures. The exterior windows are mostly single-glazing plastic steel windows or aluminum alloy windows. The roofs are not thermally insulated. The shape factors are mostly between 0.2 and 0.4. Building exteriors fall into three main styles: fully painted finishes; a combination of tiled ground floors with painted upper levels; or complete ceramic tile cladding for all floors (Fig. 3.25).

3.4.4 Analysis of Energy Efficiency Retrofitting Scale

According to data from the *China Statistical Yearbook* (1996–2016), the *China Statistical Yearbook on Construction* (2017–2020), and the Statistical Yearbook (1996–2020) of different first-level administrative regions, the existing floor area is composed of area stock, as-built area, and demolition area. Based on statistical data from the Ministry of Housing and Urban-Rural Development, the demolition area of urban residential buildings is about 0.5% of the area stock. The annual existing building floor area across different climate zones is shown in Fig. 3.26. In 2020, the total floor area of existing urban residential buildings was approximately 27.07 billion m^2, including about 3.25 billion m^2 in severe cold zones, roughly 7.92 billion m^2 in cold zones, around 11.48 billion m^2 in hot summer and cold winter zones,

City A (B subregion of the cold zone)

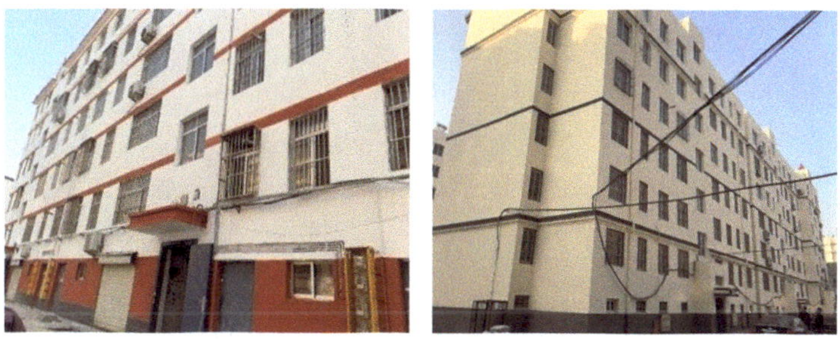

City B (B subregion of the cold zone)

City C (A subregion of the cold zone)

City D (C subregion of the cold zone)

Fig. 3.24 Field research on clean heating pilot cities

Fig. 3.25 Field research photos

nearly 3.41 billion m^2 in hot summer and warm winter zones, and approximately 1.01 billion m^2 in mild zones.

In addition, urban residential buildings in different climate zones adhere to different building energy efficiency design standards, as shown in Table 3.6. Existing residential buildings are classified by the standard implementation year. Northern regions of China implemented the *Civil Building Energy Efficiency Design Standard* (JGJ 26-86) in 1986, but the implementation level was low at the initial stage. Therefore, the focus is primarily on estimating the residential buildings complying with energy efficiency standards JGJ 26-1995 and JGJ 26-2010. The industry standards for mild zones were published later than those for other climate zones, and the corresponding province- and city-level design standards were also introduced at

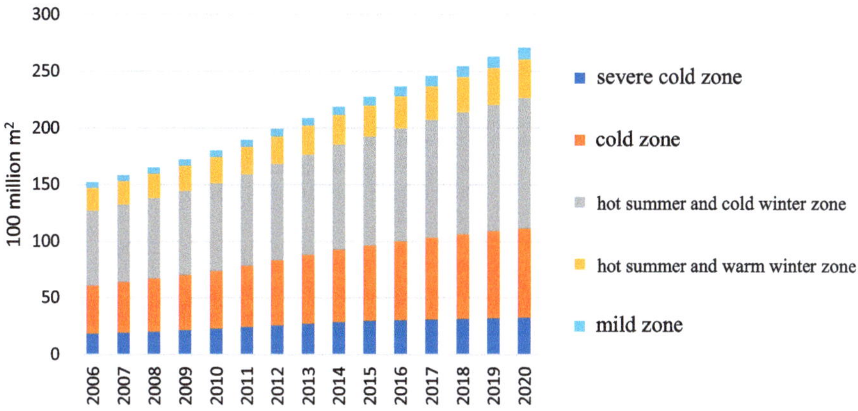

Fig. 3.26 Floor area of existing urban residential buildings in different climate zones

varying times. Therefore, the floor area of existing residential buildings complying with energy efficiency standards in mild zones was estimated by reference to the implementation year of the standard JGJ 75-2012 for hot summer and warm winter zones. Considering that it takes about 3 years for energy efficiency design standards to be implemented and converted into as-built areas, the estimated floor area of existing urban residential buildings with different energy conservation rates by 2020 is listed in Table 3.7. According to *The 13th Five-Year Plan for Building Energy Efficiency and Green Building Development* and *The 14th Five-Year Plan for Building Energy Efficiency and Green Building Development*, 990 million m^2 of existing residential buildings in severe cold and cold areas completed energy efficiency retrofitting during the 12th Five-Year Plan period and 514 million m^2 completed retrofitting during the 13th Five-Year Plan period; additionally, 70.9 million m^2 of existing residential buildings in hot summer and cold winter areas completed energy efficiency retrofitting.

As a result, the floor area of existing urban residential buildings below the second-stage energy-efficient standard is approximately 9.36 billion m^2, including about 0.59 billion m^2 in severe cold zones, around 1.45 billion m^2 in cold zones, roughly 5.15 billion m^2 in hot summer and cold winter zones, nearly 1.77 billion m^2 in hot summer and warm winter zones, and approximately 0.4 billion m^2 in mild zones.

Table 3.6 Urban residential building energy efficiency design standards in different climate zones

Energy efficiency standard	Climate zone	Description	Implementation year
Initial-stage energy efficiency standard	Severe cold and cold zones	Civil Building Energy Efficiency Design Standard (Residential Buildings with Heating Demand) (JGJ 26-86)	1986
Second-stage energy efficiency standard	Severe cold and cold zones	Energy Conservation Design Standard for Residential Buildings (Heating Residential Buildings Part) (JGJ 26-1995)	1996
	Hot summer and cold winter zone	Design Standard for Energy Efficiency of Residential Buildings in Hot Summer and Cold Winter Zone (JGJ 134-2010)	2010
	Hot summer and warm winter zone	Design Standard for Energy Efficiency of Residential Buildings in Hot Summer and Warm Winter Zone (JGJ 75-2012)	2013
Third-stage energy efficiency standard	Severe cold and cold zones	Design Standard for Energy Efficiency of Residential Buildings in Severe Cold and Cold Zones (JGJ 26-2010)	2010
	Hot summer and cold winter, hot summer and warm winter, and mild zones	General Code for Energy Efficiency and Renewable Energy Application in Buildings (GB 55015-2021)	2022
Fourth-stage energy efficiency standard	Severe cold and cold zones	Design Standard for Energy Efficiency of Residential Buildings in Severe Cold and Cold Zones (JGJ 26-2018)	2019

Table 3.7 Floor area of existing urban residential buildings with different energy conservation rates in 2020

Energy efficiency standard	Climate zones	Implementation year	As-built area calculation year	Existing floor area (100 million m^2)
Second-stage energy efficiency standard	Severe cold zones	1996~2009	1999~2012	18.8
	Cold zones	1996~2009	1999~2012	40.4
	Hot summer and cold winter zones	2010~2020	2013~2020	63.3
	Hot summer and warm winter zones	2013~2018	2016~2020	16.4
	Mild zones	2013~2018	2016~2020	6.1
	Subtotal			145.1
Third-stage energy efficiency standard	Severe cold zones	2010~2020	2013~2020	7.8
	Cold zones	2010~2020	2013~2020	24.3
	Subtotal			32.1
	Total			177.2

Chapter 4
Full Electrification and Transition in Electricity Consumption Pattern of Urban Residential Buildings

4.1 Urban Residential Building Development Concept in the Context of Energy Revolution

With the escalating severity of global climate change and the growing prominence of energy mix transformation, the building sector, as a primary domain of energy consumption, faces unprecedented challenges and opportunities. Against the backdrop of the energy revolution, the transition from traditional energy sources to clean and renewable energy has triggered a transformation in energy production and consumption patterns, while also presenting new requirements for the building energy field. In the context of addressing global climate change and achieving carbon peaking and carbon neutrality goals, the energy consumption patterns and development concepts of urban residential buildings require urgent transformation and upgrades. Buildings, as a crucial component of the power grid demand side, urgently require breakthroughs in key technological bottlenecks to achieve friendly interaction with power grids and release their potential as the demand-side regulation means. Traditional buildings, as "energy consumers", struggle to align with the current development of the energy system. The development of the new building power system necessitates the transition of an urban residential building from a mere energy consumer to an integrated energy system incorporating energy production, storage, consumption and regulation functions. This transition swims with the tide of full building electrification and can address the phase-out trend of fossil energy and the challenge of maximizing renewable energy utilization during the low-carbon transition of the building sector.

Building Energy Research Center of Tsinghua University, *Decarbonizing Urban Residential Buildings*, https://doi.org/10.1007/978-981-95-8200-6_4

4.1.1 Full Electrification of Urban Residential Buildings

The terminal energy consumption forms in traditional urban residential buildings include fossil energy, such as gas and fuel oil, and electricity. The core of full electrification lies in transforming the energy used in traditional urban residential buildings into cleaner electricity. Sharing the same goal with the low-carbon and clean transformation of power grids, the full electrification of buildings can not only effectively reduce carbon emissions but also better integrate with renewable energy utilization, thereby constructing a low-carbon and clean building energy system. The past two decades have witnessed significant progress in building electrification, including efficient refrigeration and air conditioning, heat pump water heaters, and heat pump heating systems, which have become important directions for the energy efficiency and low-carbon development of residential buildings.

However, the attainment of carbon neutrality goals in the building sector cannot rely entirely on cleaner electricity. Although electricity plays an important role in the future terminal energy consumption of buildings, the building sector still needs to advance low-carbon and zero-carbon development in all respects of the construction industry, including green building design, the research, development, and use of low-carbon building materials, and the application of building energy efficiency technologies. Moreover, the regulation and management capability of the building energy system needs to be improved. Through automated means and artificial intelligence technologies, effective system control can be implemented to avoid unnecessary waste of energy when not in use and increase the utilization efficiency of energy when in use. By integrating full electrification with the application of low-carbon and zero-carbon technologies in the construction industry, the renewable energy utilization rate and energy system efficiency can be improved, thus effectively driving the low-carbon transition of the construction industry and contributing to the achievement of dual carbon goals.

Amid the prevailing trend of building electrification, the phase-out of gas has been an important issue under discussion in the urban residential building field. Gas is usually used in such fields as heating, cooking, and domestic hot water. However, due to its nature as fossil energy, it not only causes substantial direct carbon emissions but also poses certain safety hazards. Many countries around the world have begun phasing out gas facilities in urban residential buildings to promote the electrification of household kitchen appliances and the application of other low-carbon technologies. In response to the demand for low-carbon development in China's building sector, the phase-out of gas should also become a mainstream trend. Furthermore, it is challenging to effectively promote hydrogen energy application in urban residential buildings due to its high production, storage and transportation costs, coupled with immature and unstable security technologies. From this perspective, achieving full electrification of buildings and enhancing the regulation, operation, and maintenance capabilities of building power systems will be the core path for future low-carbon development in the building sector.

4.1.2 "Production-Storage-Consumption-Regulation" Integrated Building Energy System Architecture

Traditional residential buildings, as "consumers" of energy, mainly function to fully meet the environment conditioning and use demands of indoor users, covering such aspects as heating, cooling, domestic hot water, and lighting. Therefore, the design of traditional residential buildings focuses on energy consumption control and energy efficiency optimization to reduce the energy intensity and total energy consumption of buildings. However, with the transition of the global power system towards renewable energy, the installed capacity of wind and solar PV power units accounted for over 40% of China's total, while their electricity generation reached nearly 20% of the nation's total, in 2024. In the future energy system, the installed capacity and electricity generation of wind and solar PV power sources will further increase and play a dominant role. However, the volatility of wind and PV power generation with meteorological conditions poses huge challenges to power grid stability and supply-demand matching. As a result, the demand-side building sector is required to shift its role from a mere energy "consumer" to an energy participant integrating "production-storage-consumption-regulation".

In the "production-storage-consumption-regulation" integrated building energy system architecture, the following measures should be taken. First, in terms of energy production, such equipment as solar PV panels and wind turbine generators should be installed in residential buildings/compounds to harness natural resources for energy production, thereby reducing reliance on external energy supplies (mainly the external power grid). Second, energy storage resources in urban residential buildings, for example, water heater storage tanks, the thermal inertia of buildings for storing heating and cooling energy, and electric vehicle charging systems, should be fully utilized to achieve load shifting and increase both local and grid-wide renewable energy consumption rates. Furthermore, buildings should perform their fundamental function of energy consumption to maximize energy use efficiency and minimize total energy consumption, while simultaneously meeting the requirements of and optimizing indoor environment conditioning. At last, the inherent regulating capacity of the building energy system, including the thermal inertia of buildings and the inertia of the energy system, should be fully leveraged to enable flexible regulation of energy use and effectively promote renewable energy consumption. These measures will not only contribute to the efficient utilization of the energy in the system but also enable the energy system to play a positive role in the broader power system to alleviate the grid stress during supply and demand regulation, thereby making the energy system more stable and flexible.

4.1.3 New Changes in Building Electricity Consumption Mode

In a traditional power system, the power balance mechanism mainly adheres to a "generation following load" balance mode. The power system achieves power balance control through the "generation following load" approach, where end-use electricity serves as passive physical terminals. For grid dispatching, due to a relatively low proportion of new energy power generation, the uncertain volatility of renewable electricity has a minor impact on the balance and stable operation of the grid. Moreover, as the power grid has sufficient backup capacity, the traditional "generation following load" balance mode can ensure the safe and stable operation of the grid. However, with the large-scale integration of renewable energy, greater demands will be placed on the regulation capacity of the power system. The stochastic volatility of renewable energy exhibits significant uncertainty, which profoundly affects the supply-demand balance and the safe and stable operation of the power system. If the traditional "generation following load" balance mode is maintained without exploring flexible load regulation means on the demand side, it will inevitably lead to large-scale wind and solar curtailments, while significantly diminishing the regulation capacity of the power system.

In this process, the electricity consumption side requires positive changes. In the transformation of the building's role from a mere energy consumer to the "production-storage-consumption-regulation" integrated architecture, friendly interaction between buildings and the power grid should be established on the premise of meeting the indoor environment conditioning requirements. Indoor environment conditioning involves such functions as heating, cooling, ventilation, and lighting to meet the demand for a comfortable and healthy indoor environment. In the friendly interaction between buildings and the power grid, the building energy production and consumption are regulated to assist the power grid in supply and demand matching and increase the renewable energy utilization rate. The two goals need to be coordinated. The thermal inertia of buildings and other equivalent energy storage resources should be fully leveraged to enhance the flexible regulation capacity on the building energy consumption side, while trying not to affect or to minimize the impact on indoor satisfaction.

4.1.4 Electric Vehicle and Charging Point

The vigorous development of electric vehicles is crucial to the achievement of the dual carbon goals, as it can significantly reduce direct carbon emissions while providing important energy storage resources for the new power system through vehicle-grid interaction. Driven by the dual carbon goals, the coupling relationship between the two terminal energy consumption sectors, i.e. vehicles and buildings, is being progressively strengthened: In terms of vehicle behavior, urban private vehicles (accounting for about 80% of the total number of urban vehicles) are parked in building parking lots for more than 80% of the time on average, and also for charging purposes. Regarding power grid capacity expansion, the connection of vehicle charging points to building distribution systems reduces the investment in power grid capacity expansion.

Residential buildings are the primary scenarios for the interaction between vehicles and building energy. The behaviors of urban private vehicles are particularly closely related to residential buildings. Private electric vehicles in a typical city are parked in parking lots of residential buildings for about 39%–67% of the time on average. This long-time and stable occupancy characteristic provides adequate temporal and spatial conditions for the energy interaction between vehicles and residential buildings. Residential buildings play an important role in the electric vehicle charging mode. Generally, vehicle owners connect to charging points after returning to compounds in the evening. If uncoordinated charging, namely charging immediately after connection, is adopted, the electric vehicle charging load will significantly overlap with the peak energy consumption of residential users.

By rationally deploying intelligent coordinated charging points in parking lots of residential buildings, electric vehicles will not be charged during the evening peak hours of building electricity consumption; instead, they will be charged with the valley electricity at night to meet the travel demand of the next day. Based on this coordinated charging mode, vehicle charging points can utilize the idle-hour redundant power distribution of residential buildings, thereby reducing the capacity expansion investment incurred from charging points. For buildings, this mode alleviates the pressure on distribution systems during peak electricity consumption hours; for electric vehicles, their charging demands are met while the battery life can be extended through coordinated charging management. On the whole, this mode is conducive to relieving the capacity expansion pressure on the distribution network and fully leverages the flexibility of the power demand side to provide the power system with flexible regulation capabilities.

Based on intelligent coordinated charging points, bidirectional charging points will be further developed. Their energy storage and regulation functions will be fully utilized through the bidirectional charging and discharging technology, enabling charging during the valley period of power load and discharging during the peak period. The average annual mileage of private vehicles is about 12,000 km, and the corresponding electricity consumption generally does not exceed 2000 kWh/year. Based on an average battery capacity of 60 kWh per vehicle, and assuming each charging session provides 50 kWh, a vehicle would require only 40 charging sessions per year. If the vehicle charges and discharges once per day, the loss is calculated based on an average charge and discharge efficiency of 92%, an average of 7 kWh of electricity per day is used for driving, and another 7 kWh is lost during the charging and discharging process, then in the bidirectional charging and discharging mode, there is still 36 kWh of electricity left each day that can be used for energy storage and regulation. By 2050, China is expected to have around 300 million non-commercial passenger vehicles, which could provide about 10.8 billion kWh of energy storage capacity every single day while meeting daily driving needs, thereby playing a significant regulation role.

4.2 Load-Side Autonomous Response Regulation in Power System Source-Load Interaction

4.2.1 Source-Load Interaction Demand of New Power System

The development of a new power system is the most important task for China to achieve the energy revolution with the goal of energy transition. The power system will shift from the current high thermal power penetration to high wind and solar PV power penetrations. By 2060, the installed capacity (proportion) and annual electricity generation (proportion) of wind and solar PV power in China will increase to 7.01 billion kWh (85%) and 11.9 PWh (69.2%), respectively (Zhou et al. 2021). Figure 4.1 illustrates the variation patterns of wind power, solar PV power, and electrical load over 5 consecutive days. Although the electricity provided by wind and solar PV power systems is comparable to the total electricity demand, there are notable differences between the two in terms of time scale, for instance, the wind and solar PV power output is more than three times the electrical load when the sunlight intensity is high at noon, while the power provided by power supplies at night is far less than the electrical load. As shown in the figure, throughout the third day which is cloudy with calm wind, the electrical load is notably higher than the wind and solar PV power output. In this case, to address the mismatch between the two, it is necessary to rely on electricity regulation resources that can be centrally dispatched, including thermal and hydroelectric power, which can serve as spinning reserves, and energy storage resources such as pumped storage, compressed air energy storage, and chemical energy storage. However, in the future, with a substantial cut in thermal power generation capacity and the limited proportions of other dispatchable power sources, the installed capacity of centrally dispatchable power sources will be reduced to less than 25% of the total installed capacity, which will

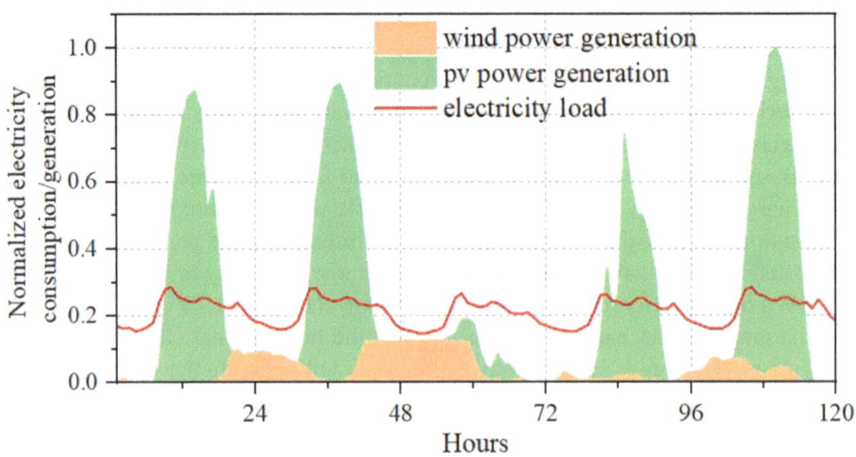

Fig. 4.1 Mismatch of changes between wind and solar PV power and electrical load within a week

be inadequate to fill the load gap in Fig. 4.1. Furthermore, energy storage resources that can be centrally allocated, with daily storage capacity far from easy to reach 30% of total daily electricity consumption, and consumption power that can hardly exceed 20% of the total installed capacity of wind and solar PV power, will be insufficient to consume the surplus electricity in Fig. 4.1 and fill the electricity gap in Fig. 4.1. Therefore, it is essential to dispatch energy storage and flexible electricity resources distributed at electricity consumption terminals to solve this problem. By changing the electricity consumption mode and the terminal energy storage, the electrical load demand curve will be made as close as possible to the power supply curve. Through joint efforts of the supply and demand sides, the two curves will finally coincide to meet the power requirements.

In a southern province of China, for example, when a high proportion of wind and solar PV power is connected to the power supply side, there is a mismatch between renewable energy and electrical load. To address this mismatch, daily energy storage capacity is required. The curve of daily energy storage capacity demand versus cumulative days is shown in Fig. 4.2. Among them, the centralized chemical energy storage is fully utilized every day for charging and discharging, which can provide a storage capacity that is approximately equivalent to 24 million kWh. The pumped storage basically operates once a day as well to provide a storage capacity of 21 million kWh. Next, this province can utilize waste heat from nuclear power plants to generate steam through heat pumps in industrial production and store heat in tanks to make heat pumps regulate electricity peaks simultaneously. According to the predicted industrial steam demand of this province, the heat-power coordination of this industrial steam can provide a storage capacity equivalent to 14 million kWh. For the remaining energy storage capacity demand, most of it can be provided by electric vehicles in this province. This province will have 4.5 million non-commercial electric vehicles. As long as one-third of them are connected to the power grid through

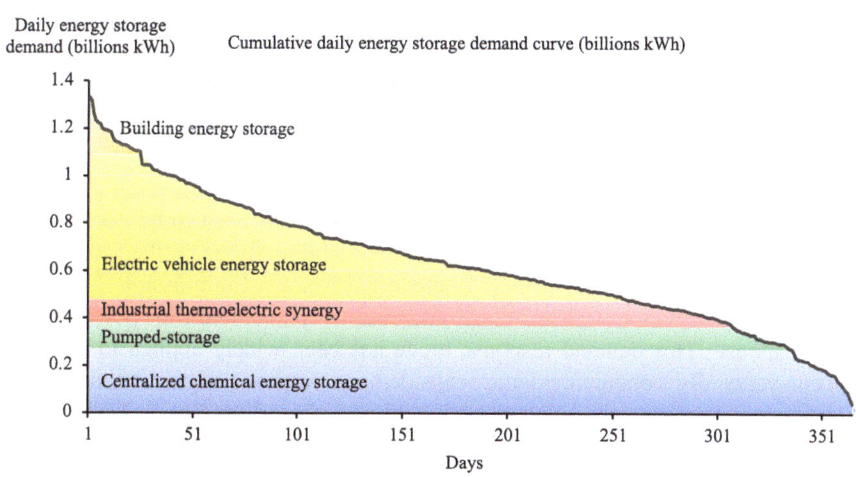

Fig. 4.2 Daily energy storage capacity demand of high wind and solar PV power penetrations in a southern province of China

charging points at the same time, they can meet the demand for short-time electricity peak regulation. The last 10% of energy storage demand requires the mobilization of the adjustable flexible energy consumption resources within buildings.

4.2.2 Source-Load Interaction Mode of Current Power System

The real-time regulation of a power system at present is as follows: Based on day-ahead prediction and optimization, the spinning reserve capacity that can be mobilized at any time is dispatched without delay upon the occurrence of a deviation, thus dynamically balancing the power supply side and the electrical load (Wang et al. 2023). For a regional power grid, the objects to be regulated are mainly hundreds of adjustable power supplies and centralized energy storage facilities. However, independent electricity consumption terminals with distributed energy storage and flexible electricity consumption capabilities, such as charging points and building electrical appliances (Jiang 2021; Liu et al. 2023a), numbering hundreds of thousands to millions, can hardly be mobilized in real time through electricity dispatch. As a result, it is difficult to make these potential flexible electricity resources regulate their electricity storage and consumption statuses dynamically according to the grid supply-demand balance conditions. To mobilize the flexible electricity consumption capabilities of electricity consumption terminals, the power sector has attempted relevant measures, which mainly include load aggregator, demand response, and time-of-use tariff.

The "load aggregator" mode involves a third-party load integrator, which integrates these flexible electricity loads, making it an equivalent flexibly adjustable electrical load that can be incorporated into the dispatching platform of the power system. This essentially shifts the problem of how to regulate a huge number of distributed loads in real time onto the load aggregator, rather than resolving it substantially. It is also highly challenging for the load aggregator to find an effective method to timely coordinate the electrical loads under its governance in response to the real-time changes in the supply and demand relationship, thus cooperating with the power grid in real-time regulation.

"Demand response" involves contracting with loads that have flexible electricity consumption capabilities in advance. When power loads need to be decreased or increased, these contract consumers are mobilized directly or notified to regulate their electric power (National Development and Reform Commission 2023), and financial compensation is given to consumers based on their real-time power regulation response. This demand response mode has been widely applied in China in recent years. By conducting large-scale demand response regulation dozens of times annually during the summer peak load periods and the Spring Festival load valley periods, the State Grid has effectively mitigated the imbalance between power supply and demand and also gained billions in RMB of economic benefits. However, the demand response regulation merely addresses the short-term imbalance between supply and demand and deals with abnormal load changes. The calculated load-side power regulation quantity is the difference between the power predicted under normal working conditions and the actual electric power. Therefore, this mode is only suitable for addressing temporary abnormal phenomena and cannot be incentivized to solve daily energy storage and flexible electricity regulation required at electricity consumption terminals after the connection of large-scale wind and solar PV power. The fundamental issue here is how to evaluate the ultimate consumers' regulation effects. Should the evaluation be based on the regulated power variation or the electric power at the specific time? If it is based on the power variation after regulation, how should the original electric power of consumers be identified and predicted when there is no response? If such regulation becomes a daily norm, how should the original electric power on the customer side be defined?

In the meantime, China has implemented a "time-of-use tariff" since the 1990s. The day is divided into different power load periods, which are peak periods with high tariffs, shoulder periods with medium tariffs, and valley periods with low tariffs respectively. This mode is primarily designed to change the electricity consumption behaviors of ultimate consumers through tariff, thereby reducing the peak-valley difference in urban power load, and it has indeed played a role in peak load shaving. For instance, influenced by the time-of-use tariff, the daily power load peaks occur after midnight in some provinces (e.g., Liaoning and Hainan) of China. This is because the electricity consumption behaviors of some industrial loads and electric vehicle charging loads are changed for low tariffs. However, the daily changes of wind and solar PV power do not strictly follow a fixed schedule, and overcast, calm wind, sunny, and windy weather conditions in a day may lead to radically different electricity output, thus requiring entirely distinct regulation actions from the terminals. Therefore, the fixed-time "time-of-use tariff" mode is unlikely to be an effective incentive signal for addressing the issues of wind and solar PV power consumption.

Implementing a dynamic real-time price system, where electricity prices act as an incentive signal, is feasible in principle; this is the approach adopted by many Western countries currently (Mehrtash et al. 2023). However, China's electricity tariffs have served many functions, such as special low-price electricity to ensure people's livelihoods and support agriculture, and various additional tariffs to support the construction of the power system. Further loading the task of flexible regulation at the power-consuming terminals onto the electricity tariffs, which are already entangled with various subsidies and surcharges, makes it difficult to highlight their incentive function for regulating electricity consumption. Moreover, determining dynamic tariffs for different scenarios in a way that promotes flexible regulation while maintaining fairness in tariff levels remains a formidable challenge.

4.2.3 Load-Side Autonomous Response Regulation

On the load side, relevant research has shown that a building and its distribution network can provide tremendous energy storage and flexible electricity resources through electric vehicles connected to charging points (Liu et al. 2023b). However, these resources primarily serve their designated functions, instead of being designed specifically to serve the power grid. They must strike a balance between fulfilling their functions and assisting the grid in peak regulation, namely engaging in the peak load shaving of the power grid whenever possible without compromising the fulfillment of their functions. This requires the study of appropriate mechanisms to mobilize various power-consuming terminals so that their energy storage and flexible energy consumption functions can be fully utilized whenever possible.

As demonstrated in demand response practices, precise control is unnecessary for massive ultimate consumers with flexible electricity consumption capabilities. Instead, a certain "call" with an incentive mechanism can effectively change their consumption behaviors. They are encouraged to decrease or increase electric power as directed by the "call", thus alleviating the pressure on the power supply side for unified dispatch and regulation and achieving the final regulation with minimal spinning reserve capacity. In the future, when uncertain wind and solar PV power become dominant energy sources and the regulation required to match their real-time changes becomes routine, if these changes can be balanced by the load-side energy storage and flexible electricity resources mobilized through an incentive mechanism, the regulation pressure on the power supply side can be significantly relieved. Thus, the power system's real-time regulation tasks caused by the random changes of large-scale wind and solar PV power can be jointly accomplished by the energy storage and flexible electricity resources distributed across millions of power-consuming terminals and the grid's directly dispatchable spinning reserve capacity.

However, to achieve such regulation, the power system must issue real-time regulation signals. Such signals shall fully reflect the real-time supply-demand imbalance in the power system and align the resultant regulation behaviors of power-consuming terminals with the load regulation direction required by the power system at that time. Moreover, they should serve as "incentive" signals that can effectively incentivize the terminals to participate actively in real-time regulation. This will be the key to enabling the power system, where a massive number of regulating entities make real-time regulation toward the same objective, to achieve effective control and regulation.

In conclusion, to address the mismatch between the high penetration of renewable energy, such as wind and solar PV power, and electrical load, it is urgent to mobilize flexible electricity resources distributed on the electricity consumption side. Current measures, including load aggregators, demand response, and time-of-use tariff, cannot fully resolve the supply-demand imbalances that change in real time. There is a need to establish regulating signals that guide flexible energy usage on the electricity consumption side in real time. Against this background, the power system has attempted to adopt the "dynamic carbon emission responsibility factor for electricity" (hereinafter referred to as "C_r") as a dynamic signal to incentivize power-consuming terminals to engage in "bottom-up" autonomous response regulation (as illustrated in Fig. 4.3), thereby participating in grid interaction. Section 4.3.3 will focus on this method and its application in residential buildings.

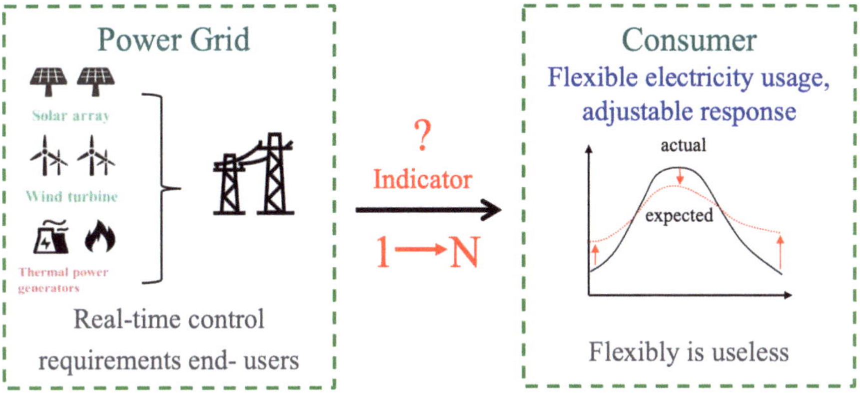

Fig. 4.3 Load-side "bottom-up" autonomous response regulation

4.3 Dynamic Carbon Emission Responsibility Factor for Electricity and Its Application in Residential Buildings

4.3.1 Introduction to C_r

The dynamic carbon emission responsibility factor for electricity (C_r) represents the carbon emission responsibility that power consumers should bear per kilowatt-hour of electricity consumed, which is measured in $kgCO_2/kWh$. The annual average value obtained by time-integrating this factor weighted by electric power over the entire year accurately equals the real total CO_2 emissions from all thermal power plants in the region, resulting from both local per-kWh electricity production and imported electricity from other regions. However, C_r at each moment does not equal the immediate real carbon emissions per kilowatt-hour, primarily because coal-fired power plants will always account for a certain proportion of total electricity generation in China's current and future carbon neutrality era. From a fundamental perspective, achieving ideal flexible regulation with coal-fired units is difficult. As a result, when wind and solar PV power generation are at high output and sufficient to meet the immediate load demand, some coal-fired units still have to operate at their minimum power output. Actually, they remain in "hot standby" mode, so they can quickly ramp up to meet the electrical load demand when wind and solar PV power output drop and can no longer meet the demand. If the power generation from coal-fired units operating at their minimum load plus wind and solar PV power generation exceeds the immediate power load, curtailment will be inevitable. This is the so-called "wind and solar curtailments". In this case, the carbon emissions from power generation by coal-fired units are not for meeting the immediate electricity demand but to ensure the power supply for future peak load periods or wind power valley periods. Therefore, the corresponding carbon emissions should not be attributed to the power consumers during that period, but rather allocated to those consuming power later when thermal power units ramp up to high output to meet demand. From this, the calculation principle of C_r is derived. The immediate C_r is determined based on the operational status of all thermal power units at every moment. When thermal power units operate at high output, C_r is higher than the immediate actual carbon emission per kilowatt-hour of thermal power. This guides the user-ends to minimize electric power as much as possible and release stored energy to reduce the amount of electricity drawn from the grid, thereby alleviating the grid's peak regulation pressure. When thermal power units bank fire and maintain operation at low power, C_r is lower than the immediate actual carbon emission per kilowatt-hour of thermal power. This guides the user-ends to increase their power draw from the grid, store energy, and consume more electricity, thereby reducing the load reduction pressure of thermal power plants and minimizing curtailment.

C_r is a time-varying variable calculated and published in real time by the power dispatch department based on the operational status of all units within the power supply area. The specific calculation method can be found in reference (Jiang et al. 2024). Take a power supply area in northern China as an example. Figure 4.4 shows the variations of C_r over a continuous week in spring, summer, autumn, and winter, respectively. As can be observed in the figure, C_r fluctuates widely on a daily basis, with an annual fluctuation range between 0.1 and 1.7 $kgCO_2/kWh$ and a high-low difference of up to 17 times, demonstrating sufficient sensitivity to incentivize user-end response regulation. In terms of variation patterns, C_r exhibits different fluctuation trends across seasons, with similar intraday variation patterns within the same season. In spring (April), for example, C_r generally exhibits "two peaks and two valleys" within a day. The "two peaks" refer to the morning peak before PV generation begins and the evening peak after PV generation ends, while the "two valleys" are the early morning and midday valleys. The midday valley is caused because the power system can hardly consume the surplus PV energy during the transitional

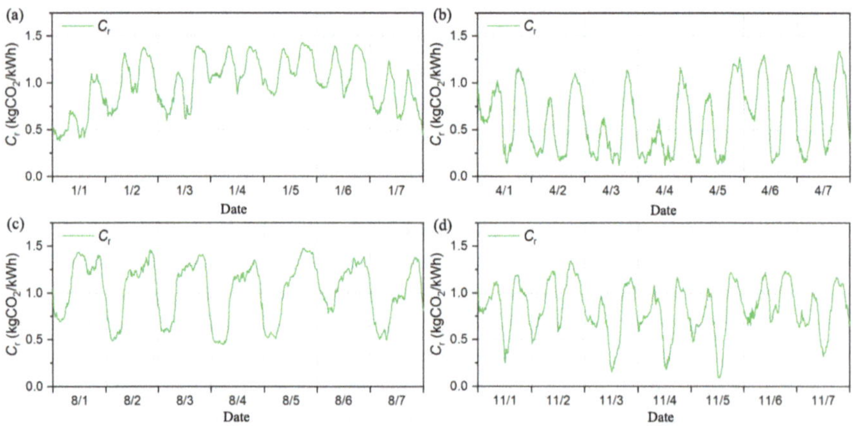

Fig. 4.4 Variations of typical weekly electricity carbon emission responsibility factors in a power supply area in northern China: (**a**) January; (**b**) April; (**c**) August; (**d**) November

season with relatively high PV power generation output. The early morning valley occurs due to the operation of thermal power units in banking fire mode despite the relatively low electrical load at night. In summer (August), C_r exhibits only "one peak and one valley" within a day. This is because the electricity consumption for air conditioning increases significantly in summer, while the existing wind and solar PV power account for a very low proportion, leaving thermal power units to shoulder nearly the entire load demand. This demonstrates that C_r is a dynamic signal that can reflect the supply-demand relationship in the power system, thereby mobilizing massive flexible regulation resources distributed on the load side to participate in power system regulation. During peak periods when thermal power units ramp up to full capacity to meet grid demand, C_r is relatively high, guiding flexible terminals to reduce load and thus alleviating the peak pressure on thermal power. When thermal power units bank fire and maintain a low load, C_r is relatively low, allowing flexible terminals to increase load and thus reducing the curtailment pressure.

4.3.2 Signal Transmission

C_r signals can be transmitted to user ends through power line carriers and public networks. For independently controlled end equipment, existing power facilities are utilized to acquire C_r power line carrier signals through smart electricity meters, eliminating the need for additional communication equipment. C_r signals can be received while the plug is energized to get electricity. Figure 4.5a demonstrates real-time C_r signal transmission and display based on the State Grid's smart electricity meter architecture. In addition, carbon emission responsibilities can be measured and calculated dynamically for electricity usage, as shown in Fig. 4.5b. For end equipment under centralized control and management, such as users of buildings with BAS systems, C_r signals can also be acquired through public networks. At present, building automation management companies, regional energy management systems, and electrical equipment aggregator platforms can acquire C_r signals through this method and uniformly regulate terminal equipment according to their demand.

The power sector calculates and releases C_r signals in real time based on the operational status of generator units. At present, the release interval is set at 15 min. In addition, considering the volatile nature of wind and solar PV power, regional power system dispatch departments will formulate the next day's generation schedule for generator units based on the forecasts of the next day's power load and renewable energy sources such as wind and solar PV power, thus calculating the next day's C_r forecast curve. The power system will issue the next day's 96-point forecast curve at 23:00 day-ahead as a reference signal for load-side terminals to respond to grid regulation. This curve is provided solely as a reference for power consumers, and carbon emission responsibilities shall still be calculated based on the actual C_r values. Figure 4.6 illustrates the comparison between forecast and real-time values of the C_r signal. Although the numerical values are slightly different, the

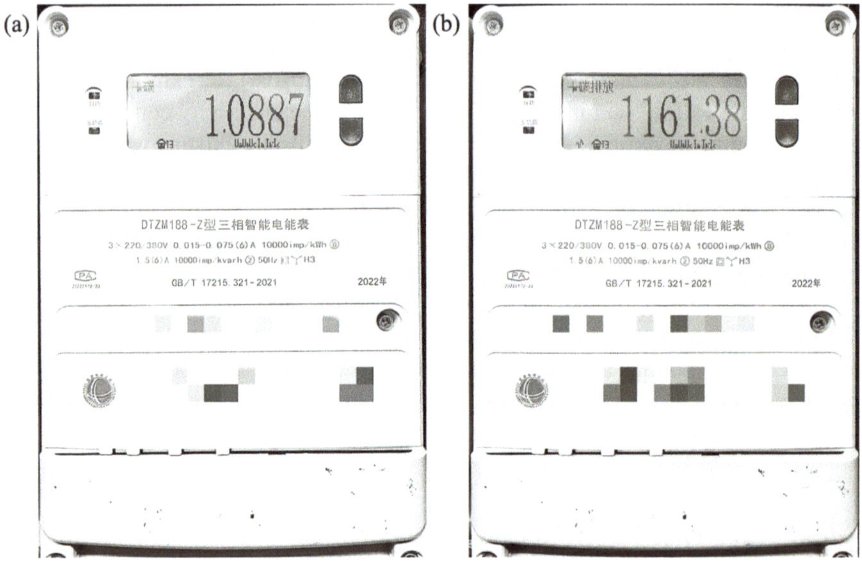

Fig. 4.5 Signal transmission based on smart energy meters: (**a**) Real-time Cr signal transmission and display; (**b**) Carbon emission responsibility measurement and calculation

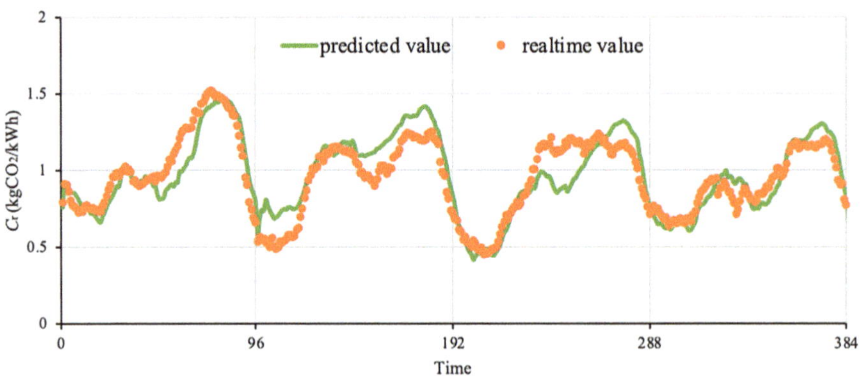

Fig. 4.6 C_r forecast values versus real-time values (4-day data)

variation trends are almost identical. Therefore, the C_r forecast values are sufficient to guide consumers in reasonably arranging electricity usage time for production/daily life based on the C_r variation trend, thereby truly achieving green electricity consumption by adjusting their electricity usage behaviors.

4.3.3 C_r Response

While ensuring basic functionalities, buildings can change their load profiles by leveraging flexible potential through optimized operation scheduling and power adjustment of electric equipment. The C_r-based response regulation enables electric peak regulation and renewable energy consumption. The electrical loads of residential buildings can be classified by their regulation characteristics into three types: energy-storable, time-shiftable, and adjustable loads.

1. Energy-storable load

   Water heaters, ice-storage multi-split air conditioners, and other electrical equipment that can store cold or heat generally only need to be turned on and operated for a certain period within the required time based on the temperature inside the storage space. Appropriately advancing or delaying their start time does not affect their performance. Therefore, by studying the variation patterns of the C_r signal, it is possible to identify the periods within a day when electricity consumption needs to be increased or minimized, as well as the required continuous operation duration and on-off ratio of connected equipment. Then, the scheduling strategy can be made in a day to avoid operation in high C_r periods (electricity shortage) and shift electricity consumption to low C_r periods (electricity surplus) as much as possible.
2. Time-shiftable load

   Time-shiftable electrical equipment, such as refrigerators, washing machines, dishwashers, and robot vacuum cleaners, can reduce loads through the energy-saving mode in non-emergency situations. Such equipment can be turned on and operated in C_r valley periods instead of C_r peak periods to realize low-carbon electricity consumption.
3. Adjustable load

   Adjustable loads can adjust their electric power based on C_r signals. For instance, air conditioners are typical adjustable loads in residential buildings, where both building envelopes and furniture have certain cold and heat storage capabilities. Consequently, short-term turn-off or output power adjustment of air conditioners will not have a significant impact on the indoor ambient temperature, and by such methods as controlling the turn-on and turn-off of air conditioners, changing the compressor frequency of variable frequency air conditioners, and changing the indoor temperature setpoints, flexible control of loads can be realized without no or minimal impact on the user comfort. Figure 4.7 shows the response process of a split air conditioner's autonomous regulation based on the C_r signal. The air conditioner performs the regulation actions of corresponding thresholds based on the values of the C_r signals received continuously.

Adjustable loads like luminaires need to set zoned and tiered adjustment ranges based on functional criticality and service assurance requirements. When C_r exceeds a predefined threshold, the luminaires will consume electricity orderly based on C_r by appropriately reducing illuminance levels in indoor spaces with lower service

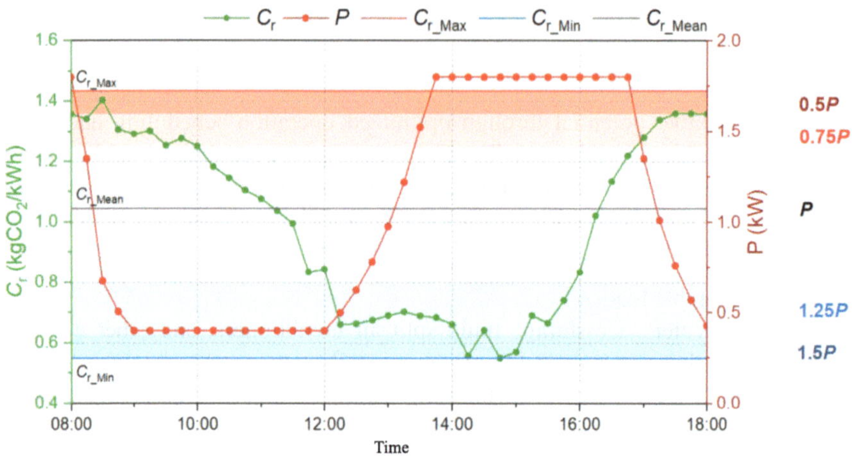

Fig. 4.7 A split air conditioner performing autonomous response regulation based on the C_r signal

assurance requirements. However, given the heightened human sensitivity to luminance variations, it is imperative to develop time-phased and zone-specific granular control strategies for lighting equipment, with comprehensive integration of comfort requirements.

A simple user-defined interface/adjustment knob can be set for all the above power-consuming terminals, enabling users to select the depth S of the equipment's participation in C_r response within the range of 0–1. The S value represents varying degrees of participation in regulation. "0" signifies no involvement in the C_r response regulation and operating under the existing mode; "1" indicates deep response, namely making adjustments based on C_r as much as possible. In this way, each electric equipment responds to C_r regulation through the manufacturer's modification of the product control strategy. Moreover, the regulation depth for each set of equipment can be changed at any time by the direct user via the adjustment knob, thus balancing the requirement to fulfill the building operation functions with the demand to assist the power grid in regulation. The above description of the control strategy only involves the simple principle and theory. However, the details of each product's control strategy are related to its adjustment characteristics and need to be researched and developed separately by the manufacturer, and its difference in performance can be a competitive differentiator among similar products. A good control strategy does not affect the functions and performance of products; in addition, it is highly performable in flexible regulation of electric power, thus increasing system flexibility or reducing the demand for battery capacity with the same system flexibility, which is also the economic benefit that can be achieved through control strategy optimization.

Currently, the Beijing Municipal Administrative Center has chosen a residential compound for demonstrating C_r-based response regulation in household scenarios. Meanwhile, in the context of the energy transition, multiple domestic building electrical equipment manufacturers are actively researching and developing intelligent building electrical equipment that can respond to C_r. The goal is to provide a batch of building electrical equipment that can perform intelligent regulation based on C_r on the market. The planned products include air conditioners, fans, water pumps, black/white household appliances, and luminaires. The hope is that a complete suite of intelligent electrical equipment can be provided to meet the demand for user-side response regulation.

4.3.4 Calculation of Power Consumers' Carbon Emission Responsibilities

The carbon emission responsibility borne by power consumers (T_c) is the time integral of the product of the electric power P_T and C_r at each moment:

$$T_c = \int P_T(\tau) \cdot C_r(\tau) \cdot d\tau \tag{4.1}$$

where, T_c denotes the total carbon emission responsibility that should be borne by a power-consuming terminal for consuming electricity with a power of P_T over a period, measured in tCO_2; C_r represents the time-dependent dynamic carbon emission responsibility factor for electricity, measured in $kgCO_2/kWh$. As C_r is always a positive value, the power-consuming terminal's carbon emission responsibility T_c is also invariably positive, namely, any electricity consumption entails corresponding carbon emission responsibility.

Take a typical residential electricity consumption curve as an example. Under the premise of maintaining the same electricity consumption, the electricity consumption curves before and after participation in the C_r signal adjustment are shown in Fig. 4.8a, while the equivalent energy storage charging/discharging behavior and C_r curve for that day are illustrated in Fig. 4.8b. According to the figure, the typical residential electricity consumption (black curve) reaches its peak at around 19:00, which lasts until approximately 22:00 and is about twice the daytime load. The lowest electricity consumption is observed in the early morning. The carbon emission responsibility per kilowatt-hour under this electricity consumption mode is calculated based on C_r, and the result is 0.71 $kgCO_2/kWh$. After the response based on the C_r signal (red curve), indoor flexible resources, such as water heaters, air conditioners, and robot vacuum cleaners (assuming the maximum power of equivalent energy storage is 50% of the base load and the capacity is 50% of daily electricity consumption), should time their electricity consumption to avoid C_r peak periods and target low C_r periods. Under the premise of unchanged daily electricity consumption, the carbon emission responsibility per kilowatt-hour after response is 0.60 $kgCO_2/kWh$, indicating a 15% decrease.

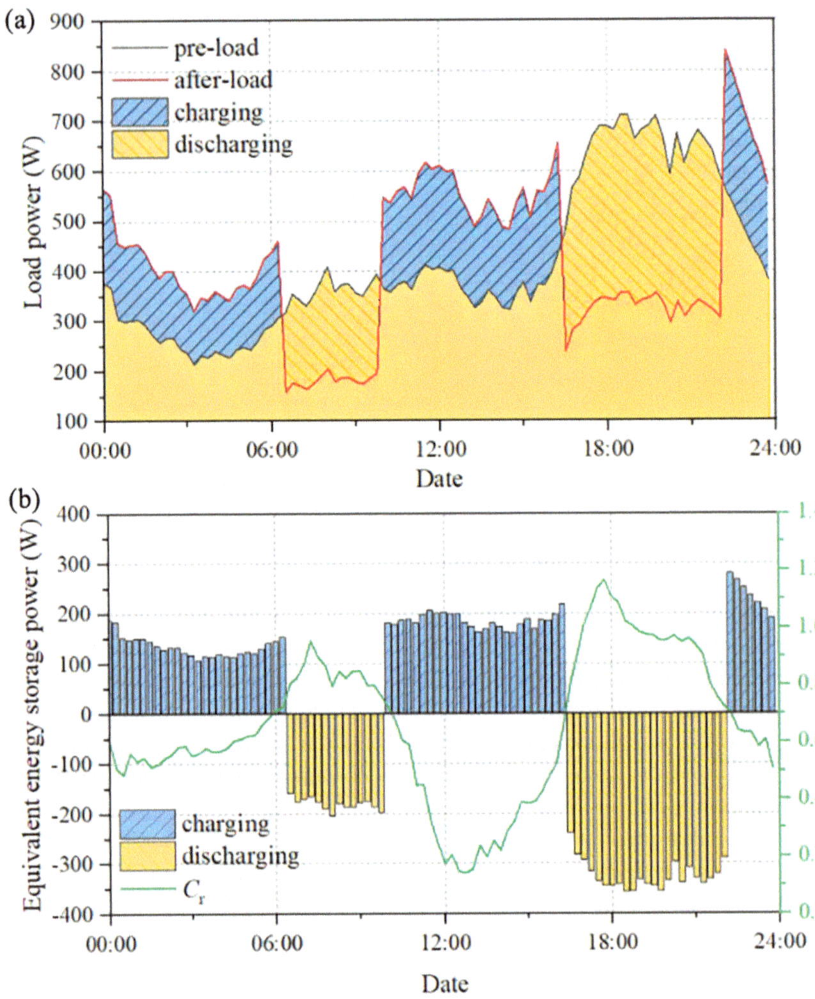

Fig. 4.8 Typical residential electricity consumption curve and response effect

Chapter 5
Flexible Resource Potential in Future Development of Urban Residential Buildings

5.1 Full Electrification Feasibility Analysis

As a crucial component of urban energy consumption, urban residential buildings cover multiple facets of energy consumption, including heating, cooling, cooking, domestic hot water, and lighting. With the advancement of urbanization and the improvement of residents' living standards, the total energy consumption of urban residential buildings continues to increase, while the energy mix is also constantly evolving. Guided by the dual carbon goals, optimizing the energy use structure in buildings and promoting low-carbon transition in the building sector have become the industry consensus, with increased electrification standing as one of the core paths.

5.1.1 Energy Use Structure in Residential Buildings

In 2022, the building operation energy consumption of urban residential buildings amounted to 450 million tons of coal equivalent (TCE), and their carbon emissions accounted for 38.5% of the national total building operation carbon emissions (China Building Energy Efficiency Association, Building Energy Consumption and Carbon Emission Data Committee 2024). These carbon emissions primarily originate from two sources. One is the direct combustion of fossil energy, such as coal, natural gas, and liquefied petroleum gas, and the other is the indirect emissions through power and central heating systems. As coal-fired power generation still accounts for a significant proportion of China's electric power structure, building electricity consumption generates substantial indirect carbon emissions. Therefore, reducing fossil energy consumption and increasing electrification while maintaining building operation comfort have become important goals.

Building Energy Research Center of Tsinghua University, *Decarbonizing Urban Residential Buildings*, https://doi.org/10.1007/978-981-95-8200-6_5

The electrification rate indicates the proportion of electric energy in the terminal secondary energy consumption within the building energy use structure. It is used to measure the level of electrification in buildings. With the development of electrification technology, increasing the electrification rate facilitates reducing carbon emissions and energy consumption in buildings. Currently, the calculation method of electrification rate also incorporates the secondary energy consumption from heat supply into the base number, underestimating the electrification rate. Take heating power as an example. When a heat pump is used for electricity-to-heat conversion, the heating power in the base number should be divided by the heat pump's conversion efficiency to obtain a reasonable electrification rate calculation result. In other words, beyond central heating in urban residential buildings of northern China, the priority of increasing electrification lies in replacing thermal demand based on conventional fossil energy, such as coal and natural gas, with electric alternatives. From the perspective of carbon emissions, direct carbon emissions in the building sector can be reduced by replacing coal and gas with electricity; moreover, indirect carbon emissions from building electricity consumption will gradually decline with the low-carbon development of China's power system.

Urban residential buildings exhibit certain regional differences in energy consumption. In northern China, the promotion of the "coal-to-gas" and "coal-to-electricity" policies has driven the gradual shift of heating methods toward cleaner alternatives. The proportion of heating energy consumption in the national total building energy consumption has been declining year by year. However, the share of gas heating remains significant, and the energy mix has not achieved full decarbonization. In southern urban areas, energy use in residential buildings is mainly attributed to air conditioning for cooling and the operation of electrical appliances, with relatively low demand for winter heating. However, due to the impact of climate change, southern regions are witnessing an increasing number of low-temperature winters, causing heating demands to rise annually. This trend has placed greater pressure on winter power loads. In response, there is a growing push towards adopting high-efficiency electric heating solutions like air source heat pumps.

Regarding the energy use structure in urban residential buildings, the penetration rate of electrical appliances such as induction cookers and electric cookware in residents' daily life continues to rise. Some energy-inefficient gas appliances are being replaced by electric alternatives. However, natural gas still plays a relatively important role in cooking, domestic hot water, and heating. Despite the gradual promotion of high-efficiency electrification technologies such as electric cookware and heat pump water heaters, the overall market acceptance still needs further improvement.

5.1.2 Electrification Pathways for Residential Building Energy Use

1. Key electrification technologies

   The electrification of urban residential buildings primarily covers cooking, hot water, and heating. In recent years, as electrification technologies continue to advance, efficient electrical energy utilization methods such as heat pumps, induction cookers, and building-integrated photovoltaics (BIPV) have become increasingly ubiquitous, providing reliable technical support for building electrification.

Heating electrification is a vital component of building electrification. Northern urban heating still relies on central heating methods to provide zero-carbon heat (including renewable energy heating, industrial waste heat, nuclear power waste heat, and other clean heating methods). For scattered heating in northern China and heating in southern China, decentralized electrified heating should be adopted, such as replacing traditional gas-fired boilers, direct electric heating and other inefficient electric heating technologies with efficient heat pump technologies. At present, low-temperature air source heat pumps can operate stably even in northern China with lower outdoor temperatures, and their energy efficiency ratios can reach above 3.0, that is, they can deliver over 3 kWh of heat for every 1 kWh of electricity consumed, thus significantly reducing heating costs (see Table 5.1).

In urban domestic water heaters, electric water heaters account for 52.9%, followed by gas water heaters at 35.1%, while solar water heaters have a utilization rate of 9.5%. Gas and inefficient electricity consumption remain the dominant methods for providing domestic hot water, while decentralized heat pump water heaters have emerged as an efficient alternative. Compared to traditional electric water heaters, heat pump water heaters can achieve an energy efficiency ratio of 3.0~4.0, thereby substantially lowering electricity consumption. Although the initial investment in heat pump water heaters is higher than that in other types of water heaters, the operating costs of heat pump water heaters can be dramatically reduced given their energy efficiency and China's domestic tariff policy (see Table 5.2).

In cooking, various methods such as steaming, boiling, frying, stir-frying, and deep-frying can now be largely electrified. However, influenced by China's cooking habits and culture, users still favor utilizing gas cooking appliances. The thermal efficiency of traditional gas stoves is about 55%~63%, while that of electric cookware like induction cookers available on the market can reach up to 90%, significantly reducing energy costs. However, constrained by the uneven heating of induction cookers and the habit of open-flame cooking, the transition to electricity is proceeding slowly. Electric flame stoves, as a new type of cook-

Table 5.1 Comparison of operating costs for different heating methods

Heating method	Operating cost (RMB/m^2)		
	Beijing	Shenyang	Harbin
Coal-fired boiler	33.84	40.83	42.09
Household gas-fired boiler	31.93	38.77	40.01
Direct electric heating	105.3	128.31	132.47
Air source heat pump (quasi-two-stage compression)	29.45	41.61	48.91
Air source heat pump (latent heat storage)	30.67	40.97	47.39

Table 5.2 Comparison of household energy consumption and operating costs for various types of water heaters

Type	Gas water heater	Electric water heater	Heat pump water heater
Annual hot water consumption (L)	45 L/(人 days) × 3 人 × 365 days = 49,275 L 45 L/(person days) × 3 persons × 365 days = 49,275 L		
Thermal efficiency	90%	95%	250%
Annual energy consumption	224 m^3	2118 kWh	847 kWh
Energy price	RMB 3.5/m^3	RMB 0.5/kWh	RMB 0.5/kWh
Annual energy cost	RMB 784	RMB 1059	RMB 424
Equipment price	RMB 1500	RMB 1200	3000

Note: The calculation is based on the current average water supply temperature of 10 °C and the water heater outlet temperature of 45 °C

ing equipment, combine the advantages of gas stoves and induction cookers. By employing low-temperature plasma technology, they efficiently convert electric energy into heat energy. They not only replicate the flame effect of traditional gas stoves but also maintain green, safe, and environmentally friendly characteristics. In the future, as electric cookware technology continues to upgrade, the electrification of cooking is poised to accelerate further.

2. Electrification pathways

With the continued progress of electrification technologies for heating, domestic hot water, and cooking, coupled with strong national policy support, electrification is gradually becoming a crucial pathway to improve energy efficiency and reduce carbon emissions. To ensure smooth electrification, the key lies in enhancing the electrical load capacity of buildings and effectively replacing traditional energy sources.

For existing buildings, the design often fails to account for the load changes that may arise after full electrification is achieved. Special attention should be given to distribution capacity to meet the increased power demand after electrification. For new buildings, electrification requirements should be considered during the planning and design phases to optimize the distribution system design. This involves increasing the capacity of the distribution system and rationally allocating distribution lines to ensure the system can effectively support the growing loads resulting from electrification in the future.

The substitution of traditional fuel consumption in urban residential buildings with electricity is the main pathway for electrification, particularly in heating, hot water, and cooking. The low-carbon transition of the power system is the key to supporting building electrification. By combining these two approaches, traditional energy consumption such as coal and natural gas will be gradually replaced with green electricity in the building sector, thus significantly reducing building carbon emissions.

5.1.3 Carbon Reduction Potential of Full Electrification

In the current electric power structure of China, the proportion of thermal power is over 60%. There is a concern that the full electrification of buildings may increase indirect carbon emissions from electricity, thereby exacerbating the carbon reduction burden. However, this concern overlooks two core facts. First, electrification technologies can significantly offset the increase in carbon emissions from electricity by improving energy efficiency. Second, direct fuel substitution in the building sector can result in immediate emission reduction benefits. In the case of gas, for example, urban gas consumption in 2024 was about 141.3 billion m^3, and the carbon emission intensity resulting from direct combustion of gas was 2.75 $kgCO_2/m^3$ (calculated based on the carbon emission factor of methane combustion). If gas is replaced with efficient electrification equipment, net emission reductions can still be achieved even based on the current electric power structure (the national average carbon dioxide emission factor of electricity in 2023 was 0.54 $kgCO_2/kWh$). For instance, in heating, the carbon emission intensity of a gas-fired boiler with a thermal efficiency of 90% per unit of heat supply is 0.31 $kgCO_2/kWh$. By contrast, an air source heat pump operating at a COP of 3.0 for the same heat supply demonstrates carbon emissions at 0.18 $kgCO_2/kWh$, achieving an emission reduction of approximately 42%. In cooking, taking Shanghai as an example, gas cooking generates carbon emissions of 1.99 $kgCO_2/h$. By contrast, electric cooking produces only 1.48 $kgCO_2/h$, resulting in a reduction of about 25% in carbon emissions. These calculations show that full electrification is not "trading one high-carbon solution for another", but rather achieving net reductions in direct carbon emissions through the energy efficiency advantages of technology. The core objective of full building electrification is clear and quantifiable, that is, to replace the annual 141.3 billion m^3 of natural gas consumption in urban areas and eliminate its direct carbon emissions (about 388 million tons of CO_2/year).

If full electrification is realized, carbon emissions from building operations will simultaneously reach net zero with upgrades to the distribution system (such as increasing per-household capacity to 8–10 kW) and growth in the proportion of green electricity. In particular, it should be noted that the transition of the energy system and the innovation in the building terminal energy consumption methods shall be dynamically coordinated. Currently, non-fossil energy accounts for nearly 60% of the installed capacity in China's power system, but some still argue that switching from gas to electricity should wait until a completely zero-carbon power system is established. This perception has neglected the temporal alignment in the implementation pathway of the energy revolution. It takes 20–30 years for the power system to achieve deep decarbonization. Delaying terminal retrofitting until the power system is fully decarbonized would not only forfeit current annual opportunities to reduce hundreds of millions of tons of direct emissions but also lead to delays in updating infrastructure and reshaping consumer behaviors. Just as the promotion of new energy vehicles and the construction of charging networks interact as both cause and effect, the electrification of buildings and the low-carbon development of the power grid must advance in parallel. During this process, carbon emissions may exhibit a curve of rising first and then falling, but this represents the inevitable transition cost of system reconfiguration. What truly demands foresight is recognizing that lifestyle transformations are often more challenging than technological iterations. The shift from gas stoves to induction cookers transcends mere equipment replacement—it constitutes a systemic evolution encompassing cooking culture and space design.

5.2 Photovoltaic Utilization in Urban Residential Compounds

With the continuous deepening of the global energy transition, distributed PV power generation, as an important component of clean energy, has been widely promoted around the world. Given their significant proportion in urban land occupancy coupled with concentrated and stable electricity demand, residential compounds are progressively emerging as a key scenario for distributed PV applications. Meanwhile, the electricity consumption characteristics of residential compounds, including the temporal variations in electrical load curves, not only determine the technical feasibility of PV system integration but also directly affect the self-consumption capacity of PV power generation and the overall economic performance of the system. This section will analyze the PV installation potential and electricity consumption characteristics of residential compounds, explore the matching relationship between them, and further evaluate the operation model and economic viability of PV systems, thereby providing the theoretical and practical basis for PV applications in residential compounds.

5.2.1 Potential for PV Installation in Residential Compounds

Distributed PV systems, as an important form of renewable energy, possess the unique advantage of being seamlessly integrated with existing buildings and infrastructure without the need for additional installation space. This approach not only enables efficient utilization of architectural space but also significantly reduces power transmission costs, thereby raising energy efficiency.

As shown in Fig. 5.1, residential land accounts for a relatively high proportion of the total land area in China's major cities (Beijing, Shanghai, Guangzhou, Harbin, Wuhan, and Kunming). Among them, the proportion of residential land is 35.2% in Beijing, 31.0% in Wuhan, 28.0% in Shanghai, 27.5% in Kunming, 26.2% in Harbin, and 19.6% in Guangzhou. This demonstrates that residential land, as a vital category of urban land use, constitutes high-quality resources for the deployment of distributed PV systems. By further combining the spatial distribution of urban land use types shown in Fig. 5.1, it can be observed that residential land in each city not only occupies a large share of urban areas but also is distributed in a relatively concentrated manner. Installing PV systems on existing building facilities, especially the rooftops of residential buildings can significantly elevate the spatial potential of PV power generation. This deployment method aligns with the concept of green and low-carbon development as it avoids occupying agricultural land, ecological land, and other land resources.

It should be pointed out that, although most residential buildings have some vacant roof space, the actual area available for installing PV modules is only 50% to 70% of the total vacant roof space (i.e., an effective utilization rate of 0.5–0.7). In response to this situation, the installation mode of PV modules must be more flexible. Modular design and customized arrangements should be adopted to fully adapt to the specific conditions of different rooftops, thereby ensuring optimal power generation benefits within limited space. The current technological path recommends maintaining the effective utilization rate between 0.5 and 0.7 and using this range as

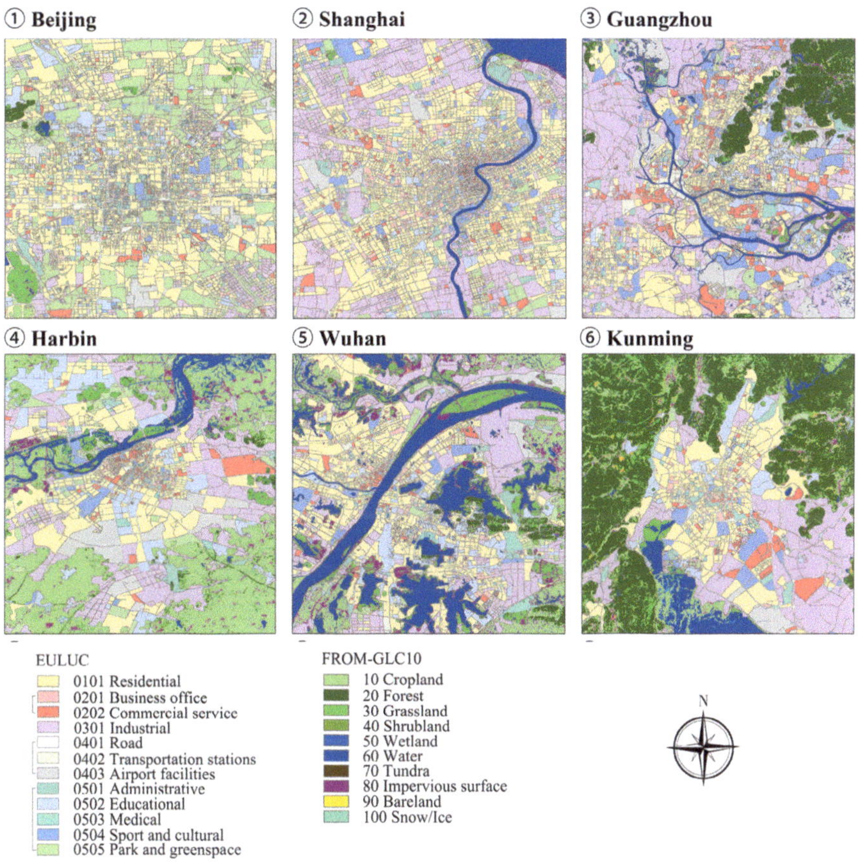

Fig. 5.1 Land use classification in major cities of China in 2018

the reference standard for installation design and evaluation. Meanwhile, the rooftop utilization efficiency can be further elevated by using lightweight modules, optimized racking designs, and intelligent management means in the future, thus unlocking greater space and higher power generation potential for PV system deployment in residential compounds.

In recent years, residential PV systems have been increasingly recognized and applied in practice nationwide (Fig. 5.2). The Wuhan Hongtuli Residential Compound PV Power Generation Project represents a typical case of PV application in high-rise residential buildings. This compound has installed 60 kWp of thin-film solar cell modules on the rooftops of four high-rise residential buildings, which can generate approximately 63,000 kWh of electricity per year, primarily to meet the lighting needs of the underground garages. The Shanghai Chongming Hemu Jiayuan Residential Compound Rooftop PV Project demonstrates another practice model of residential PV. This compound has installed 2.7 MW_p of modules on the rooftops of 88 residential buildings. The project follows a model where rooftop space is shared and all generated electricity is fed into the grid. Its annual total electricity generation reaches 2.83 million kWh, and the generated revenue is used for compound management and maintenance, benefiting over 3600 households. In addition, the Chisan Residential Compound in Hongkou District, Shanghai has installed PV systems on the rooftops of three residential buildings, and the generated electricity is used preferentially for lighting works in public areas of the community. These projects have explored the technological paths for PV systems in existing residential communities, providing feasible solutions for energy retrofitting of established compounds. The abovementioned cases fully reflect the flexibility and feasibility of residential PV systems. Whether in high-rise residential buildings, centralized communities, or existing building retrofits, distributed PV can be tailored to local conditions to deliver clean power for communities, thereby facilitating urban low-carbon development.

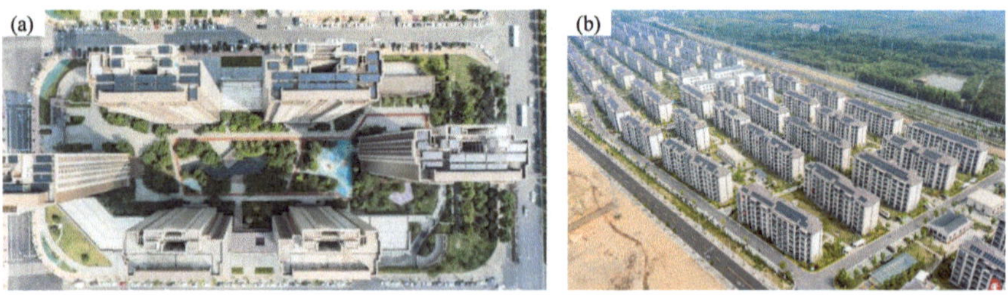

Fig. 5.2 Cases of PV projects in residential compounds. (**a**) Wuhan Hongtuli Compound PV Power Generation Project; (**b**) Shanghai Chongming Hemu Jiayuan Residential Compound PV Project

5.2.2 *Electricity Consumption Characteristics of Residential Compounds*

Residential compounds constitute a significant component of urban energy consumption, exhibiting diverse and complex electricity consumption characteristics. Figure 5.3 illustrates the variation in daily electricity consumption of 600 residential buildings in a typical year. To ensure comparability of electricity consumption data among residential buildings of different scales, the daily electricity consumption has been normalized by dividing it by its mean value. As can be seen, the electricity consumption of individual buildings exhibits significant fluctuations over time. Moreover, there are considerable individual differences in the electricity consumption behaviors of different buildings in the time dimension, as reflected in the high dispersion of the raw curve in (Fig 5.3). However, when the electricity consumption of all residential buildings is averaged, a smoother trend of variation emerges and reveals a clear seasonal pattern: Electricity consumption in winter and summer is significantly higher than that in spring and autumn, and the latter remains relatively low and stable.

The normalized electricity consumption curves on typical winter and summer days were further analyzed by selecting January 4 (winter) and July 4 (summer) as representative days (see Fig. 5.3). The hourly electricity consumption curve on a typical winter day exhibits a double-peak characteristic. The electricity consumption peaks in the morning (6:00–9:00) and evening (18:00–22:00) respectively, which is closely related to residents' daily routine. By contrast, the electricity consumption curve on a typical summer day exhibits a different characteristic. The overall load gradually increases from noon (at 12:00), peaks in the afternoon (at 18:00), and then slowly declines in the evening. This reflects the marked impact of air conditioning and refrigeration equipment under high-temperature conditions.

To have an in-depth understanding of the electricity consumption characteristics of residential buildings, a cluster analysis was made on the annual data to examine

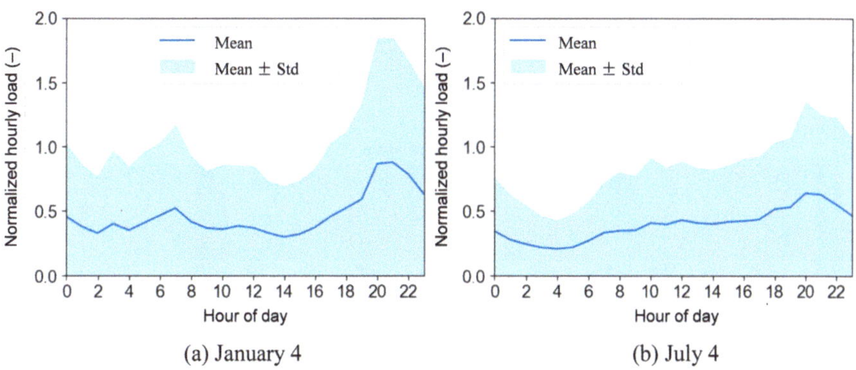

Fig. 5.3 Variation in hourly electricity consumption of 600 residential buildings (normalized by dividing it by its mean value). (**a**) January 4; (**b**) July 4

intraday electricity consumption modes. The hourly electricity consumption curves throughout the year were grouped by date, and the curve within each date was normalized by dividing it by its mean value to retain only shape information. Then, the normalized intraday curves were clustered using the k-means method, and the number of clusters was set to 2. The results are shown in Fig. 5.4. Through cluster analysis, it was found that the intraday electricity consumption modes of residential compounds could be classified into two types: The first type (Cluster 1) exhibits a typical load curve characterized by a gradual increase from noon to afternoon (12:00–18:00) followed by a slow decline in the evening. The second type (Cluster 2) demonstrates a curve characterized by double peaks which occur in the morning (6:00–9:00) and evening (18:00–22:00) respectively. This classification not only visually demonstrates the electricity consumption behavior characteristics of different residential buildings but also provides a basis for subdividing the residential electricity consumption modes. Figure 5.5 illustrates the distribution pattern of cluster labels within the annual range. The results indicate that the electricity consumption mode of Cluster 1 is mainly concentrated in summer, while that of Cluster 2 is predominantly in winter. This distribution pattern is closely related to the impact of seasonal climate changes on residential electricity consumption behaviors.

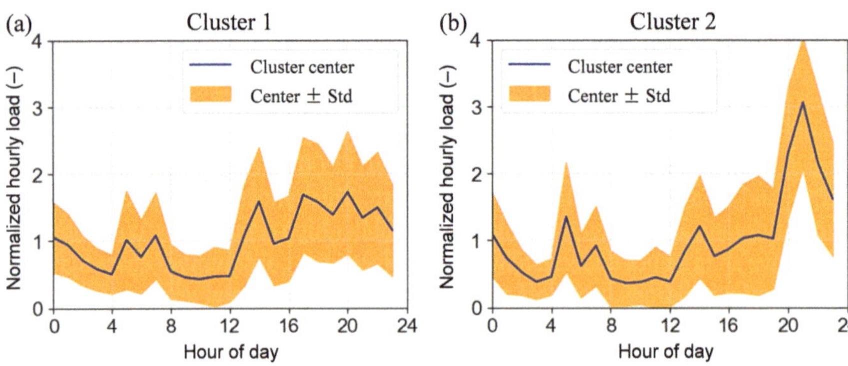

Fig. 5.4 Results of k-means clustering of the normalized intra-day electricity consumption curves for buildings

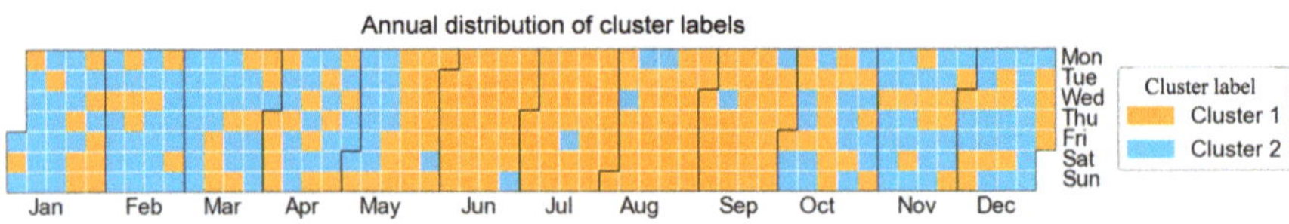

Fig. 5.5 Annual distribution of k-means clustering labels for normalized intra-day electricity consumption curves for buildings

5.2.3 *Electricity Generation and Consumption Matching and PV Consumption*

1. **Relationship between total electricity generation and consumption in residential buildings**

   The annual electricity consumption per storey of residential buildings and the annual electricity generation from rooftop PV are estimated. According to the *Standard for Energy Consumption of Building* (GB/T 51161-2016), the comprehensive annual electricity consumption per household in Beijing's residential buildings is approximately 2700 kWh. Given the number of households per storey, the annual electricity consumption per storey can be further calculated. Meanwhile, the annual electricity generation of the rooftop PV system can be calculated, by referencing the parameter of typical PV modules, that is, the capacity per square meter of PV modules is about 180 W_p, and Beijing's weather data and integrating the rooftop area of residential buildings and the proportion of area covered by PV panels.

   Based on the above data, the proportion of annual PV electricity generation in building electricity consumption in residential buildings of different storeys is obtained, as shown in Fig. 5.6. The results show that in a low-rise residential building (e.g., 4–6 storeys), the electricity generation of rooftop PV can largely satisfy the building's electricity demand, with coverage rates reaching 60% to

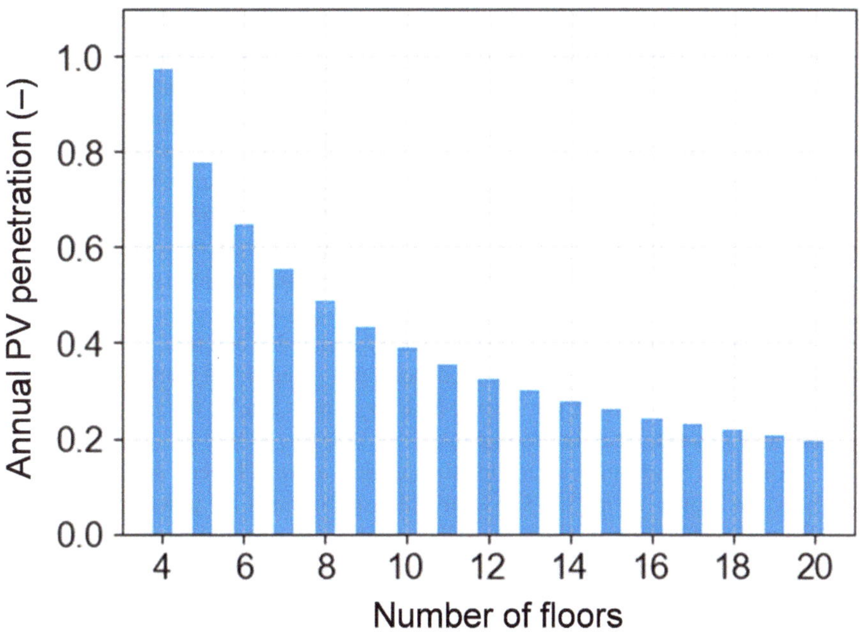

Fig. 5.6 Proportion of annual PV electricity generation in building electricity consumption in residential buildings of different storeys

100%. However, as the number of storeys increases, the proportion of PV power generation in electricity consumption gradually decreases due to the relatively fixed roof area and increased electricity consumption of users. At the 20th storey, it drops to a relatively low level of about 20%. Therefore, PV power generation in low-rise residential buildings is expected to achieve a higher self-sufficiency ratio. However, considering the inherent intermittency and volatility of PV power generation, it is still essential to make an in-depth analysis of the matching between electricity generation and consumption curves on a time scale, thus determining the actual utilization of PV power generation.

2. **Temporal matching relationship between residential electricity generation and consumption**

The temporal matching between PV power generation and building electricity consumption is crucial for improving the utilization rate of distributed PV systems. By analyzing the temporal variation characteristics of PV electricity generation and demand in residential buildings with different numbers of storeys, the matching relationship between generation and consumption can be revealed. The PV self-consumption and self-sufficiency ratios are used to quantify it. The following elaboration will be structured around three perspectives: daily variations throughout the year, hourly variations during a typical week, and matching metrics.

Figure 5.7 illustrates the daily electricity generation and consumption statuses of residential buildings with different storey counts (6-storey and 13-storey) throughout the year. As can be observed from the figure, the daily variation trend of PV power generation and the buildings' electricity demand exhibit certain differences on the whole. PV electricity generation is significantly affected by seasonal variations in solar radiation intensity. It is relatively high in summer and low in winter. By comparison, electricity consumption in buildings peaks in both summer and winter. This is primarily driven by the cooling demand in summer and the electricity demand of auxiliary equipment for heating in winter. Consequently, on an annual scale, there is an obvious mismatch in the shapes between PV power generation and building electricity consumption. Quantitatively, for a 6-storey residential building, the proportion of PV electricity generation in the building's annual electricity consumption is approximately 65%; for a 13-storey residential building, this proportion drops to 30% as a result of the significant increase in total electricity demand. Therefore, in terms of the daily total over the entire year, the buildings' electricity demand curve can largely cover the PV power curve, and PV power can be fully utilized if the intraday mismatch can be resolved.

The hourly variation characteristics during a typical week are further analyzed, as shown in Fig. 5.8. The temporal distribution of PV power generation is distinctly different from the peak-valley pattern of residential electricity demand. PV power generation is mainly concentrated at noon, while residential electricity consumption typically peaks in the morning and evening. Due to this temporal misalignment throughout the day, the temporal mismatch between electricity generation and consumption has restricted the immediate consumption and local utilization of PV power generation.

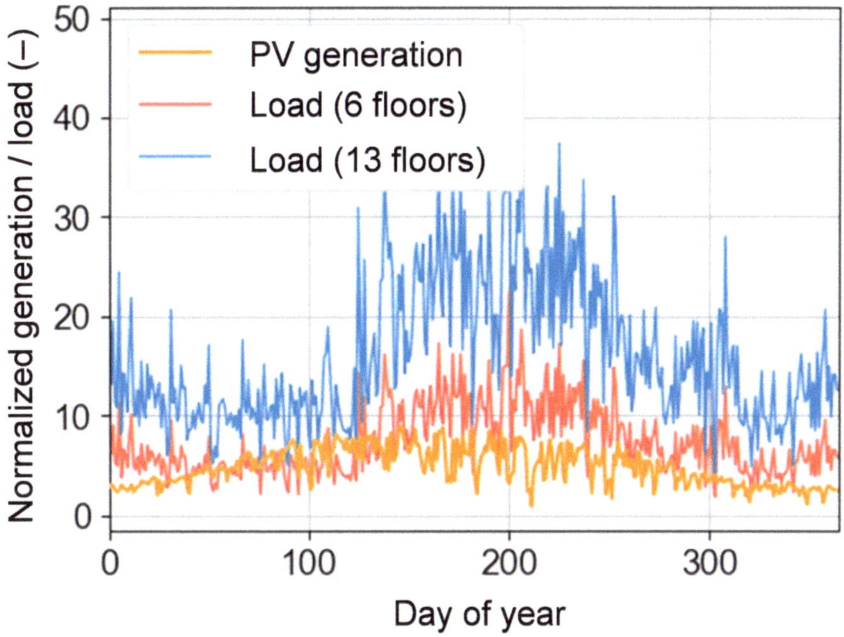

Fig. 5.7 Variations in daily electricity generation and consumption in residential buildings with different storey counts

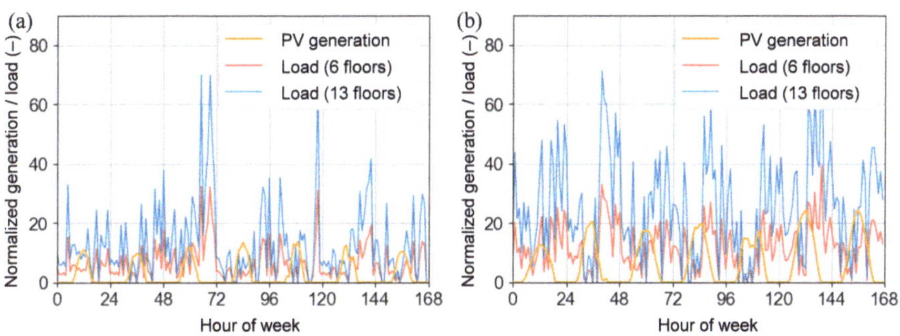

(a) Typical week in January (b) Typical week in July

Fig. 5.8 Variations in hourly electricity generation and consumption in residential buildings with different storey counts in typical weeks. (**a**) Typical week in January; (**b**) Typical week in July

To quantitatively evaluate the utilization efficiency of the PV system, the system's self-consumption and self-sufficiency ratios were calculated. Among them, the self-consumption ratio measures the proportion of PV power generation directly consumed by the building's electricity demand, while the self-sufficiency ratio reflects the proportion of the building's electricity demand met by PV power generation. Analysis results reveal significant differences in PV utilization efficiency among residential buildings with different storey counts. For a 6-storey residential building, PV power generation accounts for 65% of the building's total electricity consumption, but the self-consumption ratio is only 40%. This indicates that despite the high proportion of PV electricity generation, it is not fully utilized by the building due to the mismatch between intraday electricity generation and demand. By comparison, for 13-storey and 20-storey residential buildings, the proportion of PV power generation decreases to 30% and 19% respectively, but the self-consumption ratio rises to 58% and 66% accordingly due to increased electricity demand. Despite this, PV systems in high-rise residential buildings are still restricted by the temporal mismatch between power generation and consumption peaks, and the consumption of PV power generation is still insufficient.

In general, the electricity consumption peaks in residential buildings are concentrated in the morning and evening, and this is prominently not matched with the intraday distribution of PV power generation. For a 13-storey residential building, the self-consumption ratio can only reach 58%. The introduction of energy storage systems or load regulation technologies is the key to achieving higher PV utilization rates.

5.2.4 Operation and Economic Viability of PV Systems

1. **Connection mode and operation model of rooftop PV systems**

   Driven by the current dual carbon goals, rooftop PV systems for residential buildings have become an important application scenario for distributed PV development. Unlike industrial and commercial buildings, residential PV investors primarily consist of property management companies, whose focus is on achieving economic returns through PV systems while improving community service quality and reducing overall energy consumption. For residential buildings, common PV connection modes include full self-consumption (Mode 1), self-consumption with surplus feed-in (Mode 2), and full feed-in (Mode 3). Based on practical operating experience and policy characteristics of different regions, the following provides a systematic analysis of the three grid-connection modes in terms of system configuration, applicable scenarios, advantages and disadvantages, and implementation barriers.

   Mode 1: Full self-consumption. This mode requires that all electricity generated by the PV system be consumed by the electricity demand of the residential building, and unused electricity is not allowed to be fed back into the grid. This mode applies to residential compounds with stable electricity demand, especially those featuring high electrical loads for public facilities, such as underground parking lot lighting, elevator operation, security systems, and public area illumination. The advantage of this mode lies in avoiding complex interactions with the power grid while helping compound property management companies cut electricity costs. However, its main limitation resides in the temporal mismatch between electrical load and PV power generation when PV electricity generation accounts for a large proportion. During the peak hours of PV power generation in the daytime, many residents are away from home, resulting in a lower electrical load.

Mode 2: Self-consumption with surplus feed-in. The PV system prioritizes the electricity demand within the compound, and the surplus electricity is fed into the grid through a bidirectional electricity meter to generate revenue from electricity sales. This mode not only enables the self-consumed portion of PV electricity to fully reflect the value of high time-of-use tariffs but also generates additional revenue through surplus electricity feed-in when the PV electricity generation cannot be fully consumed. This mode imposes higher requirements on power grid management, necessitating detailed evaluation and analysis of regional distribution network capacity for reverse power transmission.

Mode 3: Full feed-in. This mode requires that all electricity generated by the PV system be directly connected to the grid, and the property management company earns revenue from electricity sales. In this mode, the PV system, equivalent to an independent power generation unit, is unrelated to the actual electricity consumption of residential buildings. Its financial model is simple with stable returns, making it suitable for scenarios where the residential compound has low electrical loads and abundant rooftop PV resources. For residential buildings, the major advantage of adopting the full feed-in mode lies in the simplicity of system design and operation. Property management companies do not need to consider the matching between electrical load and PV power generation in the compounds, and can instead obtain long-term stable revenue through fixed feed-in tariffs. However, its main limitation stems from the fact that the fixed feed-in tariffs are often set too low. For example, Beijing's fixed feed-in tariff is RMB 0.36/kWh, significantly lower than the time-of-use tariff. This causes the full feed-in mode to be less economically viable than the self-consumption with surplus feed-in mode. More importantly, with the increase in PV installations, feed-in tariffs are continuously being reduced across China (as low as RMB 0.20/kWh in some regions, and even down to RMB 0.10/kWh in parts of northwest China), significantly dampening the enthusiasm for connecting PV to grid across the country. On the other hand, the large-scale grid connection of PV also places considerable pressure on the power grid, increasing the difficulty of grid regulation and consumption. This shows that the full feed-in mode is not sustainable.

The roof of a residential building is generally considered a part common to all owners. As stipulated in Article 271 of the *Civil Code of the People's Republic of China*, a unit owner has the ownership over an exclusive unit of a building, such as a dwelling space or a space used for operating businesses, and has the right to co-own and jointly manage the common space other than the unit. Therefore, the investment and operation of a rooftop PV system require collective decision-making by the owners, such as approval from the owners' assembly. According to Article 278 of the *Civil Code of the People's Republic of China*, changing the intended use of the co-owned space or making use of the co-owned space to engage in business activities shall be subject to the unit owners' joint decision, and the quorum for this matter shall be two thirds or more of the exclusive units both by area and by number of unit owners. Below is an introduction to common investment and operation models for residential PV systems.

Joint investment by owners: Under this model, all owners collectively invest in the installation of a PV system, with income allocated based on their capital contribution ratios or other mutually agreed terms. The advantage of this model lies in its full embodiment of the owners' collective rights and interests, but its implementation necessitates a high level of collaboration and mutual trust among stakeholders. Decision-making efficiency may be somewhat limited due to the larger number of owners involved.

Investment led by the property management company: The property management company serves as the investment entity and is responsible for the construction, maintenance, and operation of the PV system. Owners benefit through reduced utility electricity charges, lower property management fees, or direct participation in income sharing. This model simplifies the owners' participation process and improves project implementation efficiency. However, it requires clearly defining the profit distribution mechanism between the property management company and owners through contracts to ensure transparency and fairness.

Investment by a third-party enterprise: Under this model, a professional PV investment enterprise is engaged to undertake the construction and operation. Owners profit by leasing rooftops, enjoying discounted electricity tariffs, or sharing income based on agreed proportions. The advantage of this model is that owners are not required to undertake the initial investment and subsequent operation and maintenance, but they must enter into long-term agreements with the enterprise to clearly define the rights and responsibilities of all parties. In addition, the enterprise, as the investment entity, may have higher requirements on the rate of return, thus entailing an evaluation of the specific rooftop resources and policy support.

The connection mode and operation model of a rooftop PV system in a residential building play a crucial role in promoting distributed PV. The different connection modes have their unique advantages and disadvantages, and the selection should be based on actual electrical loads, PV resources, and local policies. In terms of operation models, the model with investment led by the property management company and that with investment by a third-party enterprise have become the focus of current promotion due to their advantages in simplified management and efficient operations. However, the broader application of these models entails further improvement of policy support, optimization of the income distribution mechanism, and strengthening of cooperation between owners and investors to jointly promote the development of green and low-carbon communities.

2. **Investment income under different connection modes**

Based on the residential PV power generation and building electricity consumption curves in Beijing introduced in the Sect. 5.2.3, payback periods and PV self-consumption ratios of different PV connection modes across low-, mid-, and high-rise residential compounds were analyzed, with results shown in

Fig. 5.9. In the figure, Mode 1 represents full self-consumption, Mode 2 denotes self-consumption with surplus feed-in, and Mode 3 indicates full feed-in. As observed in the figure, the selection of PV connection mode is directly related to the PV self-consumption ratio (SCR):

- Low-rise residential compounds have lower SCR, and the ranking of their payback periods is as follows: Mode 2 has the shortest payback period, Mode 3 follows behind, and Mode 1 has the longest payback period;
- High-rise residential compounds have higher SCR, and the ranking of their payback periods is as follows: Mode 2 has the shortest payback period, Mode 1 follows behind, and Mode 3 has the longest payback period.

Mode 2 (self-consumption with surplus feed-in) demonstrates the shortest payback period among all three types of residential compounds. This is because this mode enables the value of PV power generation to be maximized: The self-consumed portion can replace higher-priced grid electricity, while the surplus electricity fed into the grid can generate additional revenue through electricity sales. However, in practice, this mode is often subject to multiple restrictions, including local grid management, and its widespread application is still challenging. As future grid-connected PV technology and policy requirements will become increasingly stringent, and the feed-in tariffs may be further lowered, continuously optimizing the economic viability of PV systems will necessitate increasing the local consumption of PV electricity generation. For example, energy storage or equivalent energy storage, flexible regulation, and other technologies will be adopted to further increase the PV SCR, shorten the payback period, and promote the efficient utilization and sustainable development of residential PV systems.

It should be noted that the payback periods shown in the figure were calculated based on the total investment and total income of PV systems, without considering the actual income distribution in multi-owner scenarios. In a real project, the investment and operating income of a PV system often need to be

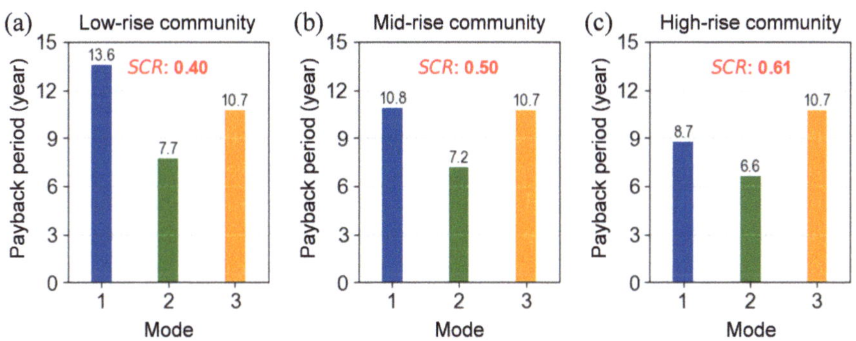

Fig. 5.9 PV SCR of different residential compounds in Beijing and payback periods of PV system in different connection modes (Mode 1: full self-consumption; Mode 2: self-consumption with surplus feed-in; Mode 3: full feed-in)

distributed among owners, and the property management company or a third-party enterprise, and different distribution modes can have remarkable impacts on the payback periods for individual investors. Therefore, before the implementation of the project, a clear consensus on the income distribution mechanism shall be reached to safeguard the rights and interests of all parties and ensure the project's economic viability and feasibility.

5.3 Regulation Potential of Flexible Resources in Residential Buildings

5.3.1 Flexible Resources of Residential Buildings

With the full electrification of urban residential buildings and the wide application of PV systems in residential buildings, how to increase the flexibility and regulation capacity in electricity consumption of buildings has become a key issue for optimizing the energy mix and improving the PV consumption rate. There are multiple types of controllable flexible resources in residential buildings, which can be used to cut electricity demand during peak load periods of the power system and increase electricity consumption when the PV output is high, thereby achieving a power supply-demand balance. According to different energy usage types and regulation methods, the main flexible resources in residential buildings consist of the following categories:

1. Electric water heaters and heat pump water heaters

   Electric water heaters and heat pump water heaters are among the most typical adjustable loads in residential buildings. In particular, heat pump water heaters have become important equipment for improving the electrification of buildings due to their high energy efficiency ratios. The core flexible characteristic of such equipment resides in its energy storage capacity, as its heating water tank can operate in advance to store heat during periods of low tariff or abundant PV generation and suspend operation during peak periods to reduce grid burden. A single household electric water heater theoretically has an adjustable load range of 1–2 kWh. Hence, in a residential compound of 100 households, the overall regulation potential can reach 100–200 kWh. In a real regulation process, the adjustable capacity varies among equipment due to the influence of factors such as equipment parameters, water temperature, and service time. An experimental study by Newcastle University suggests that domestic heat pump water heaters can reduce operating power by an average of 0.3 kW for about 50 min during peak load periods. In addition, C_r signals can be used to direct the optimization of the water heater operation mode and enhance the carbon reduction effect. The WiFi module of a water heater can connect to the mobile application program to acquire the real-time variation data of C_r in 1 day. By predicting the variation of C_r for the next day with machine learning algorithms and combining it with users' water usage habits depicted by hot water valve opening signals, the water heater operation mode for maximizing carbon reduction can be obtained.

2. Air conditioning equipment

   Air conditioners constitute the main seasonally-oriented power loads in residential buildings. The flexibility of air conditioning equipment lies in utilizing the thermal inertia of buildings and occupants' acceptable range of comfortable temperature to regulate temperatures, thus seeking a balance between comfort and grid load. Based on a baseline indoor setpoint temperature of 26 °C, adjusting the building's indoor setpoint temperature up or down by 2 °C can provide approximately 10% to 35% of air-conditioning load flexibility, without significantly affecting indoor comfort. A cluster of 100 variable frequency air conditioners can provide power regulation capacity equivalent to 20%–30% of the total electrical load in a building group.

3. Controllable smart home appliances

   Smart home appliances, e.g., smart washing machines, dryers, and dishwashers, are also potential flexible resources in residential buildings. Such appliances are characterized by flexible operation time and difficulty in adjusting operating power. Therefore, they can be started during periods of abundant PV generation or low electricity consumption to reduce the grid pressure during peak hours, thus achieving flexible energy use. Currently, such appliances from most brands can be controlled via mobile application programs, enabling the development of personalized operation strategies.

4. Energy storage equipment

   Energy storage equipment includes lithium-ion battery, lead-acid battery and other energy storage devices installed in households, as well as electrical equipment capable of energy storage. Energy storage equipment can charge during periods of high surplus PV generation and discharge during peak load periods, thereby helping to balance the grid's supply and demand. Despite their small individual energy storage capacities, when they are collectively deployed in large-scale residential building groups, their total storage capacity and regulation capability can create significant flexible potential. In a demand response event involving 1.122 million residential households, the peak load on the power grid was reduced by a total of 151,500 kW, averaging only 135 W per household.

5.3.2 Implementation Approaches for Flexible Regulation

In practice, flexible resources should be inherently adjustable, and appropriate regulation methods should be adopted. Electric water heaters and heat pump water heaters exhibit strong thermal storage characteristics and high temporal flexibility. A water heater's set temperature and operating period can be configured via a mobile application program or smart plug, enabling it to start heating water in advance during periods of lower electricity tariffs or abundant PV power generation and automatically defer or reduce operating power during peak electricity consumption periods. This approach not only enables peak load shaving but also maximizes the self-consumption ratio of PV power, thereby reducing reliance on an external power grid. Air conditioning equipment should utilize intelligent temperature control technology to make fine adjustments based on external temperature, indoor cooling/heating loads, and grid load conditions while ensuring indoor comfort. Equipment capable of energy storage should be able to automatically perform charge or discharge operations in response to regulation signals. When working in coordination with other equipment, energy storage equipment can also serve as a buffer to facilitate short-term energy dispatch, thus providing users with more autonomy in making choices. Smart home appliances primarily reduce peak electrical loads by shifting their operation to off-peak hours.

The most critical link in the above implementation approaches for flexible regulation is how to engage users in adjustment and what basis they should follow to achieve autonomous energy consumption regulation. Unlike traditional methods that rely on load aggregators or virtual power plants to issue commands (where users become controlled objects, involving issues of participation willingness and regulation satisfaction), C_r, as a guiding signal tool, acts like a "baton". It enables users to intuitively understand the relationship between their energy consumption behavior and carbon emissions, thereby consciously adjusting and optimizing their energy consumption patterns without being subject to external mandatory dispatching. C_r, as a regulation signal for the carbon emission responsibility factor, is not merely a simple adjustment parameter. Instead, it quantifies and presents to users the impact of their energy consumption behavior on overall carbon emissions through a real-time and transparent feedback mechanism. When a user's energy consumption method causes the carbon emission indicator to exceed the predetermined target, a corresponding prompt is given through the C_r signal, allowing the user to notice the connection between "cost of energy" and carbon emissions. Conversely, when the user actively adjusts its energy consumption mode, such as delaying the startup of high-energy-consuming equipment or reasonably scheduling the operation times of equipment like water heaters and air conditioners, the C_r value corresponding to its energy consumption behavior will decrease accordingly, thus providing intuitive positive feedback.

This C_r-centered regulation method, on one hand, makes the regulation process more transparent, enabling users to see in real time the specific contribution of their energy consumption behavior to carbon emissions; on the other hand, through this autonomous mechanism, it not only avoids the potential issues of low user participation and decreased satisfaction that may be caused by external mandatory regulation but also stimulates user enthusiasm. When users become aware that each active regulation they make can directly improve the carbon emission indicator, they are often more willing to participate in the process and foster good energy-saving habits in their daily energy consumption. Furthermore, C_r-based autonomous regulation does not require complex intelligent control systems or hosting mechanisms, reducing technical and managerial barriers, which facilitates promotion and widespread adoption.

5.4 Analysis of Electricity Consumption Behavior-Based Energy Conservation Potential in Residential Buildings

The main reason for the difference in electricity consumption between different households of urban residential buildings is the difference in lifestyle. The reasons that lead to high electricity consumption include the various using behaviors of different types of household appliances, the long standby time of appliances, and different ways of using air conditioners and heating devices.

There are many kinds of household appliances, and the use behavior of owners varies as well. As people's living standards have been improving, new types of household appliances continue to emerge in urban households, which is also accompanied by new lifestyle changes. Apart from essential appliances such as televisions, refrigerators, air conditioners and washing machines, the ownership of some highly energy-consuming appliances is also gradually increasing, such as freezers, dishwashers, electric dish sterilizers, electric dryers, water dispensers, and smart toilet seats. These new appliances are the reasons for different power usage behaviors and have a significant impact on electricity consumption in urban residential housing.

To explore the reason for electricity consumption differences among various households, two typical families were selected to represent households with high electricity consumption and households with low electricity consumption. The electric power of appliances and the lifestyle of the two households were surveyed. The results show that the family with high electricity consumption, which was located in Changsha, Hunan Province, consumed 6995 kWh of electricity in 2019. The family with low electricity consumption was from Beijing and consumed 999 kWh of electricity in 2019. The basic information of the two households is shown in Table 5.3.

We collected data on electric appliances' electricity consumption and surveyed the lifestyles of the two families. The annual electricity consumption of different appliances in these two families is provided in Fig. 5.10.

By comparing the two families, it can be found that air conditioners consumed 745 kWh of electricity per annum in the family with high energy consumption, which is the largest energy consumer among all appliances in this family. In the family with low electricity consumption in Beijing, air conditioners consumed only 100 kWh a year. In addition to the issue of climate differences in different regions, one of the key contributors to power use differences is the use habits of air conditioners in different families. There were two air conditioners in the two bedrooms of family A, and they were used from June to September, while the air conditioner was not used in family B until the hot weather was unbearable for several weeks in July. In addition, family A was in Changsha city of Hunan Province without a district heating system. Therefore, the electric heater used in family A had an annual electricity consumption of 430 kWh.

Aside from air conditioning, there are also several other appliances in family A that consumed more electricity than other families, such as small electric water

Table 5.3 Basic information of the selected families

		Family A	Family B
Annual energy consumption		6995 kWh	999 kWh
Family location		Changsha, Hunan Province	Beijing
Building floor area		100 m^2	160 m^2
Family members		2 middle-aged adults +1 student	2 middle-aged adults +1 student
Frequently used electric appliances	Air conditioner	Split air conditioner*3	Split air conditioner*3
	Heating	Electric heater	District heating
	DHW	Gas water heater + small electric water heating device	Gas water heater + centralized domestic hot water supply
	Others	Refrigerator, washing machine, projector, stereo, treadmill, smart toilet seat, robot vacuum cleaner, range hood, electric cooker, etc.	Refrigerator, washing machine, television, stereo, range hood, electric cooker, electric kettle, etc.

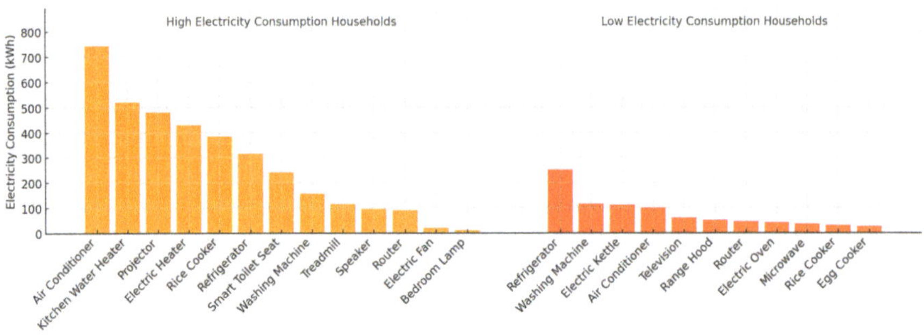

Fig. 5.10 Annual electricity consumption of key appliances in these two families (kWh)

heating devices, projectors, electric cookers and smart toilet seats. The long standby time of these devices all led to a great waste of energy.

Among all the appliances, the small electric water heating device consumed the highest amount of electricity. The DHW in family A was produced by a gas water heater and a 10 L small electric water heating device. The total monthly electricity consumption was 20 kWh in summer and 60 kWh in winter, and its annual electricity consumption for family A was 518 kWh, which was even higher than their annual electricity consumption for heating.

In addition, family A also has other high energy-consumption appliances, such as projectors, electric cookers and smart toilet seats. Each of these appliances uses at least 200 kWh every year, which is similar to the energy consumption of a refrigerator. The following passages discuss the use behavior of several high energy consumption appliances.

The combined electricity consumption of the projector and stereo in family A was 481 kWh, which was similar to TV. The projector was used for 3–5 h/day at

300 W operational power and consumed 1.3 kWh of electricity per day. The average standby power for the projector was 1.3 W, so it can be calculated that the annual electricity consumption during standby was 11.4 kWh. The total annual electricity consumption of TV in family B was 62 kWh, which was based on 256 W of operational power and 40 min of operation time on average per day. Therefore, the main reason that the projector consumed a larger amount of electricity is its higher frequency and longer use time.

The electric cooker in family A also consumed a large amount of power of 385 kWh/year, which was 11 times higher than family B's electric cooker. A 4 L electric cooker for a family of three people usually consumes 0.189 kWh of power for one meal, and the annual electricity consumption would be 140 kWh if the electric cooker is used twice a day. According to our test, family A not only used an electric cooker for normal meals but also preset the cooker for 3 h from 3:00 am to 6:00 am to cook porridge every day; each time, the cooker consumed 0.68 kWh of power (as shown in Fig. 5.11). The standby power of the electric cooker is 7.2 W, so the total standby electricity consumption of the electric cooker would be 50 kWh/year. Therefore, the difference in electricity consumption of the same household appliance could be as high as ten times depending on the ways of use. Moreover, long-time standby of heating appliances will lead to great heat dissipation and energy consumption.

Smart toilet seat was another high energy consumption appliance in family A, which consumed 241 kWh/year, equivalent to refrigerators in average families. Figure 5.12 demonstrates the typical daily power curve of the smart toilet seat in family A. There are two electricity consumption scenarios. In the first scenario, electricity is used to heat the water instantly during cleaning, the heating power is approximately 1300~1600 W, and the cleaning cycle is approximately 1 min. In the second scenario, electricity is used for heating to keep the seat warm during winter. The heating power is between 35 W~40 W during standby mode, and the average daily electricity consumption during this scenario is approximately 0.84~0.96 kWh. On a typical day, there was 1.03 kWh of power consumed by the smart toilet seat, of which 0.864 kWh of power was used to maintain the temperature of the seat,

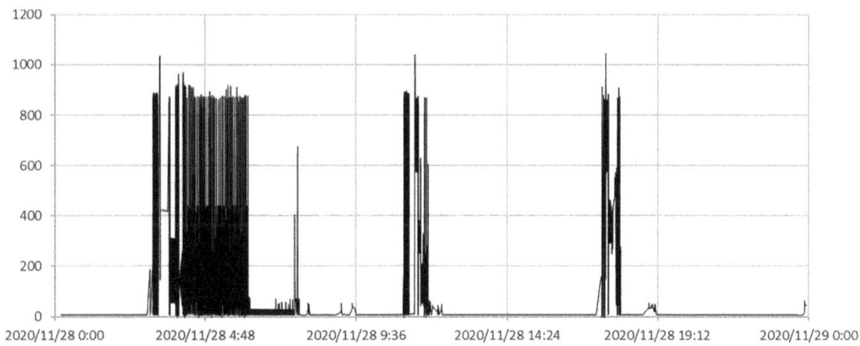

Fig. 5.11 Daily power curve of electric cooker in family A (W)

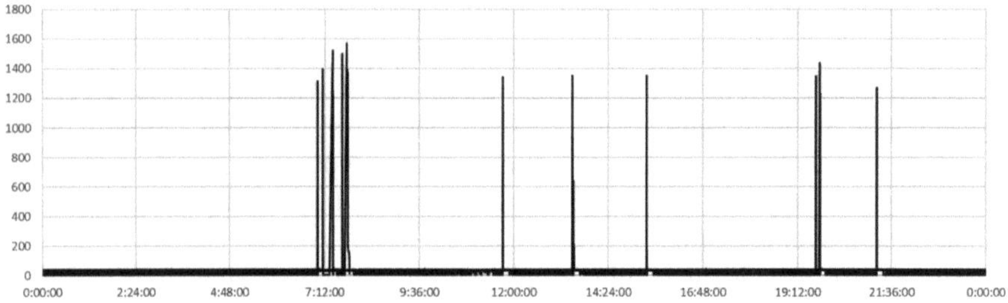

Fig. 5.12 Daily power curve of smart toilet seat (W)

accounting for 84% of the total electricity consumption, and 0.166 kWh was used for water heating, accounting for 16% of the total.

Apart from the above appliances, the annual electricity consumption of the treadmill, stereo and Wi-Fi routers of family A could not be ignored and were 111 kWh, 94 kWh and 87 kWh, respectively. The Wi-Fi router operates 24 h a day, and its average operation power was 9.9 W. Treadmill and stereo were used almost every day, and they consumed power even during the standby period. The standby power of the treadmill was 4.2 W, and the annual standby electricity consumption was 34 kWh, which was 30% of the total electricity that the treadmill consumes. The standby power of stereo was 1.8 W, and its annual standby electricity consumption was 16 kWh, accounting for 17% of the total.

Based on our analysis, we conclude that in addition to the different use behaviors towards heating devices and air conditioners, the high electricity consumption of household appliances was attributable to three main reasons. First, the long standby time and repeat heating cycle of electric water heaters, small electric water heating devices, water dispensers and smart toilet seats, which could consume large amounts of electricity each year, cannot be underestimated. Second, even though high energy consumption appliances such as electric dryers, dishwashers, electric dish sterilizers, wine cabinets, and electric ovens have not been very prevalent in urban households, energy efficiency policies need to be promulgated to lower the electricity consumption for these devices. Third, electricity consumption during the standby period, such as washing machines, TVs, etc. In family A, when appliances such as projectors, electric cookers, treadmills and stereos were not operating, they were usually in standby mode, which consumed 111 kWh of electricity in a year. Standard electricity consumption was not unacceptable for individual appliances, but the accumulation of a small amount of standby electricity for all appliances is something we cannot ignore.

These new kinds of appliances can be seen in an increasing number of Chinese households, which is one of the key reasons for the increasing electricity consumption in China. We should attach special focus on the following three types of appliances that consume high power. The detailed electricity consumption and energy efficiency potential analysis can be seen as follows:

Applications such as electric water heaters, small electric water heating devices, water dispensers, smart toilet seats, etc., which are operated in constant heating mode to maintain the water temperature or surface temperature, The purpose of constant heating is to ensure that the appliance can meet urgent hot water demand in households. According to our research, the "effective" power use for water heating only took a small share of the total amount, and the majority of the power was used to maintain the temperature of the water tank to offset the heat loss. Therefore, it can be regarded as "ineffective" power.

There are two types of appliances that require continuous heating. The first type of appliance is equipped with a water tank, such as an electric water heater, small electric water heating device, and water cooler. These appliances operate on a constant heating cycle to maintain the temperature of the water tank and stop heating when the set temperature is met. Figure 5.13 shows the power curve of the water dispenser during the 2-h standby period. The water tank of the water dispenser was heated every 20 min to sustain the temperature, which consumed 0.015 kWh of power. Based on our calculation, the average ineffective power use of the water dispenser was 46 W, which means that its daily standby electricity consumption was 0.97 kWh, higher than the daily electricity consumption of a refrigerator in an average household. The high electricity consumption of these appliances is mainly caused by poor insulation of the water tank. Therefore, we can conserve more electricity by improving the insulation performance of the tank. Additionally, one of the most popular water heating devices in the market today is the tankless water heater. It would only take approximately less than a minute to heat up the water, which is a healthier way to meet hot water demand in households while conserving almost 1 kWh/day of power.

The second type of appliance is tankless without insulation. Appliances such as smart toilet seats need to maintain the temperature of the seat at all times, and thermal insulation is unable to offset the heat radiation from the seat. Currently, smart toilet seats on the market are mainly equipped with instantaneous water heating functions due to hygiene, safety and energy efficiency requirements. The power of

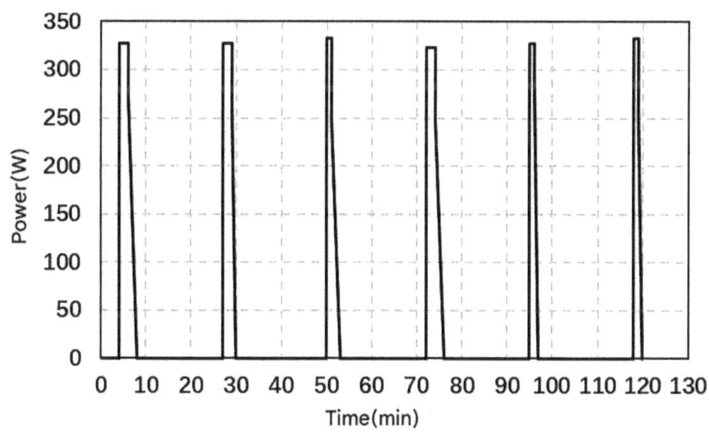

Fig. 5.13 Power curve of the water dispenser for 2 h in a real situation (W)

the electric water heater is approximately 1600 W, the power of seat heater is in the range of 40~50 W, and the power of the electric dryer is 340 W. Based on the use behavior on the electric toilet seat of family A, the total electricity consumption should be approximately 0.84~0.96 kWh to achieve 24 h of constant heating of the seat, which is 84% of the total power use, higher than that of the refrigerator. Smart control technology would be useful to reduce the high electricity consumption of these kinds of appliances. Meanwhile, some of the smart toilet seats are equipped with timers, which can be automatically turned on or off according to the user's habits. It can stop heating when no one is at home during the day and turn it on again when occupants need to use it to lower the ineffective electricity consumption for heating.

As society and the economy continue to develop and people's livelihoods improve in recent years, the number of appliances and their energy consumption are increasing. Among all the newly added appliances, dishwashers, smart toilet seats, wine cabinets, and washing machines with hot water functions are all high-energy appliances in households.

Recently, the sales volume of dishwashers has increased rapidly. The operation cycle of the dishwasher can be divided into heating and cleaning, disinfection and drying cycles. The operation duration of each cycle was between 60 and 300 min. The disinfection temperature of most of the dishwashers available on the market is approximately 70~80 °C, followed by high-temperature blow drying of the dishes. Therefore, the standard operation cycle of a dishwasher would consume 0.64~1.6 kWh of power and 300–500 kWh of power in a year, which puts the dishwasher in the high energy consumption appliance category.

The wine cabinet operates similarly to the refrigerator in a way that it also leverages a compressor to control the temperature and humidity of the atmosphere within the cabinet. Normally, the operational temperature range for wine cabinets spans from 4 to 22 °C. The amount of energy consumed is directly related to the volume of the cabinet itself. Under normal circumstances, a wine cabinet would be able to contain 30–150 bottles of wine, which would consume 0.3~0.8 kWh of power per day. For a larger wine cabinet, its electricity consumption is similar to that of a refrigerator.

The electricity consumption of using the water heating function of a washing machine is notably higher than that of using cold water. Figure 5.14 demonstrates the power curve of a single washing cycle of a washing machine in a real situation. Its workflow can be divided into three main steps, i.e. heating, washing and spin-drying. The total electricity consumption for a single washing cycle is 0.131 kWh. During the 200 s of water heating, the heating power reaches 1900 W, consuming 0.112 kWh of power, which is equivalent to 85% of the electricity consumption of a single washing cycle. As a result, the electricity consumption of using the water heating function of a washing machine with is five times higher than that of using cold water. Temperature adjustment is a commonplace function in all kinds of washing machines, and using hot water with a lower temperature or cold water conserves electricity for the washing machine.

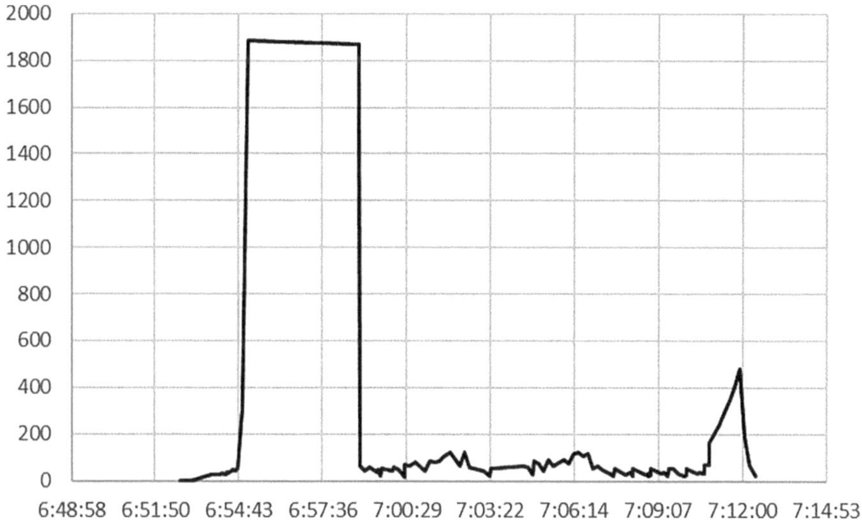

Fig. 5.14 Power curve of a single washing cycle of a washing machine (W)

For example, the yearly electricity consumption of appliances such as projectors, electric cookers, and treadmills during a long standby cycle in family A could be 111 kWh. Apart from refrigerators and Wi-Fi routers, which are switched on at all times in households, there are other appliances with long standby cycles, such as TVs, PCs, air conditioners, washing machines and electric water heaters. The standardized standby power of these kinds of appliances is between 0.5 W and 5 W, with 4~40 kWh of electricity consumption per year per single appliance.

From the previous analysis, it can be concluded that there are several main reasons contributing to the high electricity consumption of urban households in China. We need to pay special attention to the electricity consumption of household appliances with a constant heating function. Policies need to be promulgated to enhance the energy efficiency of these appliances. Smart control technologies need to be introduced to avoid unnecessary reheating and heat loss. A green and energy-saving lifestyle needs to be promoted among users to reduce the waste of energy from using these types of appliances. Policy incentives or subsidies shall not cover electric appliances such as clothes dryers, which may result in changes in lifestyle, and we should watch for the energy spikes of these high energy consumption appliances. Energy efficient technologies of all kinds shall be introduced to upgrade appliances with long standby cycles to ensure that the device can be switched to an energy saving mode during off time to reduce power use. An energy-saving lifestyle should be promoted among the populace to remind people to turn off appliances when not using.

The Chinese people have been exercising frugality since ancient times. Small and medium-sized apartment buildings have always been the mainstream in China due to land resource limitations, and the Chinese people usually try to refrain from leaving appliances in standby mode to reduce the waste of energy. Given the current economic development in China, the Chinese people are able to afford an American lifestyle with enough disposable income in their pockets. However, the majority of Chinese households are maintaining the traditional way of life with low energy consumption. It indicates that while China's current average urban income has surpassed subsistence level and is approaching moderate prosperity standards, for those who maintain average energy consumption level despite relatively high incomes, the factor constraining their energy consumption growth is not income but divergent cultural and consumption concepts.

In the context of high penetration of renewable energy, such as wind and solar PV power, into the grid in the future, power regulation and matching are more critical than energy savings. This has shifted the focus of building energy efficiency potential. The key to building energy efficiency is saving electricity when wind and solar PV power are insufficient, or using electrical appliances through storage and regulation when wind and solar PV power are abundant, whenever possible. For instance, the peak and valley values of the dynamic carbon emission responsibility factor for electricity (C_r) may differ by a factor of ten, leading to a significant difference in the carbon reduction effects of saving energy during C_r peak periods versus during C_r valley periods. In this case, rather than solely focusing on behavior-based energy conservation, it is more critical to reasonably schedule electricity consumption and enhance the capability of regulating the electrical loads of buildings. In cooking, for example, it is essential not only to reduce instantaneous electricity consumption by lowering cooker power but also to schedule cooking during periods of ample wind and solar PV power. The timing of appliance use is even more important than power reduction, thereby achieving overall superior energy conservation and carbon reduction effects.

From the previous analysis of electricity consumption of urban households, we found that there is still room for energy efficiency improvement and energy consumption reduction for all kinds of new appliances, especially for appliances with a standby mode, such as water dispensers, electric toilet seats, set-top boxes and other low-power appliances. Together with the amount of electricity consumption during the standby cycle, some of these appliances consumed similar amounts of electricity as refrigerators and other necessary devices at home. It remains an important task to introduce energy efficient standards targeting these kinds of appliances, as well as new technologies, energy saving modes and smart control mechanisms to lower electricity consumption during the standby cycle. On the other hand, subsidizing policies shall not cover these high-energy appliances. Instead, green lifestyle and energy saving behavior should be promoted among the general public to avoid the surge of electricity consumption after the growing popularity of these appliances. In the context of future high wind and solar PV power penetrations, achieving building energy efficiency and carbon reduction will require considering both the reduction of equipment power and the optimization of equipment usage timing. This approach can maximize the potential for building energy efficiency and carbon reduction.

Chapter 6
Charging Point System in Residential Compound

6.1 Background of Energy Interaction Between Electric Vehicles and Residential Buildings

6.1.1 Challenges in the Development of Electric Vehicles and Residential Community Charging Points

With the increasing electrification rate of transportation, electric vehicles, with their green and energy-efficient features, have demonstrated rapid development momentum. Among electric vehicles, non-commercial passenger vehicles account for the majority share, with an ownership ratio of about 80%. As non-commercial passenger vehicles are mainly parked and charged in residential communities, the rational layout and construction of charging points in communities are of significant importance for meeting the growing charging demands of electric vehicle owners and promoting the electrification of transportation.

However, the rapid spread of electric vehicles and residential community charging points is accompanied by numerous challenges that urgently need to be addressed. These issues not only constrain the development and spread of electric vehicles and charging points but also restrict the role of electric vehicles as energy storage and regulation resources in the energy system. Moreover, they pose risks to the safe and stable operation of urban power systems.

Insufficient charging infrastructure leads to "charging difficulties": Currently, there is a lack of a well-developed urban charging infrastructure layout, and the number and distribution of existing charging points are insufficient to meet the rapidly growing demand for electric vehicle charging. In particular, the shortage of charging equipment in the parking lots of urban residential buildings directly affects the user experience for vehicle owners, which has been a primary reason for customers' reluctance to purchase electric vehicles. Addressing this issue requires improving the layout of charging facilities in urban planning. Moderately advanced planning and charging point construction can not only meet current charging

Building Energy Research Center of Tsinghua University, *Decarbonizing Urban Residential Buildings*, https://doi.org/10.1007/978-981-95-8200-6_6

demands but also provide support for the full electrification of vehicles in the future. National and local governments have also issued relevant documents to guide the construction of charging points: As set out in the *Guiding Opinions of the General Office of the State Council on Further Building the High-quality Charging Infrastructure Network* (The General Office of the State Council of China 2023), the construction of electric vehicle charging points must follow the basic principles of “scientific layout and moderate advancement,” and it is essential to “actively promote the construction of charging infrastructure in residential areas.” According to *Beijing's 14th Five-Year Plan for Addressing Climate Change and Energy Efficiency* (Beijing Municipal Development and Reform Commission 2022), charging points and other supporting facilities shall be constructed under the principle of moderate advancement to accelerate the development of the urban charging network.

Unauthorized installation of charging points causes safety hazards: In the rapid spread of charging points, the haphazard and unauthorized installation of charging facilities has brought safety hazards. Some vehicle owners, due to the lack of standardized charging facilities in residential compounds, choose to install charging points privately, which may lead to safety issues such as electrical fires and grid instability. To avoid such hazards, clear technical specifications and regulatory requirements should be published as per relevant processes of urban infrastructure construction management. Such requirements shall cover the design, construction, maintenance, and operation management of charging points, thus ensuring the safety and reliability of charging equipment from the outset.

Uncoordinated fast and ultra-fast charging in inappropriate scenarios can have a huge impact on power grids: Uncoordinated fast and ultra-fast charging result in random fluctuations of electrical loads over a wide range, putting pressure on the safe and stable operation of power systems. The prevailing “plug-and-charge” and “fast-charging-dominant” models, following the conventional refueling pattern of fuel vehicles, generate massive instantaneous power consumption spikes that are overloading urban power grids. For example, a 600 kW ultra-fast charging point approximates the instantaneous electrical load of a 20,000-m^2 large-scale public building. Fast and ultra-fast charging modes are mainly suitable for key nodes in intercity transportation networks to meet long-distance travel needs but not for the everyday use of non-commercial electric vehicles in urban areas. Commercial vehicles typically exceed an annual mileage of 100,000 km, while non-commercial vehicles average about 12,000 km/year. Their charging modes should differ based on their distinct travel requirements. In urban contexts, especially residential areas where non-commercial vehicles constitute the most important vehicle group, the mass deployment of fast and ultra-fast charging would require a manyfold increase in the capacity of the urban low-voltage distribution network, causing huge investments for capacity expansion while making it even more challenging to utilize vehicle batteries to provide energy storage and regulation capabilities for power systems. For a residential building scenario, the focus shall be on developing coordinated slow charging, and intelligent coordinated charging points shall be rationally connected to the distribution system of the residential community. Existing distribution resources shall be fully utilized through coordinated charging to avoid large-scale grid expansion and energy storage investment.

Charging points in various urban contexts need to be rationally deployed and fully utilized: The distribution and operating modes of charging points vary significantly in the parking lots of different types of buildings. Currently, some areas have an overabundance of charging point resources while others are confronted with a shortage. Moreover, there are notable differences in the utilization rates of charging points in various settings at various times of the day. This imbalance not only wastes precious distribution resources but also hinders to some extent the spread of electric vehicles. In Beijing (Dai et al. 2025), for example, the charging loads of charging stations in typical building environments exhibit broadly consistent daily patterns, and the utilization rates of charging points are closely associated with the specific building types in which the stations are situated. The utilization rates of charging points in commercial building charging stations and standalone charging stations increase with their rated power, indicating a demand for fast charging. By contrast, the charging points in workplace charging stations and residential charging stations show the opposite trend, suggesting that slow charging can largely meet the demand (Fig. 6.1).

On the whole, the current charging point construction and the diverse and complex vehicle travel and charging modes pose huge challenges to the promotion of the energy interaction between vehicles and residential buildings. These issues involve both macro-level considerations, including urban planning, traffic management, and energy policy, and micro-level factors, such as vehicle owner behaviors

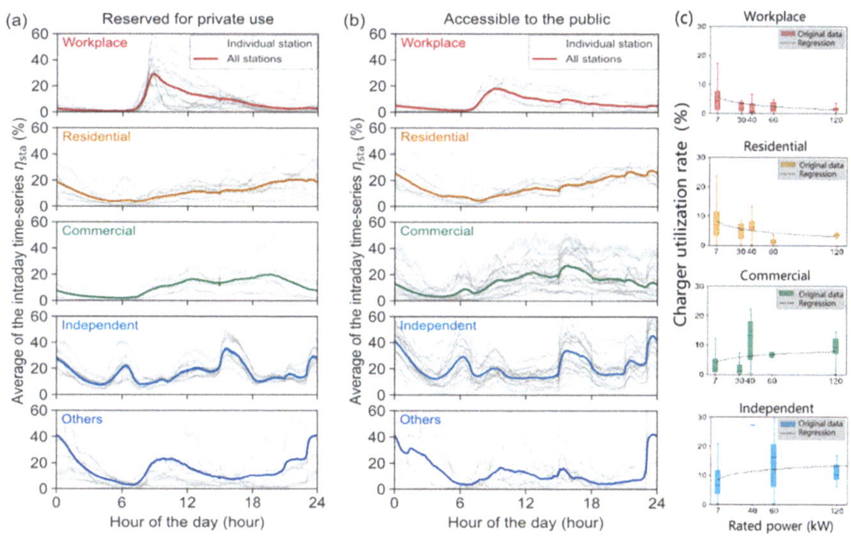

Fig. 6.1 Charging loads and charging point utilization rates in parking lots of different buildings in Beijing: (**a**) Typical daily charging loads in non-public parking lots; (**b**) Typical daily charging loads in public parking lots; (**c**) Relationship between charging point utilization rate and rated power (Dai et al. 2025)

and technology application. Only by optimizing the layout of charging points, guiding reasonable charging behavior, and improving the policy support system can these obstacles be gradually resolved, thus promoting the energy interaction between vehicles and buildings toward higher-level application practice.

6.1.2 Promoting Energy Interaction Between Electric Vehicles and Residential Buildings

Driven by the "dual carbon" goals, the coupling relationship between the two terminal energy consumption sectors, i.e. vehicles and buildings, is being progressively strengthened: In terms of vehicle behavior, urban private vehicles (accounting for about 80% of the total number of urban vehicles) are parked in building parking lots for more than 80% of their time on average, and also for charging purposes. Regarding power grid capacity expansion, the connection of vehicle charging points to residential building distribution systems reduces the investment in power grid capacity expansion. The aforementioned challenges can be effectively addressed by promoting energy interaction between electric vehicles and residential buildings.

Above all, promoting the energy interaction between vehicles and residential buildings can bring their respective resource advantages into full play, thus achieving mutual benefits for all parties involved. On one hand, electric vehicles, as distributed mobile energy storage units, have their battery capacity significantly increased in recent years. Taking private electric vehicles as an example, statistics indicate that in 2023, the average battery capacity of private passenger electric vehicles in China reached 45–50 kWh, with some models even exceeding 100 kWh (China Automotive Power Battery Industry Innovation Alliance 2025), while the average daily electricity consumption for urban private cars was only about 7 kWh (Zhang et al. 2019). This indicates that vehicle battery resources are relatively abundant and capable of providing huge energy storage and regulation potential for the energy system without influencing their daily use. Based on the production and service life data of new energy vehicles in China since 2012, the following predictions are made: China's new energy vehicle stock is expected to experience a period of rapid growth from 2020 to 2040. It is projected to approach 100 million by 2030, reach 190 million by 2035, and attain 300 million by 2040. After 2040, China's new energy vehicle market will gradually become saturated. It is estimated that China's new energy vehicle stock will reach approximately 350 million by 2050. With the rapid development of China's market and the improvement of relevant policies and regulations, the vehicle-to-grid (V2G) technology integrating electric vehicles with renewable energy will unleash tremendous potential.

On the other hand, building distribution systems do not always operate at full load. The ratio of daily peak power to rated distribution capacity for urban buildings averages only 20%–30% annually. Furthermore, the buildings operate near or at their annual peak power levels for less than 2%–3% of the time. During the extensive periods of low-load operation in buildings, there is sufficient distribution capacity available to accommodate vehicle charging needs. By combining the advantages of vehicles and buildings, rationally deploying charging points, and implementing intelligent regulation strategies, it is possible to meet vehicle mobility demands while reducing the pressure on urban power grids for capacity expansion. In addition, the mobile energy storage resources of electric vehicles are utilized to provide huge potential for distributed PV consumption and power system regulation and storage.

Residential buildings represent the main setting for energy interaction between vehicles and buildings. The behaviors of urban private vehicles are particularly closely related to residential buildings. Extensive statistical data suggest (Liu et al. 2023c) that private electric vehicles in a typical city are parked in parking lots of residential buildings for about 39%–67% of their time on average. This long-time and stable occupancy characteristic provides adequate temporal and spatial conditions for the energy interaction between vehicles and residential buildings, rendering residential buildings a main setting for electric vehicle charging, energy storage, and vehicle-to-grid charging (Fig. 6.2).

The primary challenge facing electric vehicle charging today is the prevalence of uncoordinated charging, namely, vehicle owners often choose charge time randomly based on personal vehicle usage needs and convenience rather than following a uniformly planned schedule. Uncoordinated charging may lead to greater pressure on power grids during peak periods. However, on the other side of the coin, the flexibility of electric vehicle charging constitutes a major strength as well. With the development of technologies and the improvement of charging infrastructure, electric vehicle owners can flexibly adjust their charge and discharge time based on their travel plans and grid load conditions. When vehicles are idle, they can be partially discharged in advance and then fully charged during an appropriate period. This approach not only helps maintain balance in power grids but also reduces charging costs by taking advantage of time-of-use tariff policies in some regions. The charging load in the parking lot of a residential building usually peaks in the evening, which coincides with the energy consumption peak of residential users. By rationally deploying intelligent charging points in parking lots of residential buildings, electric vehicles will not be charged, or will be discharged, during the evening peak hours of building electricity consumption; instead, they will be charged with the

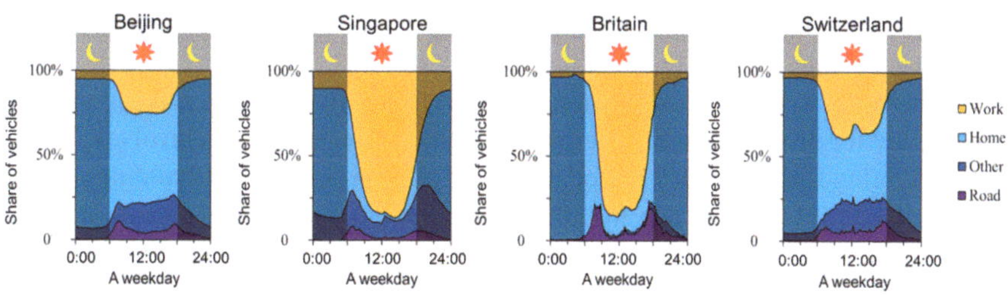

Fig. 6.2 Vehicle parking locations on weekdays in different regions (Liu et al. 2023c)

valley electricity at night to meet the travel demand of the next day. As a result, this can meet the charging demands of vehicle owners and alleviate the pressure on the distribution system during peak hours. This mode is conducive to relieving the capacity expansion pressure on the distribution network and fully leveraging the flexibility of the power demand side to provide the power system with flexible regulation capabilities.

6.2 Interaction Potential and Charging Point System in Residential Scenarios

6.2.1 Analysis of Vehicle-Building Interaction Potential in Residential Scenarios: Bidirectional Charging and Discharging Mode Holds Great Potential

In residential scenarios, the charging strategies of charging points have a significant impact on grid load distribution, and each strategy exhibits distinct characteristics in peak load shaving and energy efficiency improvement. Currently, there are three primary charging strategies, i.e., uncoordinated charging, coordinated charging, and bidirectional charging and discharging.

It is estimated that China will have approximately 300 million non-commercial passenger vehicles by 2050. If all the vehicles are electric, with an average battery capacity of about 60 kWh per vehicle, the total onboard energy storage capacity is estimated to approach 18 TWh. This amount is roughly equivalent to China's total daily electricity consumption. Non-commercial vehicles currently have an average annual mileage of 12,000 km and an electricity consumption of not more than 2000 kWh. Assuming each charge is 50 kWh, only 40 charging sessions are needed annually. Data indicates that private electric vehicles in a typical city are parked in parking lots of residential buildings for about 39%–67% of the time on average. Residential buildings serve as the main parking locations for non-commercial vehicles, and owners usually charge their vehicles when they return home.

Under uncoordinated charging conditions, vehicle batteries cannot serve as energy storage resources for residential buildings. Moreover, the large-scale random fluctuations in the electrical loads of vehicles, such as concentrated charging during peak evening hours, can place significant strain on the safe and stable operation of the power systems in residential compounds.

In the case of coordinated charging, each vehicle consumes an average of about 7 kWh/day for driving, and the energy storage function of vehicles involves shifting daily charging loads to valley periods of electricity consumption in buildings. In this way, residential buildings can install charging points with minimal or no need for capacity expansion. However, the regulating and energy storage capacity each vehicle can provide is limited, at approximately 7 kWh/day per vehicle.

In the bidirectional charging and discharging mode, if each vehicle charges and discharges once per day, the loss is calculated based on an average charge and discharge efficiency of 92%, an average of 7 kWh of electricity per day is used for driving, and another 7 kWh is lost during the charging and discharging process, so there is still 36 kWh of electricity left each day that can be used for electricity peak regulation. It is observed that if vehicles adopt bidirectional charging and discharging, they can not only meet the charging needs of vehicle owners but also help alleviate the pressure on the distribution system during peak hours by utilizing the energy storage and regulation capabilities of their batteries. This mode helps to mitigate the pressure on distribution network expansion and demonstrates huge potential for energy storage and regulation.

In a study (Li et al. 2025), electric vehicle behaviors were simulated in a Monte Carlo model, the charge and discharge power boundaries and charge and discharge energy boundaries of electric vehicles were digitally modeled based on the state of charge and travel demand data, and the shift of charging loads to the valley periods of residential loads through the adjustment of electric vehicle charge time and power by coordinated charging was simulated in the residential electricity consumption scenarios in 2025. The study results show that at an electrification rate of 30%, the average daily peak-to-valley difference can decrease by about 75% with the participation of all vehicles in bidirectional charging and discharging, and reduce by 25% in unidirectional coordinated charging, when compared to uncoordinated regulation (Fig. 6.3).

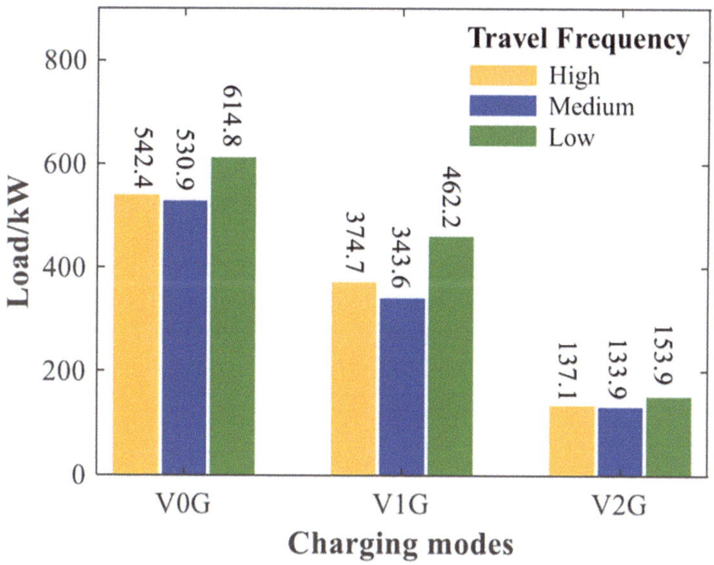

Fig. 6.3 Peak-to-valley difference of the distribution network under different charging modes

6.2.2 *Analysis of Charging Point System Integration and Regulation Modes in Residential Scenarios*

Intelligent charging points, as key components in the interaction system between vehicles and residential buildings, intelligently adjust electric vehicle charge power and time through designed intelligent charging and discharging strategies, thus enabling complementary and interactive energy usage on the user side while maintaining grid-friendly operation.

Charging point system integration mode in residential scenarios: The charging point system is deployed in the parking lots of residential compounds to achieve "one charging point per parking space." Electric vehicles are required to "connect upon parking" (i.e., connect to buildings upon parking) and are provided with "bidirectional coordinated" charging and discharging services. The figure below shows the schematic diagram of the integration mode. "One charging point per parking space" is implemented in parking lots, and no more than 50 charging points are grouped into one charging point distribution network. These charging points are powered by a uniform DC 750 V supply. The distribution network is connected to the internal distribution networks of adjacent buildings through AC/DC (the capacity of a building distribution network shall be 1.5–2 times the product of 10 kW and the number of charging points), or below the transformers in the transformer coverage area of the residential compound distribution network. The charging strategy of intelligent charging points involves adaptive regulation of their charge power by receiving busbar voltage signals (Fig. 6.4).

Charging point system operation regulation mode in residential scenarios: The AC/DC for power supply of the intelligent charging point system receives time-of-use tariff signals or "dynamic carbon emission responsibility factor" signals, and adjusts the DC busbar voltage based on these signals to regulate the charge/discharge power. The regulation method is demonstrated below by detailing the logic for regulating the charge and discharge power of charging points in a residential area based on the "dynamic carbon emission responsibility factor" (hereinafter referred to as C_r).

Calculation of C_r for a residential area: For a residential area, external connections are made through n transformers in the transformer coverage area, and each transformer is connected to a connecting line corresponding to another power

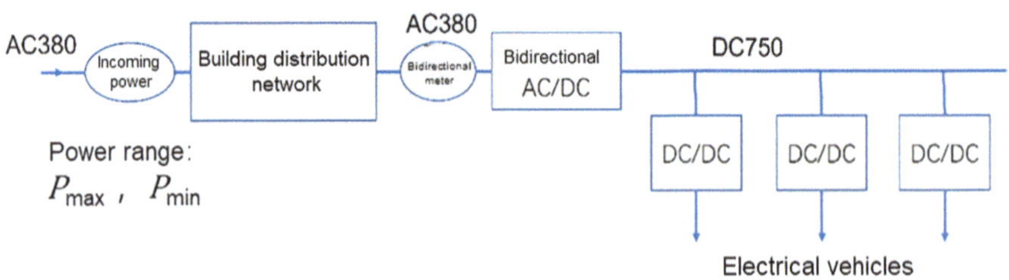

Fig. 6.4 Schematic diagram of a charging point system

supply area. In the residential area, there are direct power consumers (connected directly or through step-down transformers), power supplies (generator units, wind and solar PV power supplies and power storage units), downstream transformers in the transformer coverage, and interconnections with the power supply area of the next voltage level. The downstream transformers in the transformer coverage area may be omitted if deemed electrically equivalent to external connecting lines. When the downstream power of transformers in the transformer coverage area fails to reach 90% of the rated power and the upstream power fails to reach 85% of the rated power, the C_r for this residential area is calculated as (the sum of the power of connecting lines entering this residential area and the C_r values corresponding to these connecting lines + the carbon emission responsibility transferred from thermal power plants within this residential area + the carbon emission responsibility of centrally controllable energy storage output within this residential area) divided by (total input power of connecting lines + total thermal output power + total output power of centrally controllable energy storage within this residential area). To prevent overloading of the lines and transformers in the transformer coverage area, the C_r for this residential area should be corrected. When a connecting line or a downstream transformer in the transformer coverage area is overloaded, it is necessary to increase the carbon emission responsibility corresponding to the associated electricity consumption side. When a connecting line or an upstream transformer in the transformer coverage area is overloaded, it is necessary to reduce the carbon emission responsibility of the associated electricity consumption side.

Regulation of system busbar voltage based on C_r: AC/DC receives the C_r information issued by the power dispatch department in real time, and also collects the electric power information from the smart electricity meters at the interconnection point between the connected building distribution network and the external grid. In addition, it pre-sets the maximum input power P_{max} of the building distribution network. AC/DC determines the busbar voltage U_d of the charging system's DC distribution based on the C_r value. A higher C_r results in a lower U_d, while a lower C_r leads to a higher U_d. Simultaneously, the power at the building distribution network's entry point must be monitored in real time. When it approaches P_{max}, it should be prevented from exceeding the limit by reducing the busbar voltage immediately. When it approaches the lower limit of power (1% of P_{max}), the busbar voltage U_d should be increased to reduce discharge from charging points, thus preventing the building from feeding power back into the grid.

Intelligent charging point regulation: An increase in busbar voltage or maintaining it at a high level indicates a lower C_r value. For intelligent charging points, this means the total charge power can be increased. Conversely, when the busbar voltage decreases, it indicates a higher C_r value, and the total charge power of the charging points should be reduced accordingly. Intelligent charging points will receive such data as the busbar voltage signal U_d, the state of charge (SOC) of electric vehicles, and the maximum electric vehicle charge power P_{max} to regulate the electric vehicle charge power in real time. The system control process is shown in Fig. 6.5. The actual vehicle charge power P_{max} is positively correlated with U_d. When C_r is low, the busbar voltage rises, increasing the charging point power. P_{max} is

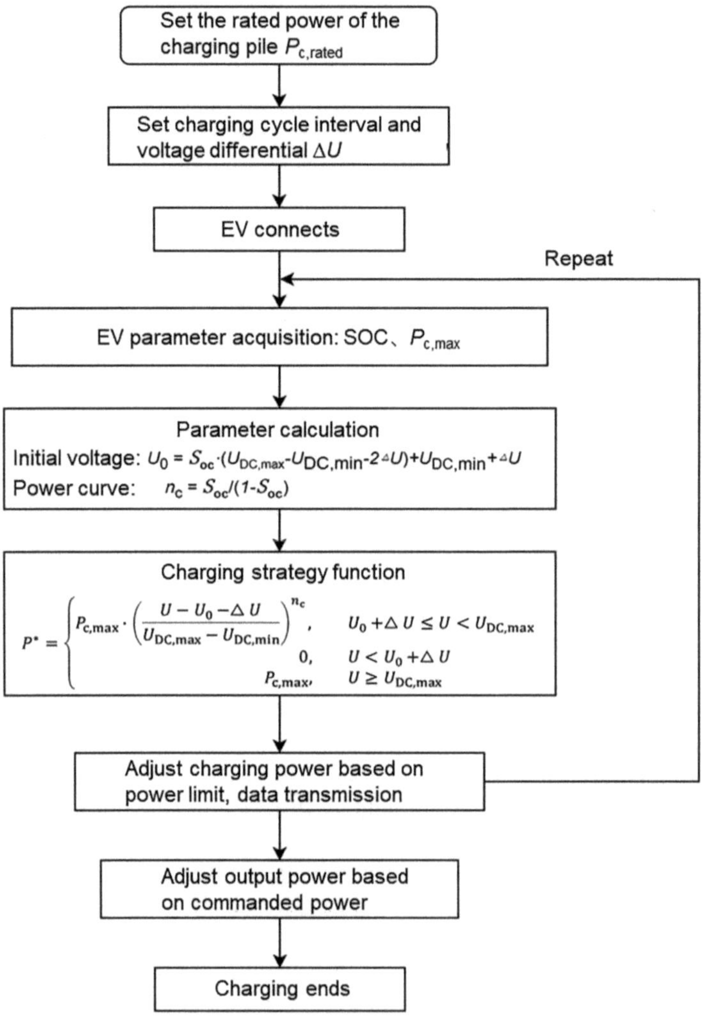

Fig. 6.5 Schematic diagram of the intelligent charging point system regulation

negatively correlated with SOC, meaning that vehicles with a lower SOC are prioritized for high-power charging under the same busbar voltage conditions.

6.2.3 *Bidirectional Charging and Discharging Mode in Residential Scenarios: Extending the Lifespan of Electric Vehicle Batteries*

In analyzing the vehicle-to-building interaction potential and charging point system modes in residential scenarios, this study explores the enormous energy storage potential enabled by the bidirectional charging and discharging mode in the interaction. In addition, the bidirectional charging and discharging mode has a positive impact on the battery lifespan. In residential scenarios, vehicles, as mobile energy storage units, are tightly integrated with building power systems. The bidirectional charging and discharging mode effectively manages vehicle battery levels during daily charging and discharging cycles through rational energy dispatch and intelligent management, thereby extending battery service life.

When integrated into the residential energy systems, electric vehicles can manage and change the calendar aging rate of batteries by adjusting the battery SOC. According to a study on the degradation mechanisms of lithium battery discharge performance (Geslin et al. 2025), controlling vehicles to charge and discharge dynamically can enhance battery lifetime compared with the conventional constant current charge and discharge. Moreover, a study on the calendar aging mechanism of lithium batteries (Lam et al. 2025) suggests that calendar aging will accelerate significantly if the batteries remain at a higher SOC level (especially near full charge, e.g., in the 90%–100% range) for a prolonged period and that it will slow down rapidly as the SOC level decreases. After the application of V2G technology (Fu et al. 2025), batteries are allowed to charge and discharge within a moderate SOC range through a rationally designed charging strategy, thus enabling significant mitigation of calendar aging. For instance, if an electric vehicle's SOC typically remains in the 90%–100% range, using V2G technology to adjust the SOC down to around 50% can significantly reduce the battery degradation rate. Therefore, by scientifically adjusting the SOC levels in coordinated charging/V2G mode, electric vehicles can, to some extent, regulate the calendar degradation rates of their batteries. A study result shows that the coordinated charging mode avoids excessively high average SOC levels caused by electric vehicle users' charging anxiety through intelligent charging. Compared to the SOC level at 81%/79% in the uncoordinated fast/slow charging mode for electric vehicles, the SOC level in the coordinated intelligent charging mode has decreased to 67%. According to the calculation results, using coordinated charging can reduce the annual battery degradation rate of electric vehicles from 7.9% to 6.8% (Figs. 6.6 and 6.7).

In addition, some research findings also demonstrate (Xu et al. 2024) that reasonably setting the charge/discharge current and frequency in vehicle-to-grid

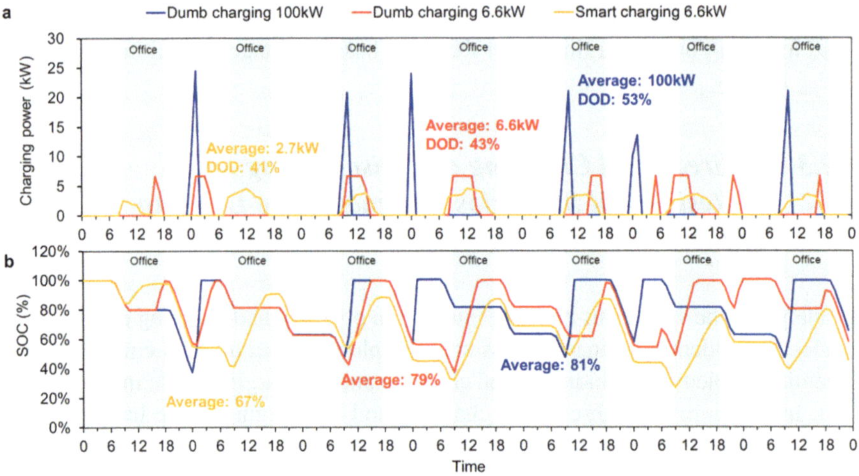

Fig. 6.6 Charging performance under different charging modes: (**a**) Charging power; (**b**) Vehicle SOC

Fig. 6.7 Impact of charging mode on battery degradation

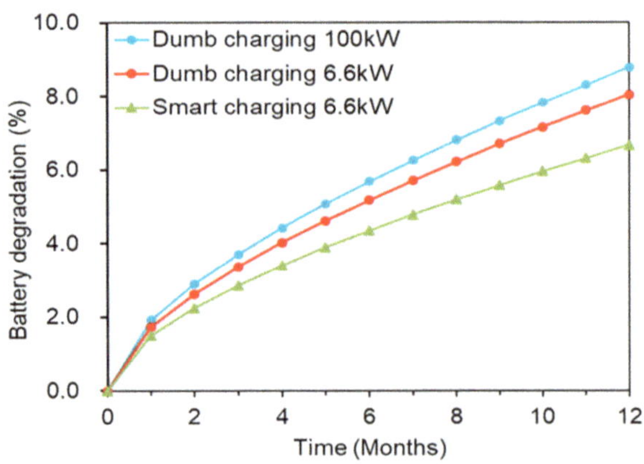

interaction can improve the growth of the solid electrolyte interface (SEI) film in batteries, thus enabling the battery degradation under charge-discharge cycling to be lower than that under pure idle conditions over the same period. By precisely regulating charge current, voltage, and SOC, the impact of adverse factors on the SEI film can be minimized, thereby inhibiting excessive growth and uneven deposition of the SEI film and maintaining its stability and compactness. This helps to extend the overall lifespan of the battery. Currently, research on mitigating battery degradation through charge-discharge cycling is still in its early stage, and further in-depth analysis is still required.

Electric vehicles participating in high-frequency vehicle-to-grid interaction can also assist in low-temperature battery heating, thereby improving their low-temperature adaptability and driving range. In low-temperature environments, the performance of lithium-ion batteries significantly declines. This decline is characterized by reduced charge efficiency, lower available capacity, and higher internal resistance. Through frequent vehicle-to-grid interaction, power batteries generate heat and warm up due to voltage polarization effects, thus restoring or achieving performance levels comparable to those at normal temperatures. Vehicles can utilize bidirectional charging for preheating before or during charging, raising the internal battery temperature to an appropriate operating range. This method shortens battery charge time, effectively reduces charge efficiency losses caused by low temperatures, and safeguards the battery against irreversible damage from low-temperature charging.

Generally, the coordinated charging strategy in residential buildings is well-suited for widespread adoption at the current stage as it has a low implementation threshold and high cost-effectiveness. In contrast, the bidirectional charging and discharging strategy holds significant potential for optimization but presents economic and technical challenges, such as battery degradation, that require further study. In future residential scenarios, combining intelligent charging strategies with electrical load management in buildings will further enhance peak load shaving effects, providing a stronger foundation for deep interaction between vehicles and buildings.

6.3 Charging Point Construction and Management Mode in Residential Scenarios

6.3.1 Charging Point Construction and Management Modes in New Residential Buildings

In new residential scenarios, intelligent coordinated slow charging points can be constructed and managed as critical infrastructure in a unified manner. Charging points must be capable of intelligent coordinated charging/discharging to maximize the regulation and storage functions of electric vehicles while minimizing excessive impact on the distribution capacity of a new residential building. Therefore, during the construction phase, charging points can be designed and purchased by developers in a unified manner to ensure their regulation function and quality; during the post-delivery operation phase, they can be maintained and managed by the property management company in a unified manner to effectively prevent potential safety hazards. Electric vehicle charging equipment incorporates specialized technologies, including power facility modifications and load management. If individual residents purchase and install this equipment on their own, it may cause overloads in the residential distribution network. Additionally, improper wiring can increase the risk of fires and electrical shocks. In contrast, unified construction and management can not only ensure professional and standardized installation but also guarantee the stable operation of the residential community's power system through centralized planning and effective coordination of resource allocation. A smart charging system can automatically schedule electric vehicle charging during periods of low grid load, thus avoiding excessive charging demand during peak hours and reducing pressure on the compound's distribution facilities. Currently, cities like Beijing (Beijing Municipal Development and Reform Commission, Beijing Municipal Commission of Urban Management 2023) and Shenzhen (Shenzhen Municipal Development and Reform Commission 2023) have proposed development plans for supercharging stations. However, constructing such stations in unsuitable urban settings, like within buildings, may cause significant problems. High-power fast charging is primarily needed in scenarios involving short vehicle stops (such as expressway charging stations) and those involving rapid recharging of operational vehicles (such as bus and taxi charging stations). The construction of ultra-fast charging stations is unsuitable for urban residential and office buildings. In urban building scenarios, the focus should be on developing coordinated slow charging. Intelligent coordinated charging points should be rationally integrated into the existing building distribution system. This approach allows for the efficient use of current distribution resources through coordinated charging, thus avoiding the need for large-scale grid expansions and significant investments in energy storage.

Furthermore, in response to the rapid growth of electric vehicles, new residential compounds can fully reserve space and power distribution conditions for charging point installation by following the principle of "one charging point per parking space," while also allowing for appropriately forward-looking construction. Compared to the current selling prices of parking spaces in residential compounds, the increase in selling price resulting from the addition of charging points is not significant, while also providing owners with the convenience of charging. Through rational planning and the early reservation of installation conditions for charging facilities, scalable deployment can be achieved as the number of vehicles grows, thus preventing the high costs and disruptions to residents' lives caused by large-scale retrofits later on. This point is also emphasized in the *Guiding Opinions of the General Office of the State Council on Further Building the High-quality Charging Infrastructure Network*: "The Employers of new residential areas must be held accountable for fulfilling their primary responsibilities. They must strictly implement the construction requirements for supporting charging infrastructure to ensure that fixed parking spaces are 100% equipped with charging infrastructure or have reserved installation conditions as stipulated to meet the requirements for direct meter connection and power supply" (The General Office of the State Council of China 2023). To further promote the implementation and improvement of the charging point management mode in new residential compounds, local governments should introduce relevant policies to standardize the construction and management of charging points and provide support for the construction and operation of charging facilities. For example, governments can encourage developers to reserve space for charging point installation in new residential compounds through subsidy policies and provide property management companies with technical support to help them establish intelligent management systems in compounds. In addition, policies should clearly define the responsibilities of compound property management companies in charging point management to ensure the safety, maintenance, and daily operation of charging facilities.

Free charging is the best choice for achieving bidirectional charging and discharging. The process of metering for bidirectional charging and discharging is extremely complex. It necessitates precise recording of electricity used for charging and discharging vehicles at different times, as well as settlement based on peak-valley tariffs. This complex metering not only requires high-precision bidirectional smart electricity meters but also necessitates a comprehensive accounting system to handle large volumes of data. According to industry data, the cost of a high-precision bidirectional metering system can be as high as several thousand yuan, and its maintenance costs should not be underestimated. Furthermore, the complex metering process may also lead to billing errors and disputes, which can further increase management and social costs. In contrast, the free charging mode can effectively avoid these issues arising from the complex metering process. By simplifying the metering process and reducing billing disputes, free charging not only lowers technical thresholds and management costs but also accelerates the widespread adoption of bidirectional charging and discharging.

For operators, vehicle charge costs are relatively low compared to parking fees. According to market research, parking fees are often a major expense for vehicle owners, while charge costs represent only a small fraction of their daily expenses. For instance, in some urban central areas, parking fees can be as high as RMB 10 per hour, and monthly rents for long-term stays typically range between RMB 300 and RMB 1000. In contrast, charge costs (calculated at a valley electricity tariff of RMB 0.30 per kWh and an average daily electricity consumption of 7 kWh) amount to only RMB 2.1 per day. Moreover, operators can earn additional income through free charging as it enables vehicles to charge at low tariffs during valley hours and return surplus electricity to the grid during peak hours. This mode not only optimizes grid operations but also generates economic value for operators. Additionally, free charging can increase user stickiness and attract more vehicle owners to opt for intelligent charging points as their daily charging solution. This economic incentive not only promotes bidirectional charging and discharging but also delivers a stable long-term user base for parking lots and charging point operators. For vehicle owners, they are more likely to participate actively in the bidirectional charging and discharging process when they enjoy free charging services. Furthermore, the technical analysis section also indicates that participating in vehicle-to-grid interaction does not negatively affect the battery life of the vehicles.

In conclusion, the charging point management mode for buildings in new residential compounds should prioritize safety and strive for the goal of "one charging point per parking space." Intelligent coordinated slow charging should be the key technical requirement. This approach should enable scientific planning and efficient resource utilization through unified construction and management by developers and property management companies. Additionally, promoting a free charging mode will help to accelerate the widespread adoption of electric vehicles.

6.3.2 *Charging Point Construction and Management Modes in Existing Residential Buildings*

In a large number of existing residential compounds, installing intelligent coordinated charging points is a solution to address the "charging difficulty." Moreover, it is also crucial for promoting the electrification of private vehicles. Therefore, the process of installing these charging points in residential compounds should be expedited. If the charge power of vehicles cannot be regulated intelligently at the charging points of an existing residential building, the random charging load from vehicles will cause the overall load to exceed the current distribution capacity limit of the compound. This could lead to the failure to install more charging points or the need for additional distribution investments. However, if intelligent bidirectional coordinated charging points are constructed, they can meet the charging needs of vehicle owners while utilizing the vehicles' regulation and storage functions to avoid exceeding the distribution capacity. In this case, installing charging points in an existing residential building is not a burden on the power system but rather beneficial. Charging points can be installed even in compounds with constrained distribution capacity. Therefore, for existing residential buildings, the key is to vigorously promote the adoption of intelligent coordinated charging points. Priority should be given to coordinated charging to fully utilize the existing distribution capacity. Where the existing distribution capacity is insufficient, appropriate expansion should be conducted, but only to meet the requirements of integrating low-power coordinated charging points.

The charging point management mode in existing residential buildings is the same as that in new buildings. In this mode, charging point operators need to reach an agreement with the owners and property management companies of residential compounds to establish charging stations as public facilities for owners' use and also take responsibility for their operation and management. To ensure the charging points are safe, reliable, and capable of intelligent coordinated charging, they shall be "constructed and operated in a unified manner" according to relevant standards.

Additionally, cooperating with the parking lots of buildings around residential compounds is also an effective way to address the "parking and charging difficulties" in existing residential compounds. In some existing old city areas, residential compounds often have a limited number of parking spaces, making parking itself an urgent issue to be addressed. In this case, the shortage of charging points within residential compounds can be alleviated by constructing public charging stations and shared charging piles in nearby office and commercial buildings, thereby meeting the parking and charging needs of vehicle owners. Co-building charging points with nearby institutions can not only improve the utilization rate of charging facilities but also promote the interconnectivity of charging networks, thus addressing concerns over charging accessibility for residents' daily mobility.

As exemplified by an office park in Beijing (Tsinghua University 2025), the park opens its parking lot to surrounding residents at night, but only allows vehicles that agree to participate in coordinated charging to enter for parking. In this way, they jointly contribute to the park's coordinated energy regulation. This not only provides parking spaces for vehicle owners from surrounding compounds but also promotes the consumption of rooftop PV power in the office park (Fig. 6.8).

In terms of policy, some cities have begun to explore and promote unified charging point construction and management in residential compounds, as well as policies and practices for sharing parking spaces and charging points with nearby office areas. For example: (1) Shanghai has explicitly endorsed shared parking space and charging point models in its policies. For instance, according to the *Procedures of Shanghai Municipality on the Construction and Management of Electric Vehicle Charging Facilities in Residential Quarters* (Shanghai Municipal People's Government 2023), parking spaces and charging points of residential quarters may be entrusted to the charging operator for unified management through negotiation, and residents are encouraged for sharing at different times. In addition, Shanghai has also promoted the addition of shared new energy motor vehicle charging points in eligible residential quarters through the renovation of old residential quarters. (2)

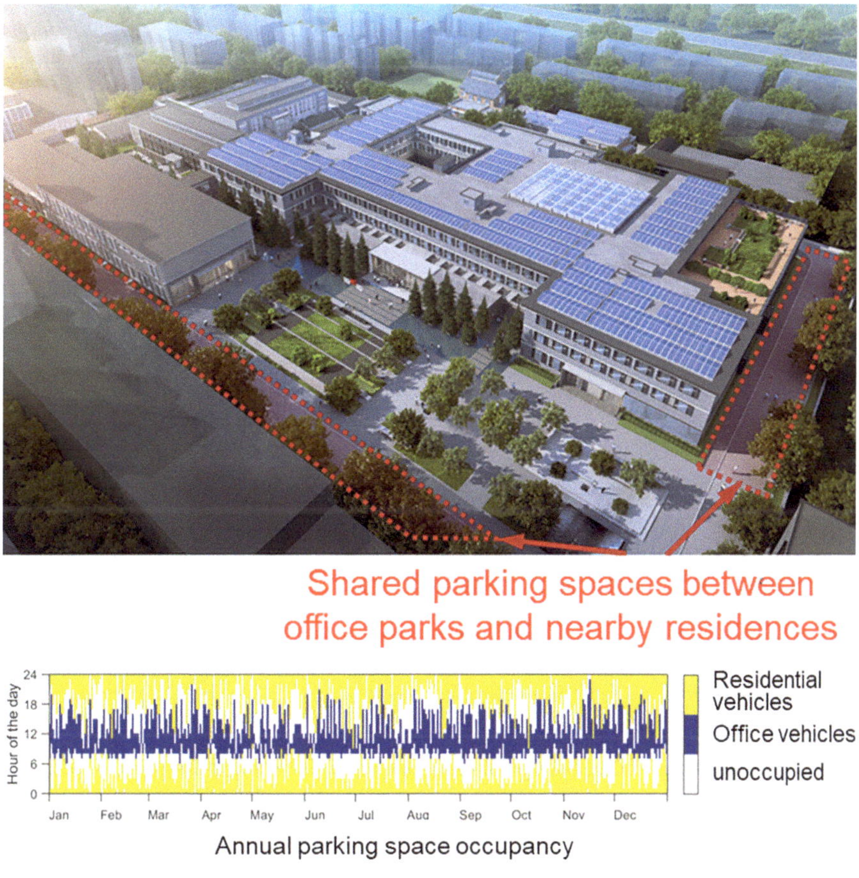

Fig. 6.8 Annual parking occupancy status of a parking space

Jinhua City, in its *Action Plan for Promoting the Construction of Charging Infrastructure to Facilitate the Spread of New Energy Vehicles in Rural Areas (2023–2025)* (Jinhua Municipal Development and Reform Commission 2023), proposes that existing residential areas may utilize the parking lots of surrounding public buildings or social public parking lots for unified construction of public charging facilities and should promote the "unified construction and operation" of parking space charging points and encourage models such as "sharing at different times" and "sharing adjacent parking spaces."

In summary, the key to the charging point management mode in an existing residential building lies in promoting the installation of intelligent coordinated charging points, thereby avoiding significant additional investment in power distribution caused by their installation. In the process of unified construction and management of intelligent charging points in residential compounds, attention should also be paid to the linkage with surrounding building parking lots and their power distribution systems, in order to provide residents with more efficient, convenient, and safe charging services.

Chapter 7
Building Electrification Technologies

7.1 Electric Cooking Appliances

Cooking, as an essential part of daily life, has a profound influence on the environment and climate change through its energy use. According to data from the International Energy Agency (IEA), carbon emissions from cooking represent about 3% of global greenhouse gas emissions and 56% of emissions from buildings. Traditional cooking appliances rely on fossil fuels, which release large amounts of greenhouse gases during combustion. In contrast, electric cooking appliances can reduce these greenhouse gas emissions significantly. As revealed by research data from Beijing, pipeline gas still makes up 74% of cooking energy in households, while electric cooking appliances account for only 13%. Therefore, to reduce carbon emissions, it is imperative to accelerate the electrification of cooking across the board.

7.1.1 Development Status of Electric Cooking Appliances

China's cooking culture boasts a long history and diverse cooking methods, including steaming, boiling, pan-frying, stir-frying, and deep-frying. Moreover, people's daily meals rely on the use of various types of cooking appliances. With the advancement of cooking electrification, a wide variety of electric cooking appliances have emerged, including electric steamers, electric cooking pots, rice cookers, induction cookers, electric frying pans, electric pressure cookers, electric chafing dishes, electric baking pans, microwave ovens, and conventional ovens. As urban lifestyles become faster-paced, many manual operations in the kitchen are progressively being replaced by personalized electric cooking appliances, such as soybean milk machines, juicers, coffee makers, egg boilers, and air fryers. Meanwhile, with the rise of smart homes, smart electric cooking appliances such as smart rice cookers

Building Energy Research Center of Tsinghua University, *Decarbonizing Urban Residential Buildings*, https://doi.org/10.1007/978-981-95-8200-6_7

and smart ovens have proliferated. They can be controlled via smartphones or voice assistants, providing people with convenient and fast cooking experiences.

The energy efficiency grades and thermal efficiency of electric cooking appliances are key factors influencing cooking energy use. To improve the efficiency of electric cooking appliances, the latest national standard titled "Minimum Allowable Values of the Energy Efficiency and Energy Efficiency Grades for Household and Similar Kitchen Appliances" has raised the existing minimum allowable values of energy efficiency. According to a survey on electric cookers (Zheng, Liu, Cui, et al. 2024), as shown in Fig. 7.1, from the perspective of market sales statistics, only 7% of products are rated at Energy Efficiency Grade 1, 14% are rated at Energy Efficiency Grade 2, the majority—68%—are rated at Energy Efficiency Grade 3, and about 11% do not meet the energy efficiency standards. Thus, it can be concluded that the overall energy efficiency level of electric cookers on the current market is relatively low, and there are some products failing to meet the energy efficiency standards. To deepen cooking electrification and minimize energy waste and environmental impact, users are encouraged to choose energy-efficient electric cooking appliances and refrain from using those failing to meet the energy efficiency standards.

Despite the growing diversity of electric cooking appliances, gas stoves remain the preferred cooking tools for many Chinese residents due to the long-term habit of open-flame cooking which is expressed in the saying "without fire there is no stove, without a stove there is no kitchen, and without a kitchen there is no home." However, gas stoves come with drawbacks, including environmental concerns,

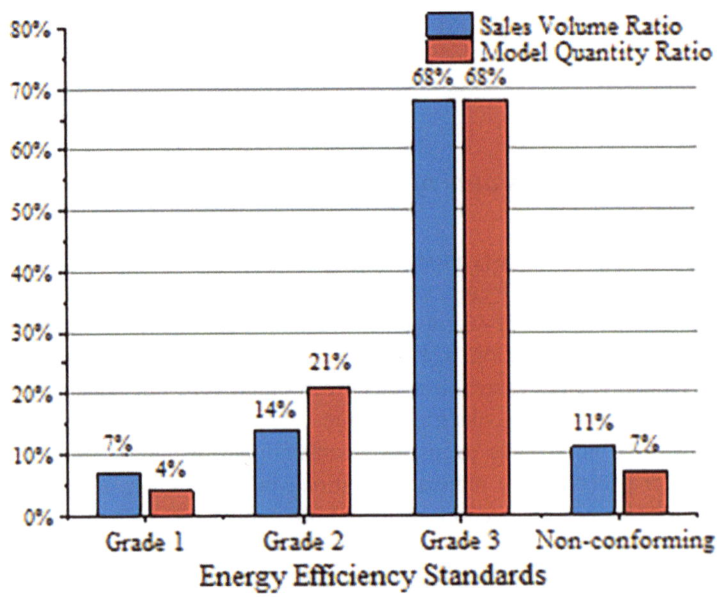

Fig. 7.1 Distribution of energy efficiency grades for electric cookers

safety risks, health hazards, and low energy efficiency. From an environmental perspective, the combustion of each cubic meter of natural gas generates 12 cubic meters of gaseous pollutants, including carbon monoxide, carbon dioxide, and nitrogen oxides. Regarding safety, gas stoves may pose potential hazards such as explosions, fires, and poisoning due to gas leaks. In terms of health, the harmful gases generated by gas stoves (such as carbon monoxide and nitrogen oxides) have a significant impact on air quality, especially in poorly ventilated kitchens. Long-term exposure to these pollutants may lead to respiratory problems, asthma, and other diseases. Concerning energy efficiency, gas stoves generally have low efficiency, approximately in the range of 55%–63%, with low thermal efficiency and large heat loss. Furthermore, gas stoves rely on gas supplies, which limit their application in remote areas.

As induction cookers are efficient, safe, environmentally friendly, and convenient, they are used by more and more users to replace gas stoves. An induction cooker operates on the electromagnetic induction principle to convert electrical energy into heat, which is then transferred to the food through the cookware via thermal conduction. With virtually no heat loss, its energy efficiency ranges from approximately 84% to 90%. Since induction cookers do not produce an open flame, they can eliminate the risk of fire and explosion, thereby offering enhanced safety. Induction cookers can be used simply by plugging them in, and they are usually equipped with a temperature control system that allows for precise adjustment of heating temperatures. Hence, they offer greater mobility and convenience. Without the need for oxygen in the heating process, induction cookers produce no exhaust gases indoors. They will generate neither direct nor indirect carbon emissions in a future energy mix with a high proportion of renewable electricity. Taking residential kitchens in Shanghai as an example (Xiong, Yu, Zhao, et al. 2023), as shown in Fig. 7.2, the use of induction cookers can reduce energy consumption by 51% and carbon emissions by 25% compared to gas stoves. If 50% of cooking methods in residential buildings in Shanghai were to be switched to induction cookers, it would result in a reduction of approximately 1.07 million tons of carbon emissions (around 9.98%) and 4.09 million tons of direct carbon emissions.

Furthermore, induction cookers offer high cost performance. Table 7.1 presents gas stoves and induction cookers available on the market. If calculated based on a gas price of RMB 4/m^3 and electricity price of RMB 0.5/kWh, the cost per unit of heat energy from electricity is 20% higher than that from gas. However, induction cookers have a significant advantage in thermal efficiency. In current products, induction cookers typically achieve 86% thermal efficiency (Grade 3), while gas stoves reach only 63% (Grade 1). As a result, the required heat input from electricity is only 73% of that from gas to achieve the same heating effect. Overall, the energy cost of induction cookers is 12% lower than that of gas stoves. Therefore, despite its higher initial investment, an induction cooker has a lower running cost compared to a gas stove. Moreover, with the improvement of induction cookers' thermal efficiency and the rise of natural gas costs in the future, the energy-saving benefits and cost advantages of induction cookers will become even more pronounced.

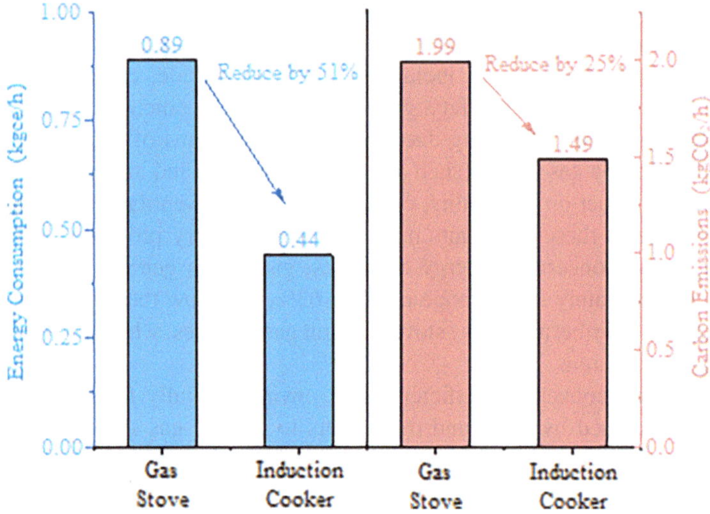

Fig. 7.2 Comparison of energy use and carbon emissions between gas stoves and induction cookers

Table 7.1 Product parameters of gas stoves and induction cookers on the market

Cooking equipment	Product	Single-burner heat load	Rated thermal efficiency/ energy efficiency grade	Price
Gas stove	A1 (dual-burner)	4.5 kW	63%/Grade 1	450 RMB
	A2 (dual-burner)	5 kW	63%/Grade 1	398 RMB
	A3 (dual-burner)	5.2 kW	67%/Grade 1	698 RMB
Induction cooker	B1 (single-burner)	2.2 kW	86%/Grade 3	359 RMB
	B2 (single-burner)	3.5 kW	86%/Grade 3	509 RMB
	B3 (dual-burner)	2.2 kW	86%/Grade 3	1399 RMB

7.1.2 New Electric Cooking Appliances

Although induction cookers have many advantages, their pace of replacing gas stoves in China has been relatively slow since their invention over 20 years ago. One key reason is that induction cookers often heat unevenly at the edges and lack the open-flame stir-frying function of traditional gas stoves. An electric flame stove, as

Fig. 7.3 Electric flame stove

illustrated in Fig. 7.3, is a new cooking appliance combining the advantages of both gas stoves and induction cookers. It can leverage low-temperature plasma technology to convert electrical energy into thermal energy. It not only replicates the flame effect of a traditional gas stove but also maintains green, safe, and environmentally friendly features. Among the current products on the market, an electric flame stove can reach an electric arc flame temperature of up to 1200 °C (compared to around 600 °C for natural gas), a power of 2500 W, and a thermal efficiency of up to 78%. Moreover, its operating cost is also lower than that of a gas stove. Electric flame stoves have short cooking times, namely, food cooks quickly. They do not compromise the fibrous structure of ingredients and are less likely to cause the loss of nutrients and moisture, thus preserving the original flavor of ingredients. Furthermore, electric flame stoves feature multi-level firepower adjustment and multiple safety protection functions, including automatic flameout protection, overheat protection, and dry-burning prevention. However, since the current price of electric flame stoves is typically four to five times that of traditional gas stoves and twice that of induction cookers, the widespread application of electric flame stoves remains constrained by economic costs. However, with technological advancements and growing concerns for health and environmental protection, electric flame stoves are expected to become a new favorite in kitchens in the future, accelerating the electrification of cooking.

With the growing demand for electrification in cooking, more and more energy-efficient and high-performance electric cooking appliances that adequately meet user needs have emerged. Thanks to the continuous upgrading of product technology, gradual reduction in prices, and the development of intelligent features, electric cooking appliances will gradually replace traditional cookware, gaining increasing recognition and application by users. All-electric kitchens will also become a favorite in the future. These technologies will further promote the comprehensive electrification of China's end-use sectors, contributing to the achievement of the carbon neutrality goal.

7.2 Electric Water Heater Equipment and Its Flexible Regulation Technology

7.2.1 Electric Water Heater

In the process of achieving zero-carbon transformation in the building sector, the comprehensive electrification of domestic hot water (DHW) production in residential households also plays a crucial role. Existing electric water heaters can be primarily categorized into two types: instant-heating type and heat-storage type. Among them, heat-storage water heaters have significantly lower heating power compared to instant-heating water heaters, resulting in relatively lower requirements for building power distribution and reduced impact on the power grid. On the other hand, the heat storage capacity of heat-storage water heaters can also serve as a method for future building demand response, promoting flexible electricity usage in buildings. Therefore, heat-storage water heaters should be the main development direction for the electrification of domestic hot water production in the future.

In recent years, electric heat pump water heaters (also known as air-source heat pump water heaters) have gradually emerged in China's water heater market. According to relevant market data, the total sale of electric heat pump water heaters in China increased from 520,000 units in 2014 to 1.23 million units in 2019, and the market size is still maintaining a rising trend. Compared to conventional heat-storage electric water heaters, electric heat pump water heaters exhibit a high energy efficiency. While electric water heaters typically have a thermal efficiency of around 95%, electric heat pump water heaters can generate 3 units of heat with just 1 unit of electricity, achieving a thermal efficiency of up to 300%. With the same water consumption, electric heat pump water heaters are far more energy-efficient than ordinary electric water heaters. Comparing the initial investment and operating costs of gas water heaters, conventional heat-storage electric water heaters, and electric heat pump water heaters, as shown in Table 7.2, it is evident that with the same water consumption, the operating costs of electric water heaters are higher than those of gas water heaters. However, the operating costs of electric heat pump water heaters are lower than those of gas water heaters. Even when accounting for the heat loss from the hot water tank of the heat pump water heater, the annual operating costs of electric heat pump water heaters can generally be comparable to those of gas water heaters.

Table 7.2 Comparison of operating costs of various types of water heaters

	Gas water heater	Electric water heater	Electric heat pump water heater
Annual water consumption, L	50 L * 365 days =18,250 L		
Thermal efficiency	90%	95%	300%
Energy consumption	82.8 m^3	784 kWh	248 kWh
Energy expense (RMB)	248.4	376.5	119.2

Currently, the equipment cost of electric heat pump water heaters remains significantly higher than that of ordinary gas water heaters and electric water heaters. However, as the market scale expands in the future, their prices will gradually decrease. From the perspective of China's zero-carbon transition and for the purpose of energy conservation and emission reduction, electric heat pump water heaters should be vigorously promoted to achieve the electrification of supplying domestic hot water.

7.2.2 Flexible Regulation Technology for Electric Water Heaters

The flexible regulation of electric water heaters can provide significant technical support for the flexible energy consumption of building power systems. Among the various technologies for flexible energy consumption in residential buildings, utilizing the storage tank of an electric water heater as flexible energy storage equipment provides a low-cost, high-potential solution for building energy regulation. Through intelligent control and thermal energy storage optimization of electric water heaters, a dynamic balance between the power system and building energy demand can be achieved. This not only enhances the energy efficiency of buildings but also provides crucial support for the stable operation of the power grid.

The core component of an electric water heater is the water storage tank, which is essentially a device with sensible heat energy storage functionality. By converting electrical energy into thermal energy, the electric water heater stores hot water in its tank, providing users with washing, bathing, and other hot water services. Unlike other household appliances, the operation of an electric water heater can be independent of real-time demand, offering flexibility in both working hours and power consumption. By preheating or delaying heating, the power demand can be shifted from peak periods to off-peak periods, achieving load transfer in the power grid. In addition, the water storage tank has excellent thermal insulation properties, enabling it to maintain stable water temperature for extended periods, and making it a low-cost short-term energy storage device.

By leveraging this property, the electric water heater can serve as a "thermal energy storage battery," participating in flexible energy regulation for residential buildings. The schematic diagram of the regulation and storage function of the electric water heater is shown in Fig. 7.4. By conducting real-time monitoring of the electric water heater valve, the user's hot water demand can be obtained, and machine learning can be used to predict the user's domestic hot water usage patterns. Through optimized control, while meeting the user's hot water demand, the electric water heater can store sufficient backup hot water during off-peak periods with lower electricity prices (e.g., nighttime) and during periods with lower dynamic carbon emission responsibility factor C_r (e.g. midday when PV power

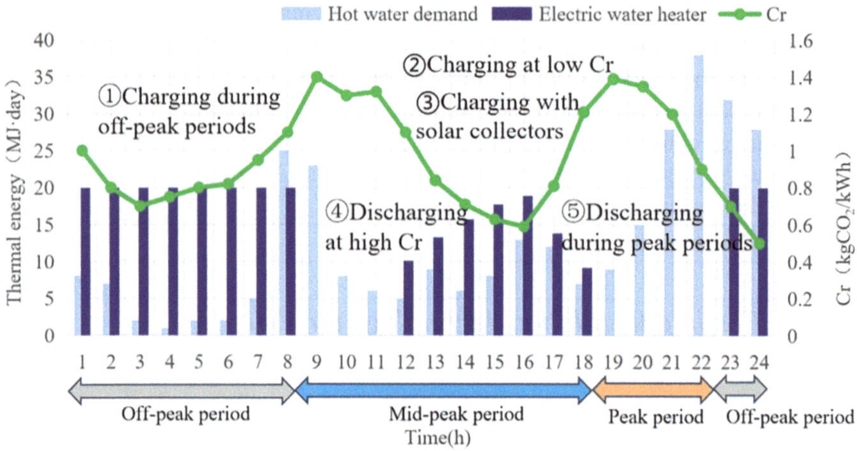

Fig. 7.4 Schematic diagram of the regulation and storage function of an electric water heater

generation is abundant). During peak electricity usage hours with higher electricity prices and periods when C_r is relatively high, the operation of the electric water heater can be paused to reduce the building's electricity demand, thereby alleviating pressure on the power grid. At the same time, to make full use of renewable resources, the flexible regulation of DHW energy consumption can be combined with solar water heaters. By utilizing a larger storage tank to store hot water generated by solar collectors during the day, supplemented by electric auxiliary heating and storage during off-peak nighttime hours, the domestic hot water demands of urban residential building households can be essentially met for an entire day. This thermal energy storage-based regulation method can not only achieve peak load shaving, but also optimize users' electricity costs and improve the utilization rate of renewable energy.

The fluctuation range of C_r throughout the year varies between 0.0 and 1.7 kgCO_2/kWh, indicating that the difference in C_r across different periods can exceed ten times. Currently, the National Energy Administration stipulates that the peak-valley tariff ratio should in principle be no less than 4:1, and the time-of-use tariff ratio across different periods is expected to exceed ten times in the future. The thermal efficiency of an electric heat pump water heater can reach as high as 300%~400%. It is evident that by fully leveraging the regulation and storage capabilities of electric water heaters and applying optimized control methods such as machine learning, DHW can be obtained at lower electricity prices and reduced C_r while meeting user demand, showcasing significant development potential.

Based on the current typical domestic hot water usage in urban residential buildings and the performance parameters of electric water heaters, it is estimated that the daily DHW demand for an urban household is approximately 60 L. The maximum hot water temperature is generally 60 °C, while the tap water supply temperature is typically 15 °C. Therefore, the daily heat requirement for DHW is 3.12 kWh.

Assuming the electric water heater has a heating power of 2 kW, completing the DHW heating process would take approximately 1.5–2 h. Based on survey data of electric water heater tanks, the overall heat transfer coefficient of the tank is typically 0.5 W/(m^2*K). With an estimated tank surface area of 1 m^2 and an indoor temperature of 25 °C, the hourly heat loss is approximately 0.0175 kWh. Assuming a hot water storage duration of 18 h, the total heat loss amounts to roughly 0.315 kWh. This involves using a hot water tank for flexible regulation of domestic hot water energy consumption, with heat loss accounting for approximately 10% of the DHW heat demand. The above estimation and analysis demonstrate that utilizing an electric water heater for flexible regulation can achieve a daily flexible regulation capacity of 3.5 kWh within an 18-h period for the users in ordinary urban residential buildings. If all 370 million urban residential building households in China participate in the flexible demand response of electric water heaters, it could achieve a flexible regulation of 1.295 billion kWh within an 18-h window every day, representing a remarkably substantial flexible energy resource. Besides, unlike the electric heat pump water heater, the electric water heater can increase the storage efficiency and store more heat for regulation by raising the water temperature, despite being limited by capacity and thermal efficiency. Through the above analysis, it can be seen that utilizing the storage tank of an electric water heater for flexible energy regulation can fully leverage the valley-load tariff, shift peak grid loads, and maintain low heat loss with minimal impact on users' hot water demand. By integrating intelligent control technology, this approach not only optimizes household electricity costs but also significantly enhances the flexibility of the building energy system, providing an effective solution for peak load shaving and load balancing in the power system.

To fully leverage the regulation function of electric water heaters, future products could be developed with integrated features such as intelligent scheduling, real-time monitoring, and remote control capabilities. These products can intelligently adjust heating time and power based on grid load and user demand, thereby reducing electricity costs for users and improving energy efficiency. It is also possible to further improve the efficiency of the electric water heater by refining the water tank function and developing a dual-tank electric water heater based on users' hot water usage habits. Among them, one water tank is designed to meet the concentrated water demand of users, featuring significant flexible regulation capacity, while another tank caters to scattered daily needs such as handwashing and cooking, requiring higher responsiveness to real-time hot water usage. In addition, it can also provide energy management services based on electric water heaters, offering users personalized water heater adjustment and storage strategies to achieve precise and efficient energy management. Through these measures, the industrialization of the electric water heater's energy storage function will be accelerated. This not only reduces energy consumption and costs on the user side but also enhances the flexible storage capacity of the power system, promoting mutual benefits for both users and the grid.

7.3 Energy Storage Multi-split Air Conditioner Technology

7.3.1 Technical Principles of Energy Storage Multi-split Air Conditioner

An energy storage multi-split air conditioner is an innovative technology that combines traditional multi-split air-conditioning system with energy storage technology. The energy storage device usually contains water or other phase change energy storage materials. Water is an excellent choice for energy storage material due to its high energy storage capacity, along with advantages such as low cost, abundant availability, and high safety. The following explanation will use water as the energy storage material.

For the energy storage multi-split air conditioner system, the energy storage device (as shown in Fig. 7.5) generally works by using an internal ice-melting mechanism. The water inside is filled once and does not participate in the system circulation, thus eliminating the need for additional devices such as water pumps or plate heat exchangers to drive the circulation of energy storage materials. This design offers great convenience for engineering applications. The heat exchange tube inside the energy storage device is generally a serpentine coil. The refrigerant flows within the heat exchange tube to exchange heat with water, thereby storing energy or extracting the energy already stored in the water.

The rated cooling energy capacity of the energy storage device can be estimated as follows. The cooling energy mainly consists of two parts: the latent heat of ice

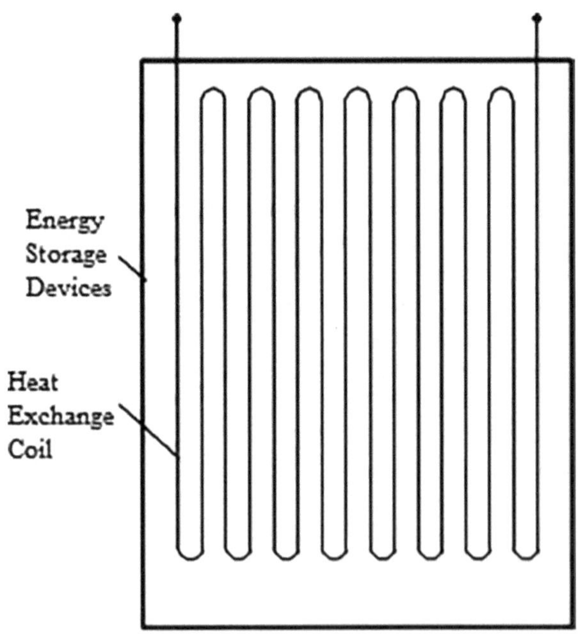

Fig. 7.5 Schematic diagram of energy storage device

and the sensible heat of water. The latent heat of ice is 335 kJ/kg, and the specific heat capacity is 4.216 kJ/(kg °C). Since the water in the accumulator may not completely freeze, a freezing rate of 90% is assumed. Therefore, for an energy storage device with a water volume of 1.5 m^3, the rated cooling energy capacity is approximately 178 kWh.

In addition, most air-conditioning equipment using external melt ice storage systems and dynamic ice-making systems do not have heat storage and release capabilities. However, energy storage multi-split air conditioners can achieve heat storage and release through a cycle of heating water into hot water and then cooling it down. The stored heat is generally used to defrost the multi-split air conditioner during winter heating. Consider an energy storage device is also loaded with 1.5 m^3 of water. Since no ice-water phase change occurs, its rated heat storage capacity only includes sensible heat, approximately 53 kWh. It can be seen that the rated heat storage capacity (53 kWh) is much smaller than the rated cooling energy capacity (178 kWh), which is why energy storage devices are generally not used to provide heat for heating but only for defrosting.

To achieve the aforesaid energy storage and release, the energy storage multi-split air conditioner must operate in at least three modes during the cooling season: cool storage mode, conventional cooling mode, and cooling release mode. During heating seasons such as winter, the energy storage multi-split air conditioner can also operate in heat storage mode, conventional heating mode, and heated defrosting mode.

1. In the cool storage mode (as shown in Fig. 7.6), the energy storage device functions as an indoor unit, with the outdoor unit providing cooling. During this process, the evaporation temperature is kept below the freezing point to ensure water can solidify into ice smoothly. The multi-split air conditioner system typically stores cooling energy during off-peak electricity hours to fully utilize valley load power and achieve valley-filling. It should be noted that due to the lower

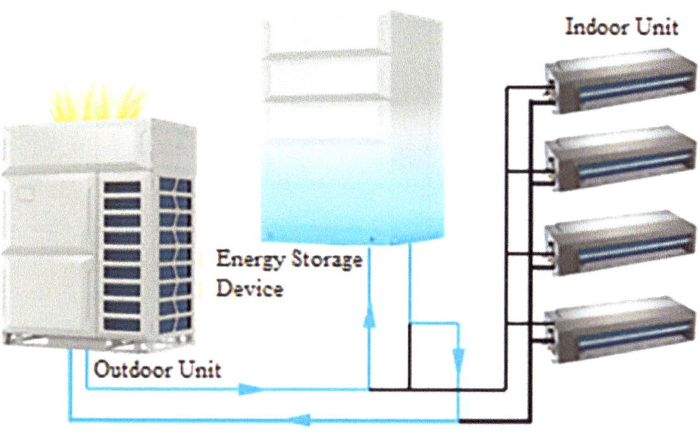

Fig. 7.6 Schematic diagram of cool storage mode

evaporation temperature during the cool storage process, the energy efficiency of the multi-split air conditioner is often lower than that in conventional cooling mode, resulting in relatively higher electricity consumption for cool storage. Considering both cooling energy storage and release, the energy storage multi-split air conditioner generally cannot reduce total electricity consumption but only shift it.

Figure 7.7 illustrates the temperature variation of water during the cool storage process. The water temperature gradually decreases to 0 °C before freezing begins. Once completely frozen, the temperature of the ice can continue to drop. Stage 1 is the sensible heat storage stage of water, stage 2 is the latent heat storage stage of ice, and stage 3 is the sensible heat storage stage of ice. The cooling energy is stored in the energy storage device as sensible heat of water, latent heat of ice, and sensible heat of ice.

2. In the conventional cooling mode (as shown in Fig. 7.8), the energy storage device does not operate. The built-in valves are closed to disconnect it from the system, and the system functions as a standard multi-split air conditioner, with the outdoor unit directly providing cooling to the indoor unit.
3. In the cooling mode, there are two types: non-refrigerant pump mode (Fig. 7.9) and refrigerant pump mode (Fig. 7.10).

   In the non-refrigerant pump mode (as shown in Fig. 7.9), the energy storage device switches the flow path to the cooling release path by controlling the integrated valve assembly. After the refrigerant flows out of the outdoor unit, it passes through the energy storage device and exchanges heat with ice/water, absorbing cooling energy before entering the indoor unit for refrigeration. During this process, the refrigerant's cooling capacity is enhanced due to the increased cooling energy it carries. Therefore, under the same indoor load conditions, the multi-split system operating in the cooling release mode requires a lower compressor output compared to a conventional multi-split system. This reduces the compressor's operating frequency, thereby lowering power

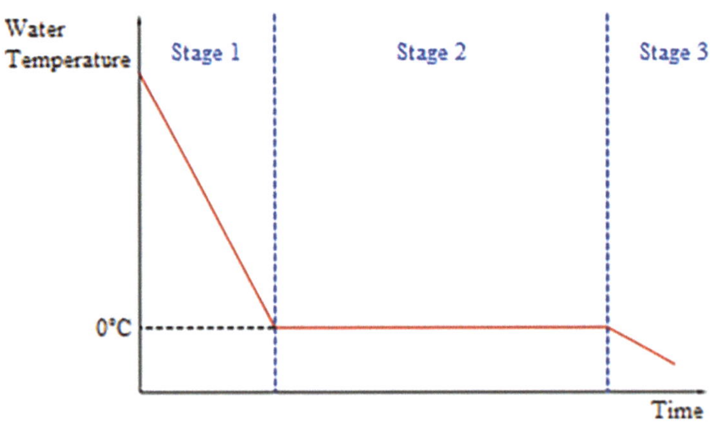

Fig. 7.7 Schematic diagram of water temperature variation during cooling energy storage process

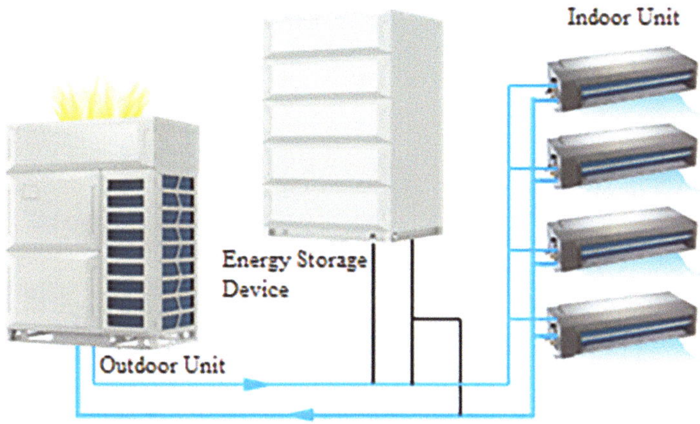

Fig. 7.8 Schematic diagram of conventional cooling mode

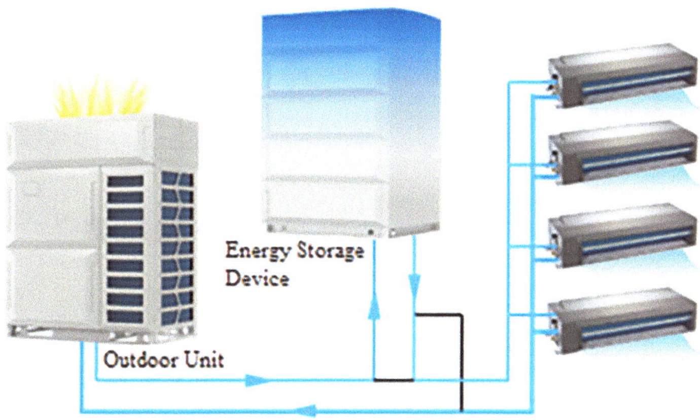

Fig. 7.9 Schematic diagram of non-refrigerant pump cooling release mode

consumption and significantly reducing the system's demand for peak-period electricity. Meanwhile, the ice in the energy storage device gradually melts, and once fully melted, the water temperature continues to rise until it reaches the same temperature as the refrigerant outlet temperature of the outdoor unit (approximately 30 °C), at which point the cooling energy is completely released.

In the refrigerant pump mode (as shown in Fig. 7.10), the outdoor unit shuts down, with both its compressor and fan deactivated. Instead, the refrigerant pump is activated, while the energy storage device controls the built-in valve combination to switch the flow path to the cooling release path. The refrigerant flows between the energy storage device and the indoor unit, exchanging heat with the energy storage material to absorb cooling energy in the energy storage device, and then evaporates in the indoor unit to release cooling energy, thereby

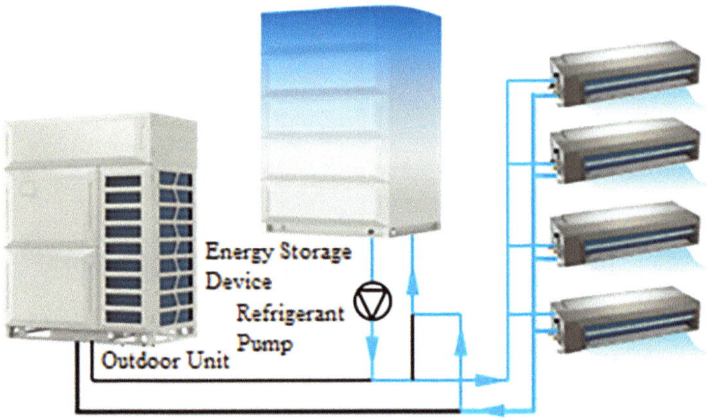

Fig. 7.10 Schematic diagram of refrigerant pump cooling release mode

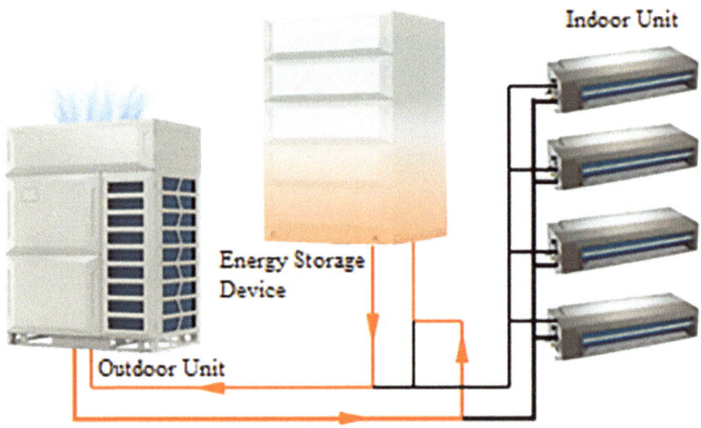

Fig. 7.11 Schematic diagram of heat storage mode

achieving the cooling function. Since the high-power consumption components, such as the compressor and fan in the outdoor unit stop working during this process, and only the refrigerant pump operates at low power consumption, the power of the multi-split air conditioner system can be reduced to 20% or less of a conventional cooling system while maintaining the same cooling capacity. This significantly decreases the demand for peak-period electricity by the multi-split air conditioner system.

4. Heat storage mode (as shown in Fig. 7.11) where the multi-split air conditioner heats the energy storage device to warm water (reaching a final temperature of 35~45 °C), thereby storing thermal energy.
5. In the conventional heating mode (as shown in Fig. 7.12), the energy storage device does not operate. The built-in valve is closed to disconnect it from the

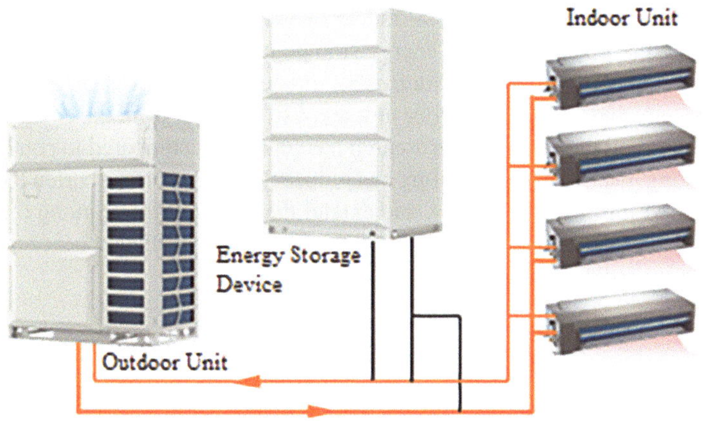

Fig. 7.12 Schematic diagram of conventional heating mode

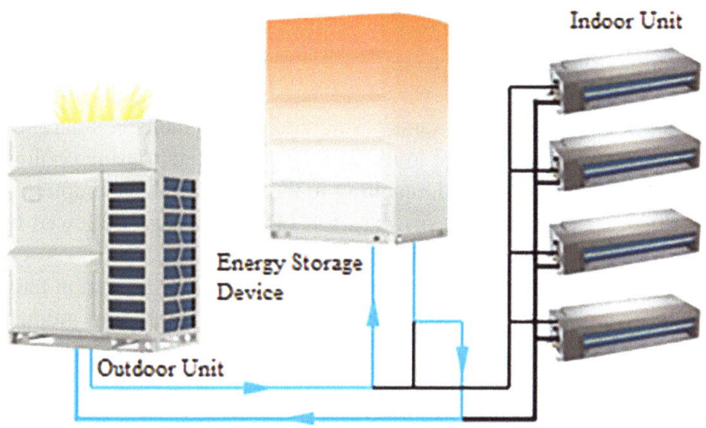

Fig. 7.13 Schematic diagram of heated defrosting mode

system, and the system functions as a conventional multi-split air conditioner, with the outdoor unit directly supplying heat to the indoor unit.

6. Heated defrosting mode is shown in Fig. 7.13. During winter heating, the multi-split air conditioner outdoor unit needs to absorb heat from the low-temperature outdoor environment to supply heat indoors, so the refrigerant temperature in the heat exchange tube of the outdoor unit is very low, often below 0 °C. When the water vapor contained in the outdoor air comes into contact with the low-temperature heat exchange tube, it gradually condenses and forms frost on the outdoor unit's heat exchanger. Once the frost layer accumulates to a certain thickness, the multi-split air conditioner needs to stop heating and instead extracts heat from the indoor side to warm the outdoor unit's heat exchanger, thereby melting the frost layer. During this process, discomfort may occur due to heat extraction from the indoor side. Additionally, when switched back to heating

mode, the pipe temperature of the indoor unit requires reheating, which prolongs the period of discomfort. The energy storage multi-split air conditioner can utilize stored heat for the heated defrosting mode by releasing heat. In this mode, the system shuts down the indoor unit and extracts heat from the energy storage device to melt the frost layer. Research has found that compared to conventional defrosting, heat-release defrosting can reduce defrosting time, shorten the recovery period for heating capacity, and extend the heating cycle, among other benefits. Generally, about 3 h of heat storage can provide sufficient heat for over 13 defrosting cycles.

7.3.2 Technical Effects of Energy Storage Multi-split Air Conditioner

To cover the cooling demand of a typical residential building with an area of 5000 m^2 for one cycle of 10 h—where the first 5 h correspond to off-peak electricity periods and the latter 5 h to peak electricity periods—an energy storage device with a water capacity of approximately 30 m^3 would be required. Taking a conventional multi-split air conditioner covering the same residential area as the benchmark, the power consumption of an energy storage multi-split air conditioner during cooling energy storage and release was evaluated, resulting in the electricity consumption schedule shown in Fig. 7.14. The process involves 8 h of cool storage, followed by 5 h of conventional cooling and 5 h of cooling energy release.

During the 8-h cool storage period, the system's target evaporation temperature dynamically changes with the water freezing process, stabilizing gradually after

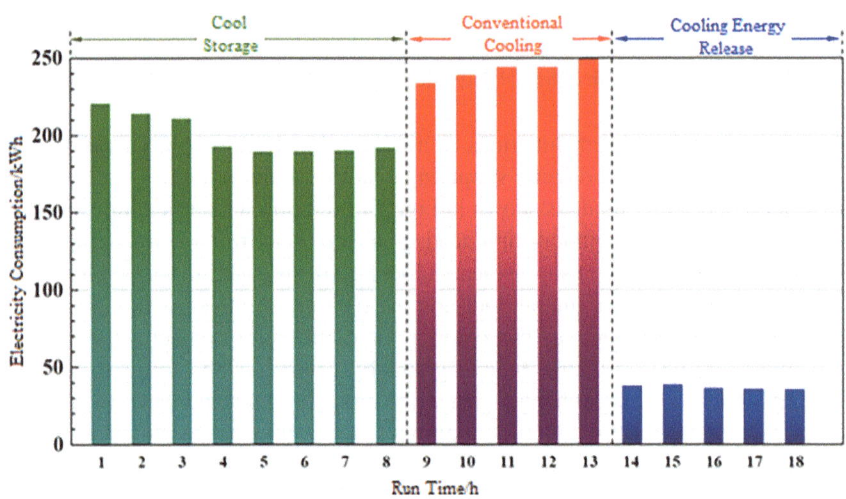

Fig. 7.14 Power consumption schedule of energy storage multi-split air conditioner

5 h. Although the cool storage capacity demand of the energy storage device is lower than that of conventional refrigeration, the unit output during cooling energy storage is practically equivalent to conventional cooling due to the requirement of providing an evaporation temperature below −8 °C for ice formation. The average power consumption during the cooling energy storage process is approximately 81% of that during conventional cooling. After the completion of cooling energy storage, the total 10-h cooling period is divided into two 5-h segments—one for conventional cooling and the other for cooling energy release. From Fig. 7.14, the difference in power consumption between conventional cooling and cooling energy release periods is clearly visible: Taking the refrigerant pump cooling release mode as an example, during cooling energy release, the average power consumption is approximately just 15% of that during conventional cooling. This means about 85% of the power consumption is shifted during the 5-h peak electricity period. By integrating the power consumption over a 10-h cooling period, the electricity consumption can be shifted by approximately 43% compared to conventional units.

7.4 Air Source Heat Pump Underfloor Heating Technology

Air source heat pump underfloor heating is a building heating method that utilizes an air source heat pump as the heat supply source and radiant floor heating as the heating terminal. Compared to traditional gas heating, the air source heat pump is driven by electricity, which can fully leverage the carbon reduction potential of renewable energy generation. Additionally, the radiant floor heating terminal possesses significant thermal storage capacity, enabling it to store the heat generated by renewable energy-powered heat pumps, and thereby promoting energy-efficient and low-carbon development in urban residential building heating. In recent years, a type of heat pump dual-supply product has emerged in the market. The product utilizes an air source heat pump as the cold and heat source, capable of simultaneously meeting the cooling and heating demands of buildings. It employs fan coils or air duct units for cooling and underfloor heating for warmth.

7.4.1 System Composition and Principles

The heat pump dual-supply unit consists of three parts: the air source heat pump unit, the fan cooling terminal, and the underfloor heating terminal. Based on the medium for heat and cold transmission, the systems are primarily categorized into three types: water cooling & water heating system, refrigerant cooling & water heating system, and refrigerant cooling & refrigerant heating system.

1. The heat pump unit of the water cooling & water heating system (as shown in Fig. 7.15) contains a refrigerant-water heat exchanger, which transfers cooling

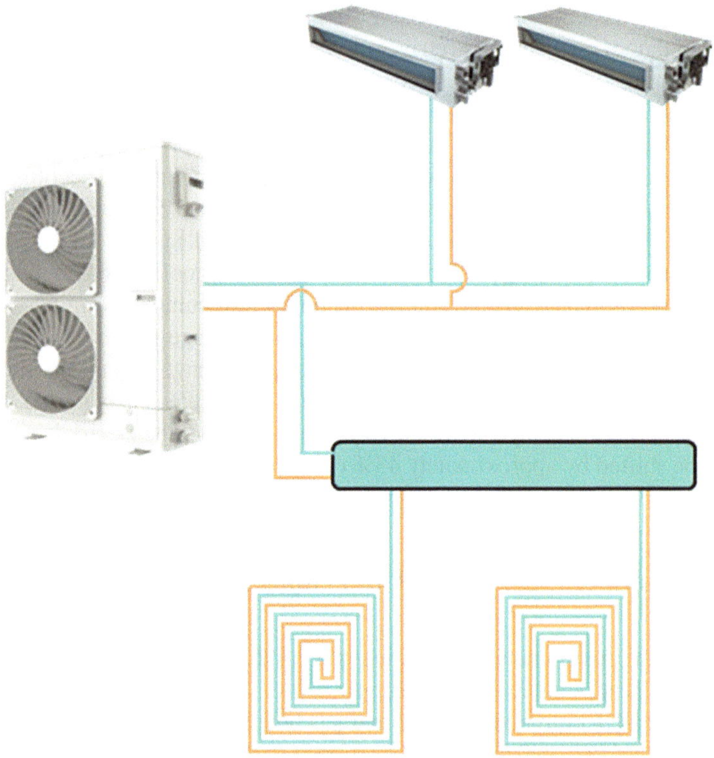

Fig. 7.15 Water cooling and water heating system

or heating energy from the refrigerant to water. The chilled water then delivers cooling energy to the fan coil for air conditioning, while the heated water transfers heat to the underground piping system for space heating. The system is relatively simple and easy to maintain, being one of the earliest heat pump dual-supply products to enter the market.

2. The refrigerant cooling and water heating system (as shown in Fig. 7.16) operates similarly to a traditional multi-split air conditioner during cooling. The direct-expansion ducted air conditioner terminal directly supplies cooling energy to the indoor space through refrigerant evaporation. During heating, the system utilizes an externally or internally installed refrigerant-water heat exchanger in the heat pump unit to transfer the heat generated by refrigerant condensation to water, which then circulates through underground pipes to provide heating. Compared to the water cooling & water heating system, the refrigerant cooling & water heating system, due to absence of heat exchange process, offers a higher energy efficiency and faster response during cooling. It is currently the leading heat pump dual-supply product in the market.
3. The refrigerant cooling and refrigerant heating system (as shown in Fig. 7.17) utilizes refrigerant for direct heat exchange at the terminal units in both cooling

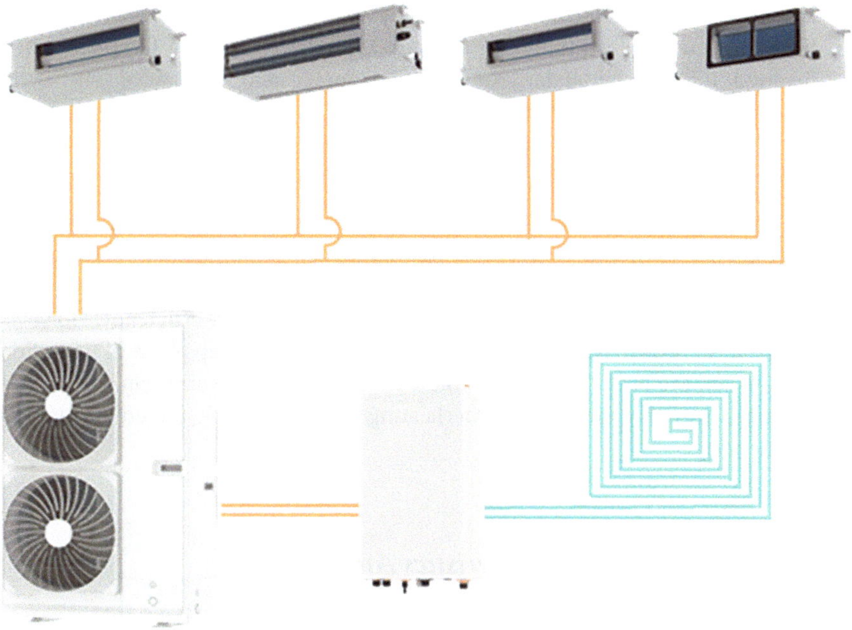

Fig. 7.16 Refrigerant cooling and water heating system

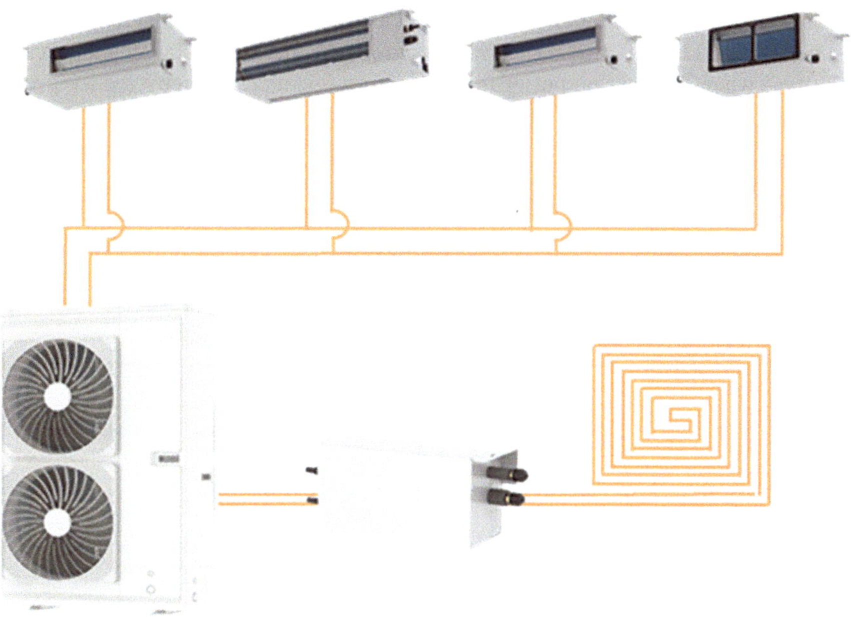

Fig. 7.17 Refrigerant cooling and refrigerant heating system

and heating modes. Specifically, it employs a direct-expansion ducted air conditioner for cooling, while during heating, the refrigerant from the main unit is diverted through an underfloor heating adapter and exchanged via direct-condensation capillary tubes embedded in the floor. Among the three types of heat pump dual-supply systems, this system boasts the highest heat exchange efficiency and energy savings, making it an emerging product in the market.

The heat pump dual-system allows for selective activation of partial indoor units during the cooling period, embodying the "partial-time, partial-space" feature. For heating mode, zonal control of underfloor heating can be achieved through underfloor heating manifolds or adapters, thereby enhancing energy efficiency. However, due to the hysteresis in room temperature rise during the startup phase of underfloor heating, to achieve better comfort, whole-house underfloor heating can also be used for heating, meaning the underfloor heating in all areas is turned on and off simultaneously.

7.4.2 Operational Characteristics Analysis

The operational characteristics of heat pump underfloor heating can be analyzed from three aspects: comfort, energy efficiency, and economic performance.

In terms of comfort, underfloor heating not only offers the aforementioned advantage of "warm feet and cool head," but also features stable room temperatures. The primary reason is that the floor has significant thermal mass, which acts as a "buffer" when the heat output of the heat pump fluctuates, thereby reducing temperature variations in the room. Figure 7.18 illustrates the room temperature variation during intermittent operation of underfloor heating. In the shutdown period, the heat stored in the floor continues to provide heating for the indoor space. In a specific project case, the room temperature drop was less than 1 °C after 10 h of shutdown. Therefore, when the air source heat pump unit undergoes defrosting during

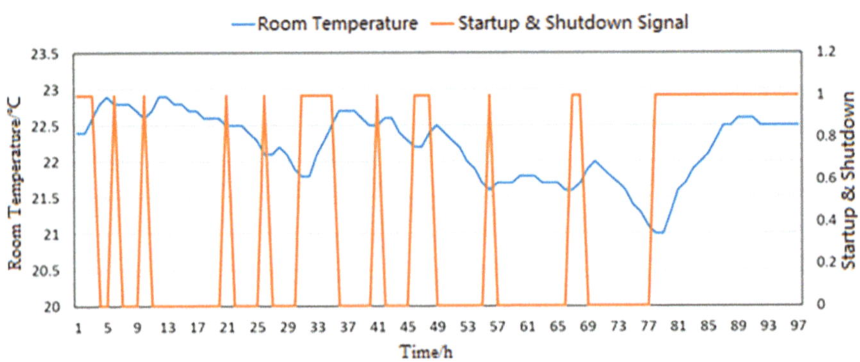

Fig. 7.18 Room temperature variation during intermittent operation of underfloor heating

shutdown, there is almost no change in indoor temperature, demonstrating improved comfort compared to fan heating (where room temperature drops during unit defrosting).

In terms of energy efficiency, underfloor heating utilizes the entire floor as a "heat exchanger." Compared to radiators and air supply terminals, it offers a larger heat transfer area. Under the same heat exchange capacity, the temperature difference required for heat transfer is smaller, meaning the heating temperature can be lower. Typically, underfloor heating operates at 35 °C, whereas radiators and air supply terminals often require temperatures as high as 50 °C. Therefore, for the heat pump unit, the underfloor heating terminal is more energy-efficient. At the same time, since underfloor heating primarily utilizes radiant heating to warm the human body, under the premise of achieving the same level of thermal comfort, the design temperature of a room with radiant heating can be 2 °C lower compared to convection heating (such as air duct-based air supply terminals). This, to some extent, reduces the building load during winter.

In terms of economic performance, heat pump underfloor heating is currently primarily used in hot summer and cold winter regions, where it can be compared with traditional gas-fired underfloor heating. Taking a building with an interior area of 100 m^2 in Zhejiang Province as an example, its gas consumption during winter (December to February) is approximately 20 m^3/day per 100 m^2, with a total heat supply of about 700 MJ/day. At a gas price of RMB 3.00 per m^3, the winter operating cost for gas underfloor heating is RMB 5400. Under the same total heat supply, with an average energy efficiency ratio (EER) of four for the air source heat pump, the daily electricity consumption for the heat pump underfloor heating is approximately 49 kWh. At an electricity cost of RMB 0.60 per kWh, the winter operating cost for the heat pump underfloor heating is RMB 2646, which is 51% lower than that of gas underfloor heating, demonstrating significant economic benefit advantages.

7.4.3 Solutions for Energy Saving and Carbon Reduction

Against the backdrop of China's "carbon peaking and carbon neutrality" policy, efforts can be made in the following three aspects to further enhance the energy-saving and carbon-reducing capabilities of heat pump underfloor heating:

1. User behavior: Underfloor heating has significant thermal storage capacity. Users can turn off the heat pump unit during periods of low heating demand based on their room temperature requirements, utilizing the stored heat in the underfloor heating system for intermittent heating. This reduces the operating time of the heat pump, achieving energy conservation. On the other hand, for certain water cooling & water heating systems and refrigerant cooling & water heating systems, users can set the water supply temperature when using underfloor heating. According to the *Technical Specification for Radiant Heating and*

Cooling (JGJ 142-2012) (as shown in Table 7.3), the heat output of underfloor heating at a 35 °C water supply temperature is reduced by 20~50% compared to that at a 40 °C water supply temperature. Therefore, by reasonably setting the water supply temperature, the heat output of the heat pump underfloor heating can be effectively reduced, preventing excessive heating and lowering the energy consumption of the heat pump.

2. Engineering construction: The actual performance of underfloor heating is closely related to material selection, installation, and commissioning during the project. As shown in Fig. 7.19, the structure of the underfloor heating terminal is complex, comprising an insulation layer, reflective film, steel mesh, heating pipes, concrete, and floor finishing layers. The heat transfer and thermal storage properties of these materials significantly impact the energy efficiency of underfloor heating. For example, tile flooring exhibits superior heat transfer performance compared to wooden flooring. To achieve the same heat transfer capacity, tile flooring requires lower water supply temperatures, effectively enhancing the operational efficiency of heat pumps. Additionally, the installation of insulation layers can substantially reduce downward heat transfer and inter-unit heat transfer in underfloor heating systems, minimizing heat dissipation. On the other hand, to improve the thermal storage capacity of underfloor heating, phase change thermal storage flooring (Zheng, Liu, Cui, et al. 2024) can be employed. Compared to traditional underfloor heating systems without phase change material (PCM), the use of PCM significantly extends the heat release duration of the flooring during shutdown periods. By integrating phase change thermal storage flooring technology with intermittent heating, the goal of peak shaving and valley filling for the power grid can be achieved, thereby reducing carbon emission.
3. Unit control: The operational efficiency of an air source heat pump unit is related to factors such as outdoor temperature and load rate. By optimizing the operation strategy of the underfloor heating system, the unit efficiency can be effectively improved, achieving energy savings. Based on the above approach, Fig. 7.20 presents a high-efficiency energy-saving technology for air source heat pumps utilizing underfloor thermal storage. By leveraging big data to establish a dynamic operational model for the underfloor heating system (Xiong, Yu, Zhao, et al. 2023), the heat pump unit can predict future building load demands using weather forecast data. Combined with the thermal storage capacity of the underfloor system, it optimizes the heat pump's operation schedule and output capacity, achieving system energy conservation while maintaining indoor thermal comfort, with an energy conservation rate exceeding 32%.

To further enhance the carbon reduction effect of the system, the air source heat pump unit can also be combined with PV power generation equipment. During a day when the outdoor temperature is higher, the efficiency of the heat pump unit improves, and it can be driven by PV electricity. At this time, the unit can not only provide heating but also store excess heat in the floor. During the night, heating can be achieved as much as possible by releasing heat from the floor, thereby reducing the energy consumption of the heat pump unit. Meanwhile, the operation of heat

Table 7.3 Heat supply of underfloor heating at different water supply temperatures

Avg. water temp (°C)	Indoor air temp (°C)	Pipe heating spacing (mm)									
		500		400		300		200		100	
		Heat supplied upward	Heat transferred downward	Heat supplied upward	Heat transferred downward	Heat supplied upward	Heat transferred downward	Heat supplied upward	Heat transferred downward	Heat supplied upward	Heat transferred downward
35	16	64.4	18.4	72.6	18.8	81.8	19.4	91.4	20	100.7	21
	18	57.7	16.7	65	17	73.2	17.4	81.7	18.1	89.9	19
	20	51	14.9	57.4	15.2	64.6	15.6	72.1	16.1	79.3	16.9
	22	44.3	13.1	49.9	13.3	56	13.7	62.5	14.2	68.7	14.9
	24	37.7	11.3	42.4	11.5	47.6	11.9	53	12.2	58.2	12.8
40	16	82.3	23.1	93	23.6	105	24.2	117.6	25.2	129.8	26.5
	18	75.5	21.4	85.3	21.8	96.2	22.4	107.7	23.3	118.8	24.4
	20	69.7	19.6	77.6	20	87.5	20.6	97.9	21.4	107.9	22.4
	22	62	17.9	69.9	18.2	78.8	18.7	88.1	19.4	97.1	20.4
	24	55.2	16.1	62.3	16.4	70.1	16.8	78.3	17.5	86.3	18.3

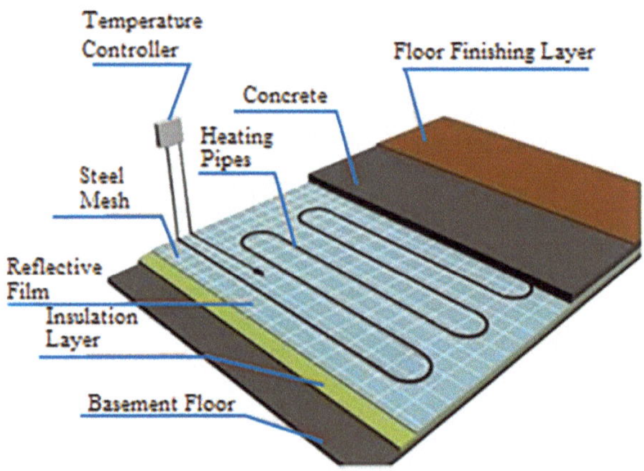

Fig. 7.19 Schematic diagram of underfloor heating terminal structure

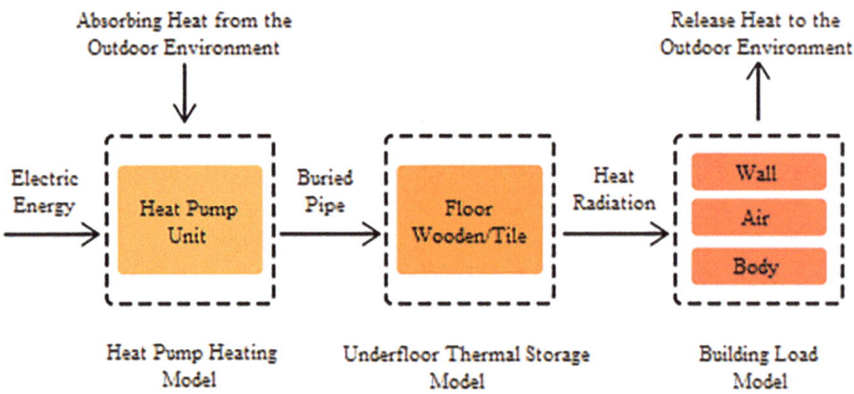

Fig. 7.20 Dynamic operation model of heat pump underfloor heating system

pump underfloor heating must consider the real-time "dynamic carbon emission responsibility factor" (C_r) (Jiang, Zhang, Zhang, et al. 2024). When the power system is in a load valley, the C_r value is lower, allowing heat pump units to increase electricity consumption for underfloor thermal storage (raising floor temperature), and thereby effectively assisting the grid in accommodating wind and solar power while reducing wind and solar curtailments. Conversely, during peak load periods of the power system when C_r values are higher, the heat pump units reduce electricity consumption and provide building heating through underfloor heat release (lowering floor temperature). For example, to minimize the electrical load of residential buildings during the evening peak hours (18:00~22:00), the heat pump unit can be shut down during this period, utilizing the floor's heat storage for heating. Relevant calculations (see Appendix) indicate that to meet the building's heating demand during the evening peak hours (4 h), the floor temperature only needs to decrease by 2 °C, effectively ensuring indoor thermal comfort when the heat pump is off.

Appendix

The thermal storage capacity of underfloor heating primarily comes from the concrete material of the building floor, with a specific heat capacity of approximately 1 kJ/(kg K) and a density of about 2200 kg/m^3. Assuming a floor area of 100 m^2 and a thickness of 50 mm, the total heat capacity of the floor is 11,000 kJ/K; meanwhile, the winter heat load per unit area of the residential building is approximately 50 W/m^2, resulting in a total load of 5000 W. During the evening peak hours (18:00~22:00), the total heat demand is 20,000 J.

Based on the aforementioned total heat capacity of the floor, a decrease of approximately 2 °C in floor heat release during heat pump shutdown is sufficient to meet the heating demand during evening peak hours. Since the floor temperature is generally above 30 °C, with a temperature difference of more than 10 °C from the room temperature, the heat supply rate for underfloor heating remains essentially unchanged during heat dissipation and cooling. This ensures the thermal comfort indoors when the heat pump is shut down during evening peak hours.

It should be noted that if the floor incorporates phase change materials (PCM) (such as calcium chloride hexahydrate, with a density comparable to concrete and a specific heat capacity approximately three times that of concrete), the thermal storage capacity of the floor increases. Under the condition of providing the same amount of heat, the required temperature drop for heat release can be effectively reduced. When the heat pump is shut down, the floor can maintain indoor thermal comfort for a longer duration.

Chapter 8
Indoor Environment Conditioning Technologies for Urban Residential Buildings

8.1 Environmental Requirements in Urban Residential Buildings

In recent years, China's rapid socio-economic development and rising living standards have significantly transformed the housing consumption concepts and demands of urban residents. Residential development priorities have shifted from pursuing "quantity" to embracing "quality," while housing construction has transitioned from focusing on "functionality" to emphasizing "comfort." As a result, the use of heating and cooling equipment to improve indoor living quality has become increasingly prevalent. However, creating desirable indoor environments in urban residential buildings now presents new challenges and dilemmas: the choice between maintaining constant temperature and humidity and allowing variable conditions; the preference for mechanical ventilation or natural ventilation; and the pursuit of an optimal balance between service quality and energy consumption. This section will explore these three core issues in depth.

8.1.1 Constant Temperature and Humidity and Variable Environments

An environment with constant temperature and humidity can provide stable thermal comfort. It protects the human body from cold and heat stimulation and eliminates the need for active heat regulation, making people feel more comfortable. However, such an environment is limited in accommodating the individualized demands of residents. This is particularly challenging in China, where the vast territory spans multiple climate zones, including tropical, temperate, and frigid zones from south to north, creating significant climatic variations. Coupled with variations in thermal adaptation across gender and age groups, a uniform constant temperature and

Building Energy Research Center of Tsinghua University, *Decarbonizing Urban Residential Buildings*, https://doi.org/10.1007/978-981-95-8200-6_8

humidity setting can hardly meet these diverse comfort requirements. For instance, the elderly prefer warmer settings, while younger individuals favor cooler ones. At the same temperature (e.g., 26 °C), different individuals may have completely opposite feelings; some find it comfortable, while others feel overheated. Such individual differences cannot be adequately addressed in centrally controlled constant temperature and humidity systems. In these systems, the indoor temperatures of all households are preset uniformly, preventing residents from adjusting indoor temperature to their actual needs. Consequently, this standardized model fails to address the true comfort demands of households.

In reality, the indoor environment requirements of residences are remarkably dynamic and individualized. These include whether there are occupants in the room and whether they are sleeping, resting, enjoying entertainment, or making noise. When in different states, people have significantly different requirements for temperature, humidity, ventilation, and even whether the window needs to be opened, and the indoor environment requirements of occupants can be related to their subjective mood. Thus, a genuinely comfortable and human-centric residence can be very difficult to create by using a whole set of automatic air conditioning, ventilation, and lighting equipment that are controlled and regulated by a highly automatic system. Since such systems can never truly know the real needs of the occupant, true comfort requires user-adjustable environmental control on demand.

From the physiological aspect, the human body has developed sophisticated heat regulation capability through long-term evolution and can adjust to the thermal changes in the natural environment. If the human body maintains good thermal regulation capability, it can easily adapt to any environment with suboptimal thermal conditions without feeling noticeable discomfort. However, overly relying on cooling and heating equipment will undermine the thermal regulation capability of the human body. Studies indicate that occupants living in mechanically air-conditioned and heated environments over extended periods have high requirements for indoor thermal comfort, and their comfortable temperature range shows a trend of shifting towards extreme temperatures. In comparison, the people living in non-conditioned environments experience frequent temperature fluctuation, and their bones, muscles, sweat glands, and physiological functions have been exercised, which, to a certain extent, enhances their thermal regulation capability and allows them to adapt to a wider ambient temperature range.

A long period of living in an environment with constant temperature and humidity may lead to "air-conditioning maladjustment syndrome", where the symptoms include sweat gland and sebaceous gland contraction, glandular occlusion, and poor blood flow. In the case of excessive difference between indoor and outdoor temperatures, a person entering/exiting a room may experience the discomfort caused by excessive thermal shock. Such temperature difference may even impact the health of the occupant, where the thermal stimulation may cause a cold, and there are other symptoms such as heatstroke, headache, hypersomnia, fatigue, and arthralgia. Therefore, staying in an environment with constant temperature and humidity will eliminate the discomfort from the extreme heat in summer and extreme cold in winter, but it will change the thermal adaptation of the human body developed in the natural environment and thus adversely impact health.

In fact, it was only a few decades between the invention of air conditioning and its wide application, while the passive thermal regulation strategies relying on building insulation, natural ventilation, and clothing substantially display the physiological ability of the human body to adapt to the natural environment. This low-carbon and eco-friendly lifestyle is still favored by most residents. The field investigations across China indicate that under the same temperature and humidity, natural ventilated environments instead of air-conditioned spaces are preferred by residents. Residents take a "use only if necessary" attitude towards air conditioning. As long as the indoor temperature is within an acceptable range, they prefer to open the window for natural ventilation or use other natural cooling equipment such as electric fans. If the indoor temperature notably exceeds the comfortable range, intermittent use of air conditioning will ease the discomfort without obviously affecting the thermal adaptation of the human body and thus is the preferred option that strikes a balance between comfort and health.

8.1.2 Mechanical Ventilation and Natural Ventilation

The ventilation method in residential buildings concerns the occupant health, thermal comfort, and building energy efficiency, and thus fulfills a very important function. It affects the indoor thermal environment and also the air quality.

Ventilation affects indoor thermal and humid environment and energy consumption in two ways: when the outdoor temperature is suitable, ventilation will effectively extract the extra heat and humidity generated indoors, reduce the operation duration of air conditioning and energy consumption; when the outdoor environment is excessively hot and humid or excessively cold, ventilation will create extra hygrothermal loads or cold loads indoors and cause the increasing of air conditioning energy consumption. Meanwhile, ventilation is also important to expel various indoor pollutants and maintain good indoor air quality. Indraft of fresh air will dilute the indoor pollutants such as CO_2 and VOC and help maintain a healthy indoor environment. However, in the event of severe outdoor pollution, e.g., hazy weather and excessive PM2.5 concentration, ventilation may bring in more hazardous particles.

Therefore, ideal ventilation technology should meet indoor environmental requirements while minimizing energy consumption. There are currently three main ventilation methods: (1) **door-window joint infiltration**: air infiltration through windows and doors is the simplest ventilation method; (2) **open-window natural ventilation**: obtaining adjustable ventilation volume through opening windows will easily realize high air change rate, e.g. 5–10 times per hour (ACH) or more; (3) **mechanical ventilation**: a intake/exhaust fan driven by electricity will accurately control the ventilation volume.

In developed countries, including countries in Europe and the United States, mechanical ventilation systems are now the mainstream means of building ventilation. This system can precisely control the ventilation volume. Due to the changed interior-exterior pressure difference, however, the building usually needs high air tightness (to prevent uncontrolled air infiltration), installation of openable windows is even prohibited (open-window ventilation is impossible), and a high level of electricity consumption is required to power the ventilation fans. The filters installed will improve the indoor environment in the event of severe outdoor pollution, but there are clearly certain limitations: under the favorable weather conditions that account for over 95% of the time of a year, the users have to continue relying on mechanical ventilation instead of natural ventilation. This will fail to further improve the indoor air quality and will also cause continuous energy consumption of the fans. In addition, without adequate maintenance, mechanical ventilation systems may become a hotbed for bacteria breeding, and consequently, smaller particles or bacteria may be brought indoors and become another pollution source. In comparison, the room air cleaners popular in the market are more advantageous. The air cleaner can be flexibly controlled by the occupants according to the outdoor pollution and their bodily feelings. In the case of a favorable outdoor environment, open-window ventilation without using cleaners is more rational.

It is worth noting that the ventilation requirements of residences are not constant, and instead dynamically vary depending on the occupant number and the pollution intensity. Taking decoration materials for example, the VOC emission will gradually decrease over time, which means that ventilation requirements are high in the early days after decoration and then gradually decline. Moreover, when the room is unoccupied, excessive ventilation is unnecessary. Even if ventilation is inadequate when the room is unoccupied, ventilation volume can be increased when the resident returns to the room to meet the air quality requirements. However, the current mechanical ventilation systems are designed based on a fixed fresh air rate standard, maintaining constant minimum fresh air rates during the operation stage, so the fixed-rate operation fails to accurately reflect the actual dynamic ventilation requirements of residents. Such design flaws often cause two problems in practical operation of mechanical ventilation systems: either excessive ventilation volume beyond the practical requirements, resulting in waste of energy, or insufficient ventilation volume failing to meet the indoor air quality requirements.

Besides, there are other problems in the use of mechanical ventilation systems abroad. For example, due to increased heating fees, many residents choose not to turn on the mechanical ventilation equipment; even if some residents turn it on, it is used for a short time every day. In this situation, the indoor pollutants cannot be effectively expelled, which will threaten human health.

Maintaining proper airtightness of the building and applying intermittent ventilation through natural infiltration, opened windows, and natural ventilators are the ventilation methods used by most residential buildings in China. The model combines proper door and window air tightness to meet the basic ventilation requirements of residences, and intermittent natural ventilation through window opening or natural ventilator for flexible control of the ventilation volume, striking a perfect balance between energy conservation and indoor air quality. In this model, occupants can control the door and window opening degree and time based on actual conditions: they can open the window for ventilation when the outdoor temperature is favorable and close the window at night or during extreme weather. When natural ventilation is inadequate, other equipment such as a natural ventilator can be used to supplement ventilation volume. In this model dominated by natural ventilation, when the outdoor temperature, humidity, and cleanness are favorable, occupants may open the windows and adequately use natural ventilation to create a desirable indoor environment, which will avoid extra electricity consumption of the fans and also reduce the cooling and heating losses caused by the ventilation. It allows the occupant greater freedom and a more comfortable environment and is a ventilation model capable of sustainable development.

From the aspect of climate, in most regions, the outdoor environment is comfortable for the human body for more than half of the year (this fact itself reflects that humankind has adapted to most climatic conditions through their evolution). Opening the window will create a comfortable enough indoor environment. Throughout human history, natural ventilation has been regarded as an important approach to create and maintain a suitable indoor environment. The physiological characteristics of the human body indicate that natural ventilation can create a wide range of thermal comfort, and moderate environmental changes are beneficial for human health and comfort. Therefore, natural ventilation is a healthy and green method for indoor environmental control and is worth popularizing.

The natural ventilation model with suitable air tightness and intermittent window opening allows occupants enough freedom to control the ventilation volume, creates a healthier and more comfortable dynamic indoor environment, and substantially reduces energy consumption. Therefore, this model should be widely applied in China.

8.1.3 Service Quality and Energy Consumption

Over 70% of the building's energy is used for indoor environmental control in buildings, e.g., temperature control, humidity control, and ventilation. The operation energy consumption mainly depends on whether the resident maintains "full control" over the indoor environment or is satisfied with "improvement", whether the indoor is maximally isolated from the outdoor or the contact with the outdoor is maximally maintained, and whether the physical parameters are strictly controlled to be the "optimum values that meet the requirements of the occupant" or the proximity to the natural environment is maximally maintained as long as the basic requirements of occupants are satisfied. For both residential buildings and office buildings, significant differences in energy consumption of the same type of building do not stem from whether advanced building energy efficiency technology is applied, but instead from the various indoor environments provided and the lifestyle of the building occupants. For residences to maintain constant indoor environmental parameters, including temperature, humidity, air quality, and noise levels, a large amount of equipment needs to keep running. Notably, in extreme weather conditions, the equipment loads will increase considerably; consequently, the energy consumption will increase sharply.

In a sense, the basic function of buildings is to provide occupants with services. Evaluation of a building's quality should focus on its service quality. Why shouldn't humans desire the best possible service quality and enjoy the best possible living environment? As shown in Fig. 8.1, the relation between the service quality of the building and the energy consumption required to achieve such service is not a linear relation.

When buildings only provide basic service functions, the required energy consumption is minimized. To improve the building system and provide the service functions that maintain the basic comfort, the energy consumption will increase to some extent. When pursuing invariant indoor environmental conditions for a long period, the required energy consumption will increase considerably. The goal is to achieve "full control" and create the "optimum" indoor environment. With this goal achieved, certain technological innovations are employed to enhance the system's efficiency and reduce energy consumption, but significant energy savings can be difficult to achieve, and the energy consumption is usually much higher than that of the "improved" indoor environment.

Under the premise that the per-capita building energy consumption essentially does not increase in China, the only possible option is to improve the indoor environment through technological innovation and thus further satisfy the requirements of the occupant. It is unreasonable to let the per-capita building energy consumption increase significantly in the name of "improving people's living standards". These are the building energy-efficiency-related issues that must be addressed. We should balance and coordinate these requirements, befittingly restrain this endless pursuit of services, and pay more attention to energy saving, resource conservation, and environmental protection. We should preferably use the natural environmental

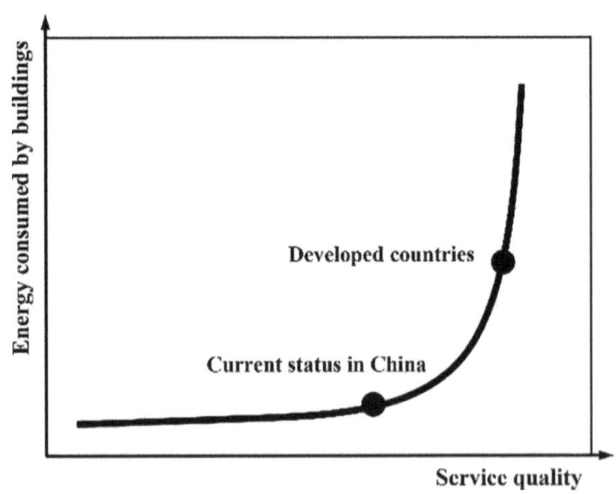

Fig. 8.1 Relation between building energy consumption level and service quality

condition instead of directly resorting to mechanical equipment and systems, to make a major difference in energy consumption; we should properly readjust our goals for indoor environment, aiming to achieve an environmental setup that covers only "partial space" or "part time" and carries "a certain degree of non-guarantee rate", but remains acceptable or tolerable to the building user or occupant. The goals should be maximally close to the natural environmental conditions available, so that the natural resources are maximally used, to minimize the reliance on mechanical systems. This is the key to creating low-energy buildings and achieving high building energy efficiency.

8.1.4 Summary

In conclusion, the residential environment we require is a habitable space that prioritizes natural conditions, which is reasonably adjustable, and that is independently controlled. From the aspect of natural environment, a suitable living environment can be created through various passive methods and the self-adjustment of the occupant (e.g., open-window ventilation and shading), and harmonization with the natural environment can be achieved through the adjustability and adaptability of the human body. If the requirements for the environment are still unsatisfied, then the mechanical systems can be the last resort. In this environmental control concept, the indoor temperature, humidity, and ventilation volume are not supposed to be kept at a certain set point. Instead, only if the results of the passive methods deviate far from the required comfort level, will the mechanical systems be deployed to improve the indoor environment. There is no "desire to control" the indoor environment. Instead, improvement is required only if discomfort is felt. In this way, energy consumption is very low.

In comparison, deploying the high-tech environmental control system used by office buildings in residential buildings, i.e., creating a constant residential indoor environment through a whole set of mechanical systems, is unnecessary and unreasonable. A number of residential projects featuring "five-constant" (constant temperature, humidity, cleanliness, air quality, and acoustic level) have been initiated in China in recent years. Such projects often pursue the so-called constant "optimum condition" and have forsaken the most important element--whether the occupant can independently control the living environment. They are designed to provide the optimum services through uniform high-tech approaches. This model neglects the interaction between human and nature and causes a high level of unnecessary energy consumption, so it is not the main development direction of residences in the future.

As stated in the report at the 19th National Congress of the Communist Party of China, socialism with Chinese characteristics has entered a new era; the principal contradiction facing Chinese society has evolved. What we now face is the contradiction between unbalanced and inadequate development and people's ever-growing needs for a better life. With socio-economic development, the people's living standard has been significantly lifted, and they have ever-growing needs for a better life.

However, the "five-constant" system, as an idealized indoor environmental control technology, is not the right path to a better life. The "five-constant" system is designed to create an absolutely stable indoor environment, but it neglects people's differentiated and dynamic requirements for the environment and may even adversely impact the health of some residents. Furthermore, the realization of such a system relies on the technical equipment that requires high energy consumption, which will aggravate energy consumption and carbon emissions and will hinder the "carbon peaking and carbon neutrality" goal. Therefore, the "five-constant" system is not the right path to people's better lives. The genuine path to people's better life is the people-oriented, healthy, comfortable, green, and low-carbon concept of living environment. We should adequately use the natural conditions and rationally use the mechanical systems to satisfy the differentiated requirements of residents, so as to provide the people with a genuinely better life.

8.2 Development of New Intelligent Building Envelope

According to statistics, the carbon emission caused by the building energy consumption in China accounts for a quarter of the carbon emission of the whole society. Of the overall building energy consumption, the cooling and heating caused by building envelope take a fairly high proportion. According to a study, the cooling or heating energy loss caused by poor performance of building envelope exceeds 30%. Therefore, developing and utilizing efficient building envelope to minimize the energy consumption and carbon emission in building operation is one of the main approaches to realizing "carbon neutrality" in the building industry. With the emergence of new materials and technologies, as well as the increasingly high requirements for intelligentization, new concepts and solutions for building envelope have been devised, to create new building envelope with high suitability and performance in engineering application. From the aspects of control method (active or passive control) and the building envelope type (transparent or opaque envelope), as shown in Fig. 8.2, this section presents a few typical technologies as case studies, to expound the reality of new intelligent building envelope and afford technical references for the professionals in this field.

8.2.1 Passive Adaptive Control of Transparent Building Envelope

(1) **Thermochromic glass**

Thermochromic glass can indirectly control its sunlight transmissivity based on its temperature, thus realizing passive energy conservation of the building. As the temperature changes, the thermochromic responsive material on the glass

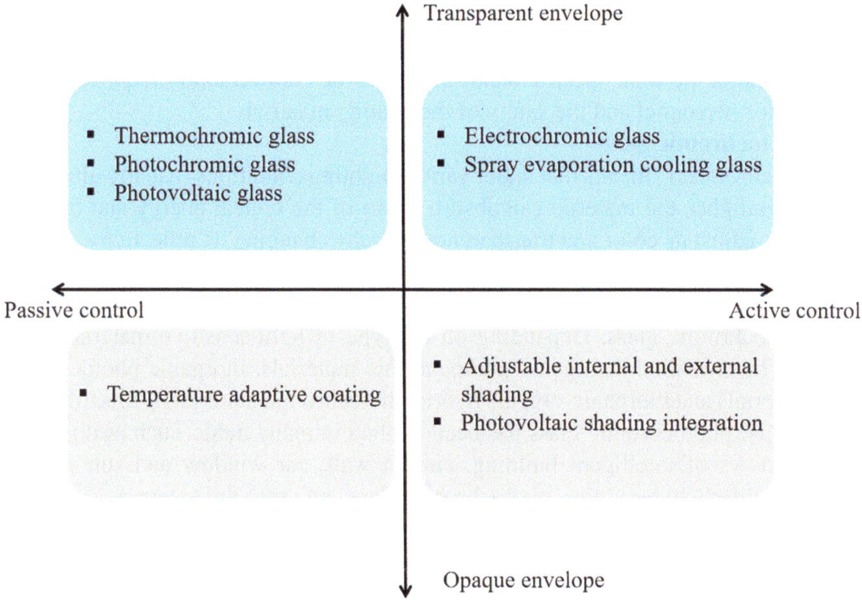

Fig. 8.2 Types of new intelligent building envelope

surface can reversibly change from uncolored state to colored state. Common thermochromic responsive materials include VO_2, ionic liquid, perovskite and hydrogel. When such materials are applied on the surface of ordinary glass, the surface temperature of the thin film changes with the weather. If the surface temperature is higher than the phase transition temperature, the material will reversibly change from insulating phase to conductive phase (colored state). Consequently, the sunlight transmissivity is changed, thereby realizing control over the solar heat in different seasons (Kalnay and Cai 2003).

In practical application, this technology still faces many challenges: the sunlight control capacity is insufficient, the visible light transmissivity is too low, and the phase transition temperature is too high. The sunlight control range is between the two states of the thermochromic material, so the light control range may have certain limitation. VO_2 is believed to be an excellent coating material for thermochromic glass. However, its natural color is brownish-yellow, and it becomes semitransparent after phase change and has low visible light transmissivity, so it will not meet the occupant's requirements for using natural light. Most thermochromic materials have high phase transition temperature, while the phase transition temperature in practical application needs to approximate room temperature.

At present, thermochromic glass is mostly used in the regions with hot summer and cold winter where sunshine duration is long in summer and short in winter. However, the high phase transition temperature of thermochromic materials has impeded their application in the regions with warm climate, and the commonly used coating material VO_2 is toxic, hazardous to human health. Future studies need to solve the aforesaid three key problems, and also need to take into consideration the solar spectral bands that meet the comfortability requirements of indoor personnel and the safety of the coating materials.

(2) **Photochromic glass**

"Photochromism" means that under various lighting conditions (usually ultraviolet or sunlight), the material can absorb a part of the optical energy and dynamically adjust its color and transparency, thereby changing its reflectivity. A light-sensitive material (e.g. silver salt, nitrogen-containing compound or organic molecule) is applied as a thin film or particles on the glass surface, to produce photochromic glass. Depending on the type of light-sensitive materials, they can be classified as organic photochromic materials, inorganic photochromic materials and inorganic-organic hybrid photochromic materials (Howard 1833).

Currently, photochromic glass has been applied in many fields, such as doors and windows of intelligent building, curtain wall, car window and sun glasses. Particularly in buildings, photochromic glass can sense the external solar radiation to effectively control the light intensity indoors and improve visual comfort; besides, it can also block heat transfer, reduce air-conditioning load and thus significantly reduce energy consumption. Nonetheless, photochromic glass has certain limitations, including slow response, limited color change range and effect, short service life, and unstable performance in extreme environments. If the properties of existing light-sensitive materials are improved and such materials are combined with the innovative materials such as PV, thermochromic and electrochromic materials, photochromic glass can become an efficient green material with multiple functions to be applied more widely and effectively (Fig. 8.3).

(3) **Photovoltaic glass**

Photovoltaic glass refers to the glass that changes between transparent and opaque states under the action of light/heat and that is used for PV power generation. In terms of structural difference, photovoltaic glass can be in two types: modularized

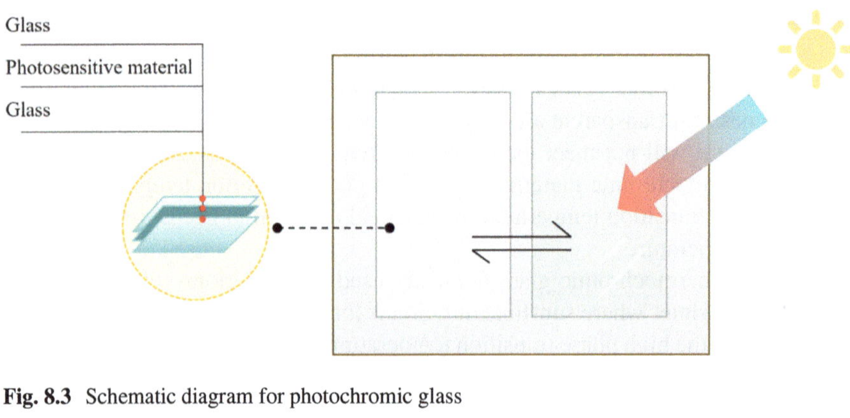

Fig. 8.3 Schematic diagram for photochromic glass

photovoltaic glass and the integrated photovoltaic glass, as shown in Fig. 8.4 (Mhoff et al. 2010).

Photovoltaic glass experiences phase change under the action of ambient temperature or illumination intensity, thereby changing its transparency and color. It can also absorb sunlight and convert it into electric energy. By changing between transparent and opaque states, photovoltaic glass can control indoor illumination without consuming extra energy and realize intelligent control of indoor temperature and illuminance, thus fulfilling the energy conservation and energy production functions. Many factors, such as material selection, photoelectric conversion efficiency, photochromic/thermochromic response speed, environmental adaptability and encapsulation method will affect the properties of photovoltaic glass.

Photovoltaic glass is suitable for building, automobile and aircraft, and it can reduce the energy consumption of lighting, heating and air-conditioning systems. Currently, photovoltaic glass faces many challenges, e.g. mismatch between phase transition temperature and practical application environment, imbalance between photoelectric conversion efficiency and photochromic/thermochromic performance, and the problems in long-term stability and encapsulation. In the future, photovoltaic glass will be designed for higher photoelectric conversion efficiency, better photochromic/thermochromic performance and better environmental adaptability.

8.2.2 Passive Adaptive Control of Opaque Building Envelope

Temperature-adaptive coating is the type of coating that can control its radiation property according to temperature variation. Its radiation parameters (e.g. shortwave absorptivity and longwave emissivity) can be dynamically adjusted according to the ambient temperature variation, so as to optimize the heat exchange of

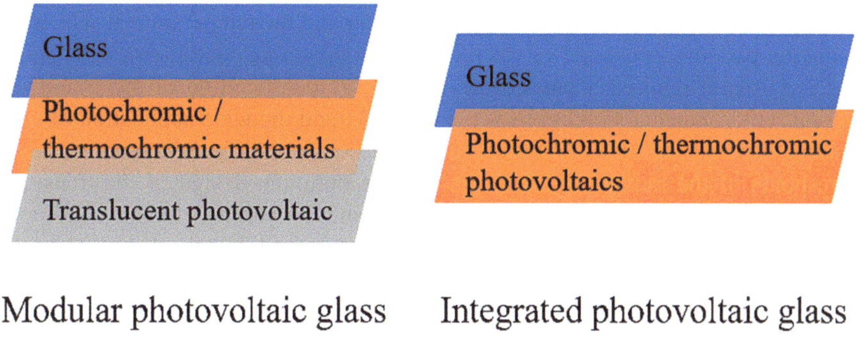

Fig. 8.4 Two types of photovoltaic glass

the outer surface of building envelope (control the heat gain/release of building envelope) and achieve more efficient energy management. As shown in Fig. 8.4, the coating needs to possess low shortwave absorptivity and high longwave emissivity when the ambient air temperature is high, in order to reduce solar radiation heat absorption, improve longwave radiation heat dissipation, and maintain a cool environment inside the building. When the ambient air temperature is low, the coating needs to have higher shortwave absorptivity and lower longwave emissivity, so as to maximize solar radiation absorption, minimize longwave radiation heat loss and maintain a warm environment inside the building (Sheng et al. 2017).

The radiation parameters of temperature-adaptive coating are variable, usually realized by changing the phase state, color or structure of the functional components in the coating. The upper and lower thresholds of the radiation parameters, as well as the temperature at which the radiation parameters change, depend on the properties of the raw material. Therefore, temperature-adaptive coatings with various radiation properties can be produced by using different raw materials according to the practical needs. This technology is particularly suitable for regions with high requirements for both cooling and heating, such as the hot summer and cold winter region. The intelligent control capability can substantially reduce the energy consumption of building cooling and heating, and also improve the thermal comfort of the living environment, so as to meet the service requirements of the building in different seasons. In addition, temperature-adaptive coating also has promising potential in creating other artificial environment, e.g. automobile, warehouse and oil tank. However, temperature-adaptive coating will, to certain extent, reduce the BIPV installation space, and the building energy conserved by such coating may not necessarily outweigh the benefit created by PV power. For example, the rural areas have large roof space, and not every room needs cooling/heating. Consequently, the energy benefit created by the PV system installed in the rural roof space is greater than the energy conservation benefit brought by the coating. In the future, application of temperature-adaptive coating needs to adequately tap the economic potential of the architectural space.

In the ongoing research, the matching between the upper/lower threshold of radiation parameters of temperature-adaptive coating and the temperature at which the radiation parameters change versus the practical operating conditions of the building is virtually ignored. Consequently the energy conservation of the coating in various environmental conditions is not optimum and the wide application of such coating has been restricted. Further research needs to specify the upper and lower thresholds of the radiation parameters and the optimum design of the transition temperature. Besides, the evaluation model suitable for various climatic conditions and buildings needs to be developed, so as to comprehensively demonstrate the energy conservation potential of the coating and ensure its optimum performance in practices.

8.2.3 *Active Intelligent Control of Transparent Building Envelope*

(1) **Electrochromic glass**

Electrochromic glass is a type of intelligent material that changes its optical properties (e.g. transparency or color) with electric current. As an effective means of energy conservation, it has been widely applied in building envelope. As shown in Fig. 8.5, it resembles the laminated glass composed of five layers. Under the action of externally applied electric field, electrochromic glass experiences electron and ion transfer that triggers oxidation reduction reaction, thus changing its luminous transmittance. When in oxidized stated, the glass is usually transparent or lightly colored; when in reduced state, the glass changes into darker color and gains the ability to absorb or reflect light. This change is reversible, i.e. after the electric field changes, the glass can switch between transparent state and opaque state without the need for mechanical motion (Oke 1973).

The properties of electrochromic glass are affected by many technical factors, where the selection of material and the properties of ionic conductor layer are especially important. The electrochemical stability of the material directly affects the color change speed, color change range and service life, while the ionic mobility of the ionic conductor layer determines the charge transfer efficiency and the overall response time. Besides, electrochromic glass is troubled by slow color change speed, high energy consumption and limited scope of application, which further limits its engineering practice.

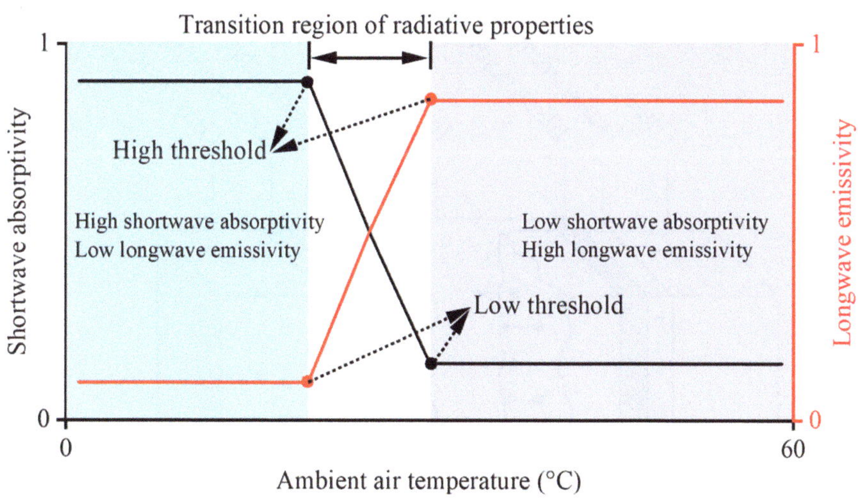

Fig. 8.5 Variation trend of the radiation parameters of temperature-adaptive coating suitable for the building

In the future, electrochromic glass will be designed to be more efficient, economical and durable. Through integration with sensors and the intelligent control system, more intelligent light control and multifunctional application can be hopefully realized. With optimized manufacturing process and increasingly mature technology, its application prospect will be broadened.

(2) **Evaporative spray cooling glass**

The structure of evaporative spray cooling glass is shown in Fig. 8.6. It is composed of roof glass, water tube, atomizing nozzle, control valve and water tank. Its basic operation process is as follows: the water in the water tank is delivered into the atomizing nozzle installed on the roof glass surface via the control valve and water tube, and is then sprayed as mist to reduce the temperature of the outer surface of roof glass. The mist spray will also block solar radiation and further reduce building energy consumption. According to a study, mist spray is better than water spray, because the former can increase the contact area on the glass surface and enhance evaporation. In addition, the water droplets can stay in the air for a longer time to block the sunlight (Yang et al. 2013). Main factors affecting evaporative spray cooling glass include type, placement (number, installation angle and spacing, etc.) and flow rate of nozzle. Evaporative spray cooling glass can be installed on the roofs of large commercial buildings in the southern regions and the transport hubs, reducing building energy consumption without affecting indoor daylighting. Currently this system still needs to improve the nozzle performance and reduce the operating energy consumption of the water pump (Fig. 8.7).

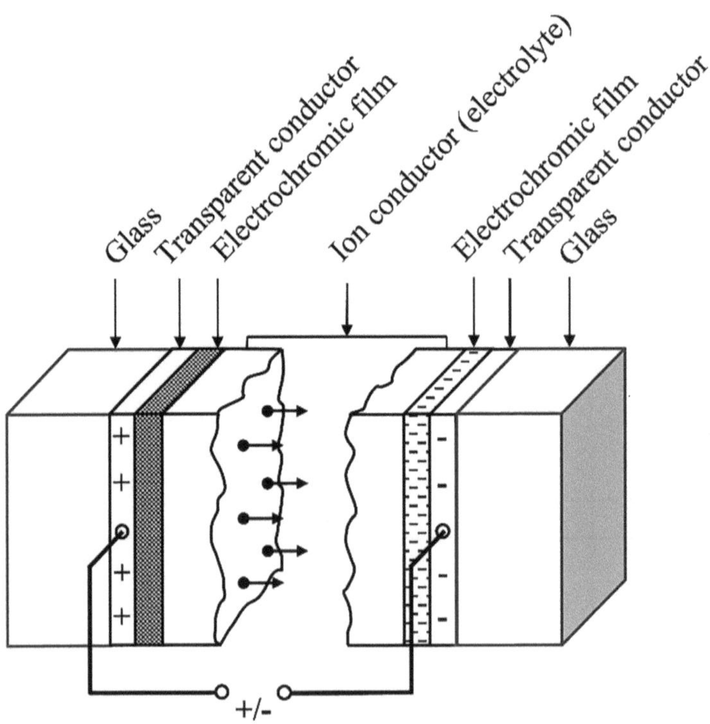

Fig. 8.6 Structure of electrochromic glass

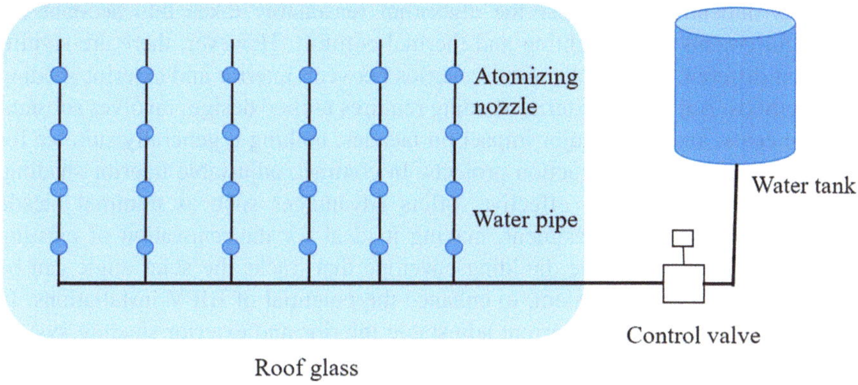

Fig. 8.7 Structure of evaporative spray cooling glass

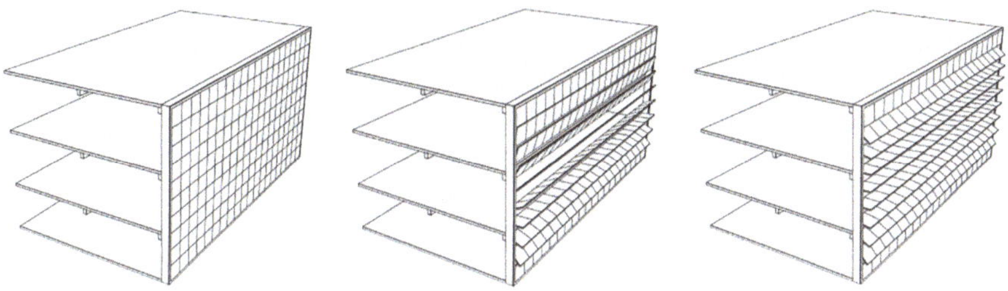

Fig. 8.8 Diagram for building envelope facade that can track the sun angle

8.2.4 Active Intelligent Control of Opaque Building Envelope

(1) **Adjustable interior and exterior shading**

Adjustable interior and exterior shading system can control the shading capacity by adjusting the angle of exterior component and the opening of the interior shading curtain. Through simulation technology and automatic shading algorithm, shading capacity is increased when there is strong sunlight, not only to reduce glare and incoming solar radiation but also to maintain a comfortable luminous environment and thermal environment indoors. The shading capacity decreases when there is weak sunlight, so as to receive more natural light (Santamouris 2015). Adjustable interior and exterior shading system has long been applied in architectural design. In 1987, Jean Nouvel applied adjustable shading surface at the Arab World Institute (Pairs, France). He used various photosensitive sensors to control the mechanical structure and adjust the structural change, in order to control the sunlight entering the building (Fig. 8.8).

In essence, the adjustable interior and exterior shading system adds intelligent control to the physical interception of solar radiation. The key to its effectiveness depends on whether the algorithm reasonably takes into account the requirements of daylighting and thermal comfort. However, there are significant differences in application scenarios between interior and exterior shading methods. Adjustable exterior shading requires unified design, involves substantial costs, and has a major impact on facades, making it generally suitable for large-scale new construction projects. In contrast, adjustable interior shading, although slightly less effective, offers advantages such as minimal facade impact and lower investment, making it ideal for the renovation of existing buildings. In the future, building envelope that tracks the solar angle can be integrated with PV system, to enhance the potential of BIPV installations. In addition, although the current adjustable interior and exterior shading system has intelligent control, such a system often operates in isolation. In the future, integrated control of multiple systems may be introduced, incorporating the shading system into the overall intelligent system of the building to create a better built environment.

(2) PV shading integration

In integrated PV shading structure, the PV module replaces traditional shading panel (Fig. 8.9). On one hand, the PV module functions as a PV panel, receiving incoming solar radiation to generate electricity. On the other hand, it serves as a shading panel, effectively blocking solar radiation from entering indoor space. This not only prevents significant temperature rises indoors during summer but also reduces glare issues caused by direct sunlight (Rosenzweig et al. 2005).

The impact factors of PV shading integration technology are as follows. (1) PV module materials: the PV cell serves as the core material in the construction of a PV module, with diverse cell types exhibiting differences in material

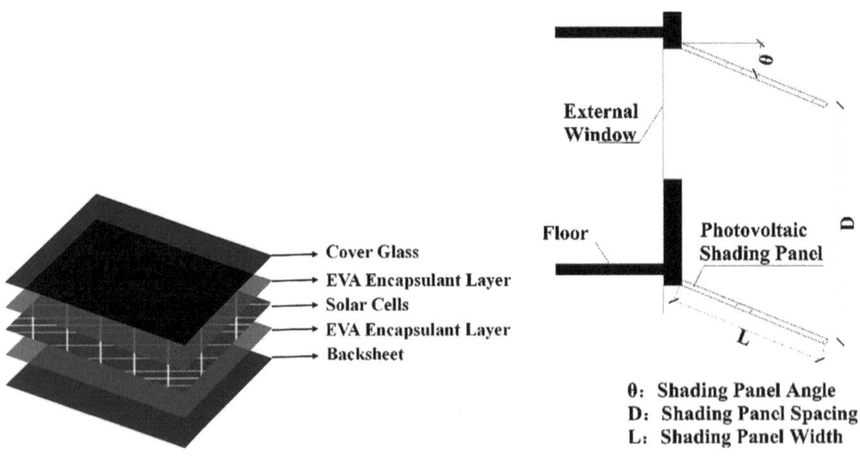

Fig. 8.9 PV module as shading panel

costs, solar energy utilization efficiency, and design complexity. (2) PV module installation: the installation position of PV module (e.g. horizontal or vertical), module tilt angle, width, and orientation affect the result of solar radiation reception, thereby influencing the electricity generation and shading performance of the module. 3) environmental conditions: for the overall climatic environment, the illumination intensity undergoes dynamic changes under different climatic conditions, thereby affecting the comprehensive performance of PV shading. For individual micro-environment, factors such as dust, rain and snow can contaminate PV modules, preventing some solar radiation from reaching the PV cells and consequently reducing power generation efficiency.

Currently, integrated PV shading structure is mostly used in building facade shading. However, with consideration of the factors such as cost and technology, this structure is primarily fixed in installation, and there is a lack of research on dynamic PV shading panels that can more efficiently utilize solar radiation and enhance shading performance. Therefore, achieving intelligent dynamic control is of great significance for promoting PV shading integration technology. At the same time, PV shading integration is also an effective means to increase the potential of BIPV installations in the future.

In summary, the advantages, disadvantages, and suitable application scenarios of new intelligent building envelope are shown in Table 8.1. As can be seen, the most significant feature of an intelligent building envelope compared to a traditional building envelope is its flexibility, which allows it to adapt to various climatic conditions, making it more suitable for regions with variable climate (such as hot summer and cold winter). Additionally, in practical application, it is essential to consider the economic feasibility of the equipment, analyze the relation between benefits and costs, as well as assess its impact on the aesthetic appeal of the building. Under the guidance of the "dual carbon" strategy, as PV costs continue to decline, the modular integration of building surface with PV in the future will be an effective means to fully exploit the potential of building exterior surface and reduce building operation energy consumption and carbon emission. This requires overcoming two challenges: (1) The service life of PV panels is only 20–25 years, less than one-third of a building's service life, so the module needs to enable flexible replacement of PV panels; (2) The heat dissipation issue of PV panels, especially when modularly integrated with the building envelope, requires particular attention as efficiency reduction caused by heating must be addressed. Future building design should preferably take into consideration additional space for PV installations, avoiding sharp or stacked exterior designs that may cause shading and reduce PV efficiency. Decorative PV modules such as PV tiles or solar windows can also be used to balance architectural aesthetics with PV installations.

Table 8.1 Characteristics of new intelligent building envelope

Control type	Transparent or not	Building envelope	Advantages	Disadvantages	Application scenarios
Passive control	Yes	Thermochromic glass	No energy consumption, added shading	Glass itself has poor transparency, and the material is toxic	Buildings in the region with hot summer and cold winter
	Yes	Photochromic glass			
	Yes	Photovoltaic glass	Added renewable electricity	Low power generation efficiency	Large French window or glass curtain wall
	No	Temperature-adaptive coating	Suitable for both summer and winter	Mismatch between the thresholds of coating's radiation parameters and the actual operating conditions	Buildings in the region with hot summer and cold winter, artificial environment such as automobile, warehouse and oil tank
Active control	Yes	Electrochromic glass	Controllable color change	Slow color change	Large French window or glass curtain wall
	Yes	Evaporative spray cooling glass	Low operating energy consumption	Potentially impact aesthetic appeal and lighting	Roof of large-space public building
	No	Adjustable interior and exterior shading	Dynamic adjustment to various needs in the day	Exterior shading requires high investment and significantly impacts the facade; the effectiveness of interior shading is limited	Exterior shading: large new projects; interior shading: old building renovation
	No	PV shading integration	Added renewable electricity	Potential obstruction; low power generation efficiency	Low-rise buildings in open areas

8.3 Retrofitting of Existing Residential Buildings

8.3.1 Basic Key Points in Energy Efficiency Retrofitting

Before building energy efficiency retrofitting, an investigation of building operation energy consumption should be conducted, and the result should be compared with current energy consumption standards. The residential buildings exceeding the constraint values of energy consumption indicators should undergo energy efficiency retrofitting. The primary reason lies in the fact that existing residential buildings, constructed in different eras, exhibit varying thermal performance in their building envelope components and differing energy efficiency in equipment and systems such as heating and air conditioning. Before formulating an energy efficiency retrofitting plan, it is essential to conduct a survey of the building operation energy consumption and make a comparison with the current national standard titled *Standard for Energy Consumption of Building* (GB/T 51161). For those exceeding the constraint value range of building energy consumption indicators, energy efficiency retrofitting may be considered.

Before building energy efficiency retrofitting, an energy efficiency diagnosis should be conducted first. Based on the diagnosis results, the scope and objectives of comprehensive or partial building energy efficiency retrofitting will be determined.

Based on the identified scope and objectives of energy efficiency retrofitting, the energy efficiency retrofitting plan and technical measures will be developed with consideration of technical reliability, operability, and economic feasibility, while maximizing the potential for energy conservation and emission reduction of existing equipment and systems. On the basis of the energy efficiency retrofitting plan, the energy efficiency retrofitting design will be completed.

After the energy efficiency retrofitting plan or design has passed the review and obtained the owner's approval, the project will be implemented; after acceptance of the energy efficiency retrofitting project, the effect of energy efficiency retrofitting shall be evaluated.

The building energy efficiency retrofitting of existing residential buildings is primarily meant for the buildings that are below the second-stage building energy efficiency level and whose main structure has service life of no <20 years after the retrofitting.

China has a vast territory, and different regions in the country have significant differences in climatic conditions and in economic and technological development. The building energy efficiency retrofitting of existing residential buildings should be carried out based on actual conditions, taking into account local geographical and climatic factors as well as economic and technological capabilities. This may involve comprehensive or partial energy saving retrofitting of the building envelope and HVAC system. The comprehensive energy efficiency retrofitting of the building envelope includes the retrofitting of exterior walls, roofs, and exterior windows, while partial energy efficiency retrofitting refers to upgrading one or several components of the building envelope based on technical and economic conditions. The comprehensive energy efficiency retrofitting of the HVAC system includes the transformation of all components such as the heat source, outdoor pipeline network, indoor heating system, or HVAC system. Partial energy efficiency retrofitting refers to the modification of only one or several of these components. In areas with favorable conditions, comprehensive energy efficiency retrofitting can be considered.

For severe cold and cold areas, the focus is on improving the thermal performance of key components of the building envelope, such as exterior walls, doors, windows, and roofs, as well as enhancing the energy efficiency of the heating system.

For the regions with hot summer and cold winter, the focus should be on energy efficiency retrofitting of the building envelope, with priority given to improving the thermal performance of exterior windows and the thermal insulation performance of roofs.

For the regions with hot summer and warm winter, priority should be given to improving the shading performance of exterior windows, the thermal insulation performance of roofs and east-west exterior walls, and improving the conditions for natural ventilation in rooms. Priority should be given to the methods that reduce the solar radiation absorption, such as reflective insulation, sunshade insulation, ventilation insulation and green insulation. Additionally, increasing the opening area of exterior windows can improve indoor ventilation.

Due to varying conditions such as climate resources, energy-saving and carbon reduction policies, energy efficiency levels and economic development in different regions, it is advisable for each locality to establish minimum energy efficiency retrofitting objectives while complying with local energy conservation and carbon reduction policies. For buildings undergoing comprehensive energy efficiency retrofitting, in addition to meeting the local energy conservation and carbon reduction policy requirements, the energy consumption of buildings in cold areas should be reduced by 30% or more compared to pre-retrofitting levels, while buildings in other climate zones should achieve a reduction of 20% or more. For buildings undergoing partial energy efficiency retrofitting, the retrofitted areas should preferably meet two or more energy-saving performance requirements, and must satisfy at least one requirement specified in the current mandatory national engineering construction standard *General Code for Energy Efficiency and Renewable Energy Application in Buildings* (GB 55015). The retrofitting mainly involves exterior walls, roofs, exterior windows, shading measures, and building equipment systems. If feasible, the retrofitting should also include floors and partition walls.

8.3.2 Exterior Doors and Windows and Sunshade

For exterior window retrofitting in severe cold, cold, and mild climate zone A, the requirements for both heat transfer coefficient and airtightness must be met, with sound insulation performance also taken into consideration. If feasible, cold area B may be given external shading measures.

In the regions with hot summer and cold winter, the retrofitting of exterior windows should not only meet the requirements for heat transfer coefficient but also meet the requirements for air tightness, openable area, visible transmittance, and solar heat gain coefficient of the exterior windows, with the sound insulation performance also taken into account.

In the regions with hot summer and warm winter, the retrofitting of exterior windows should meet the requirements for openable area and solar heat gain coefficient, while also take into consideration the sound insulation performance, visible transmittance, airtightness, and heat transfer coefficient of the exterior windows.

The Building Energy Research Center (BERC), Tsinghua University, in collaboration with the China Academy of Building Research and other institutions, conducted a comprehensive airtightness survey of over 60 existing residential buildings in the northern regions. The survey results indicate that due to poor construction quality and issues such as deformation of exterior windows, buildings constructed in China before the 1990s exhibit poor airtightness. Even with doors and windows closed, significant air leakage persists, resulting in an air change rate exceeding 1.5 times per hour. In recent years, energy-efficient doors and windows have been used in newly constructed buildings and building energy efficiency retrofitting projects, and the technology for external thermal insulation on walls has been applied, significantly improving the overall air permeability performance of buildings. The air change rate in some buildings can be less than 0.5 times/h. The results of the air permeability performance tests for residential buildings in two cities of the severe cold area in 2014 are shown in Fig. 8.10. In recent years, with the improving airtightness of the external windows of envelop structure,

In the regions with hot summer and cold winter, the airtightness of doors and windows needs improvement. Test results from the winter of 2014, as shown in Fig. 8.11, indicate that most residential buildings have an air change rate of 1–2 h^{-1}, while older buildings from the 1980s can reach 3.2 h^{-1}. Poor air tightness leads to significant air infiltration, thereby increasing heating load.

Exterior doors and windows in transparent building envelope as well as transparent curtain walls are weak links in the building envelope. In the energy efficiency standards, the heat transfer coefficient of exterior doors and windows is significantly higher than that of the roofs and exterior walls of non-transparent building envelope. Moreover, as the window-to-wall ratio increases, the thermal performance requirements for exterior doors and windows become more stringent, necessitating a lower heat transfer coefficient. The higher the energy conservation rate, the more significant the impact of exterior doors and windows on building energy consumption. In addition, poor air tightness of exterior doors and windows can lead to cold air infiltration, increasing energy loss and reducing indoor comfort. Therefore, retrofitting the exterior windows to enhance their thermal performance and airtightness is of great significance for improving indoor comfort and reducing building energy consumption. In the southern regions with hot summer and cold winter or

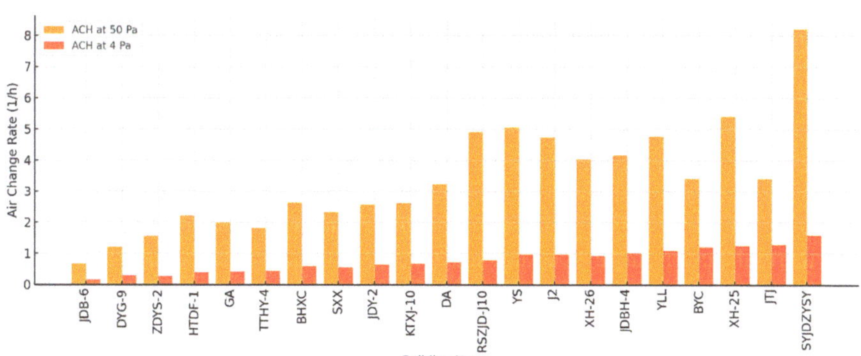

Fig. 8.10 Measured air change rate of buildings

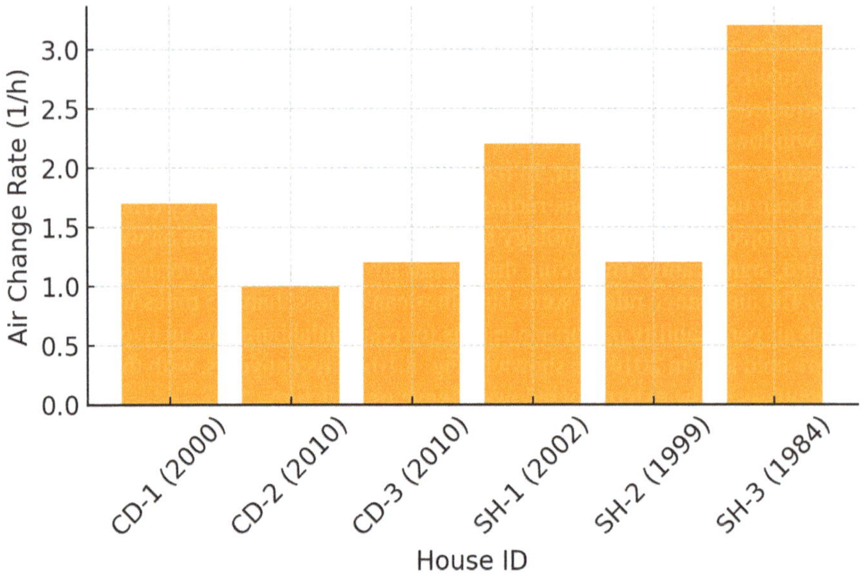

Fig. 8.11 Measured air change rate of the buildings in the regions with hot summer and cold winter (blower door method, 4 Pa)

hot summer and warm winter, enhancing the shading performance of exterior windows is one of the key focuses in energy efficiency retrofitting for exterior windows.

The retrofitting of exterior doors and windows shall be determined based on the specific conditions of the existing residential building, with comprehensive consideration of performance requirements such as safety, sound insulation, ventilation, and energy efficiency, as well as the preferences of the residents. Methods such as replacing windows or adding an additional layer of windows are advised. When the thermal performance of doors and windows in existing residential buildings fails to meet current national or local standards and the original window frames cannot be reused, the replacement of complete window may be implemented. When there is no desire to alter the original exterior window and the windowsill has sufficient width, addition of a window can be considered, i.e. adding another window unit on the exterior (or interior) side of the existing exterior window. When a window is added, the spacing between the two layers of windows shall be calculated to meet the thermal performance indicators of the windows and arranged reasonably. If the original window frames remain unchanged and only the single-pane glass is replaced with double-pane glass, or if the existing windows already have insulating glass that meets the thermal performance requirements, replacement may not be necessary. If the thermal performance does not meet the requirements, the deficiency can be compensated by appropriately enhancing the thermal performance of other building envelope components through a trade-off analysis.

The exterior doors and windows of residential buildings are often replaced during secondary decoration. Even within the same building, there can be significant variations in exterior doors and windows, including differences in profiles, glass

Table 8.2 Heat transfer coefficients of different types of glass

Type of glass	Spacing between glass panes (mm)	Heat transfer coefficient K W (m^2 k)
Single-pane glass	–	6.4
Double-pane insulating glass	6	3.4
	12	3.0
Triple-pane insulating glass	12	2.0
Double-pane low-emissivity coated insulating glass	12	1.6

configurations and other aspects. Meanwhile, it is difficult to determine the thermal performance of exterior windows through on-site testing. Therefore, a preliminary assessment of the performance of existing exterior doors and windows can be conducted by combining on-site survey results with data review. Table 8.2 provides conventional parameters for typical glass configurations, which can serve as a reference. The thermal performance of profiles can be preliminarily assessed based on factors such as material, thickness, and the year of construction or replacement.

In addition to considering the heat transfer coefficient of exterior windows, attention shall also be paid to the solar heat gain coefficient of exterior windows, which is primarily determined by the glass configuration. The solar heat gain coefficient not only accounts for the direct transmission of solar radiation through the transparent building envelope, but also accounts for the secondary heat transfer process from the exterior window components to both indoor and outdoor spaces after absorbing radiation. Compared to the shading coefficient, it provides a more comprehensive and accurate representation of indoor solar heat gain. In severe cold and cold areas, a high solar heat gain coefficient can effectively reduce the heating energy consumption in winter and improve indoor comfort; while in the regions with hot summer and warm winter or hot summer and cold winter, a lower solar heat gain coefficient helps significantly reduce summer cooling energy demand and enhance indoor comfort.

When shading is required on exterior doors and windows at the project location and shading is feasible, consideration can be given to adding fixed or movable external shading features, such as window shading panels, louvers, or roller shades. To add external shading structures or features, loads must be calculated. When shading is infeasible, the method of applying architectural window films can also be considered, as it is easily implementable and cost-effective, and yields good results.

The improvement of the thermal performance of exterior windows primarily relies on advancements in window frame construction, glass configuration, and the performance of components such as spacers. Typically, the performance of window frames can be enhanced by increasing their thickness and width or by filling them with thermal insulation materials. For glass performance, improvements can be achieved by adding more glass layers, controlling the gaps between panes, using vacuum glass, applying low-E coatings, and enhancing the performance of spacers. In recent years, with the improvement of energy efficiency standards and the promotion and application of ultra-low-energy buildings, the exterior window industry has made significant technological advancements. Examples include widely used materials such as aluminum alloy with thermal breaks, aluminum-clad wood profiles, and PVC profiles.

To meet the current energy efficiency standards in severe cold and cold areas, the broken bridge aluminum alloy profiles are typically available in 60, 65, and 70 series, with the thickness of the thermal barrier strips being around 20 mm. If triple-pane insulating glass and warm-edge spacers are used, and the profile cross-section thickness is increased to that of the 90 series, the heat transfer coefficient of the exterior window can be reduced to less than 1.0 $W/m^2 \cdot K$. Plastic profiles (PVC) are typically available in 60 and 65 series with three cavities. To ensure stability, steel reinforcements are commonly used inside the profiles. When the profile series is upgraded to 80 or above, the number of cavities is increased to no fewer than 6. Additionally, by filling the functional cavities with thermal insulation materials and using suitable glass configurations, the heat transfer coefficient can also be reduced to below 1.0 $W/m^2 \cdot K$. With the same technical approach, aluminum-clad wood profiles can also achieve excellent thermal insulation performance.

The spacer bar at the edge of the glass is another factor affecting the overall heat transfer coefficient of the window, and is the weakest link in exterior windows because it has the most heat transfer. Metal cold-edge spacer bars can create noticeable thermal bridges at the points where the glass is embedded in the frame, particularly with aluminum alloy spacers, which carry risks such as condensation and mold. The use of warm-edge spacers made of materials such as fiberglass-reinforced composites can reduce the thermal bridge coefficient, prevent mold and condensation, thereby lowering the overall heat transfer coefficient of the exterior window.

When external shading measures are not feasible, applying window films to buildings can enhance the sun-blocking performance of glass while maintaining moderate light transmission, thereby achieving thermal insulation. Building window films should generally be applied to the interior surface of the exterior window. For areas where interior installation is particularly challenging, exterior films may be used, provided that they feature an outer layer with UV resistance and enhanced weather durability.

8.4 Efficient Heat Pump Equipment and Terminals

Air source heat pumps, such as heat pump air conditioners and multi-connected air conditioning systems, are the primary cooling equipment for urban residential buildings during summer. They also serve as crucial heating facilities for buildings not connected to district heating networks and those connected to such networks but requiring heating services outside the heating season. Moreover, with the development of building electrification in the future, they will undertake more and more heating and hot water supply functions for residential buildings. Recently, heat pump systems for residential buildings have been driven by demands for higher energy efficiency, higher comfort, and the substitution of refrigerants. High-comfort terminals, variable frequency technology, pressure ratio adaptation technology, and energy storage technology were investigated to support the realization of the carbon neutrality goal in the building sector.

8.4.1 Low Ambient Temperature Air Source Heat Pump

The air source heat pump is the most common type of residential heat pump. Compared to the relatively stable indoor air temperature, the wide variations of outdoor ambient air temperature lead to significant changes in the operating conditions of air source heat pumps. Residential air source heat pumps shall be capable of adapting to changes of the outdoor environment, maintaining sufficient cooling and heating capacity while achieving high energy efficiency, particularly in terms of heating performance in low ambient temperature.

(1) **Pressure ratio adaptation technology**

The compressor is the main energy-consuming component of an air source heat pump system and also the component most affected by changes of system operating conditions. To achieve high energy efficiency of the air source heat pump under a wide range of variable working conditions throughout the year, it is necessary to realize the condition-adaptive adjustment of the compressor. Existing research indicates: During cooling operation, the pressure ratio of an air source heat pump system typically varies between 2–3, while during heating operation, it increases to 3–5 or even higher values. Therefore, compressors for residential heat pumps shall possess the capability to actively adapt to a wide range of pressure ratios.

In recent years, a novel pressure ratio adaptive compressor technology has emerged. Hefei Shengsansong Cooling and Heating Technology Co., Ltd. has developed a single-unit two-stage scroll compressor, which can be used for high-temperature hot water and steam production. Figure 8.12 shows the schematic diagram of the compressor. This compressor used two-stage scroll design, where two orbiting scroll is set as back-to-back configuration. During the operation of the compressor, gas directly enters the low-pressure stage scroll through the suction pipe. After the first stage compression, it is discharged into the internal chamber of the compressor, where it mixes with the refrigerant gas discharged from the injection loop. It then undergoes secondary compression via the high-pressure stage scroll. Finally, it is discharged out of the compressor through the exhaust pipe. It is estimated that this method can reduce the axial force of the scroll by 30% and decrease friction loss by 60%. The two-stage scroll compressor has a wider pressure ratio range. With vapor injection, the internal pressure ratio can be automatically adjusted between 3 and 7, offering a broader operating range. The lowest evaporation temperature can reach −40 °C, and the highest condensation temperature is 65 °C. Test data shows that the heat pump equipped with this compressor achieves a heating COP of 2.62 under low-temperature rated

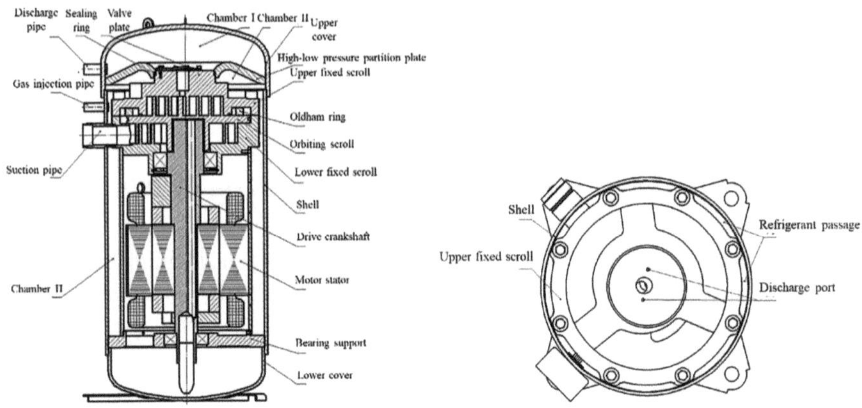

(a) Front view (b) Top view

Fig. 8.12 Schematic diagram of the singe-unit two-stage scroll compressor. (**a**) Front view. (**b**) Top view

heating conditions (ambient dry/wet-bulb temperature: −12 °C/−14 °C, outlet water temperature 41 °C), and reaches a heating COP of 3.44 under rated heating conditions (ambient dry/wet-bulb temperature: 7 °C/6 °C, outlet water temperature 45 °C). This compressor has also been utilized in the development of high-efficiency heat pump steam turbine units.

(2) **CO_2 low-temperature heat pump**

The carbon dioxide heat pump exhibits excellent heating performance and has been well applied in fields such as heat pump water heaters in the early years. However, the cost increase caused by high pressure has become a major obstacle limiting the large-scale application of CO_2 heat pumps in projects, such as space heating. However, with the enactment of national clean heating policies and the indigenization of key components, CO_2 heat pump heating has recently achieved accelerated development.

Figure 8.13 shows the diagrams of a novel CO_2 air source heat pump with a subcooler. Unlike conventional transcritical carbon dioxide heat pumps, this system incorporates an additional conventional R134a heat pump between the gas cooler outlet and hot water inlet. On one hand, this design achieves subcooling of the CO_2 at the gas cooler outlet, thereby enhancing the efficiency of the CO_2 heat pump. On the other hand, it preheats the inlet hot water, boosting the heating capacity. China Academy of Building Research conducted tests on the heat pump heating system in a public building in Qinhuangdao City and a residential building in Shijiazhuang City (Fig. 8.14). The test results indicate that under operating conditions with an average daily outdoor temperature of −10.9 to 2.5 °C, the measured daily COP ranges for unit No. 1 and unit No. 2 in the Qinhuangdao project were 2.75–3.09 and 2.76–3.15, respectively. When the outdoor temperature drops to a minimum of −18 °C, the COP of Unit 1 and

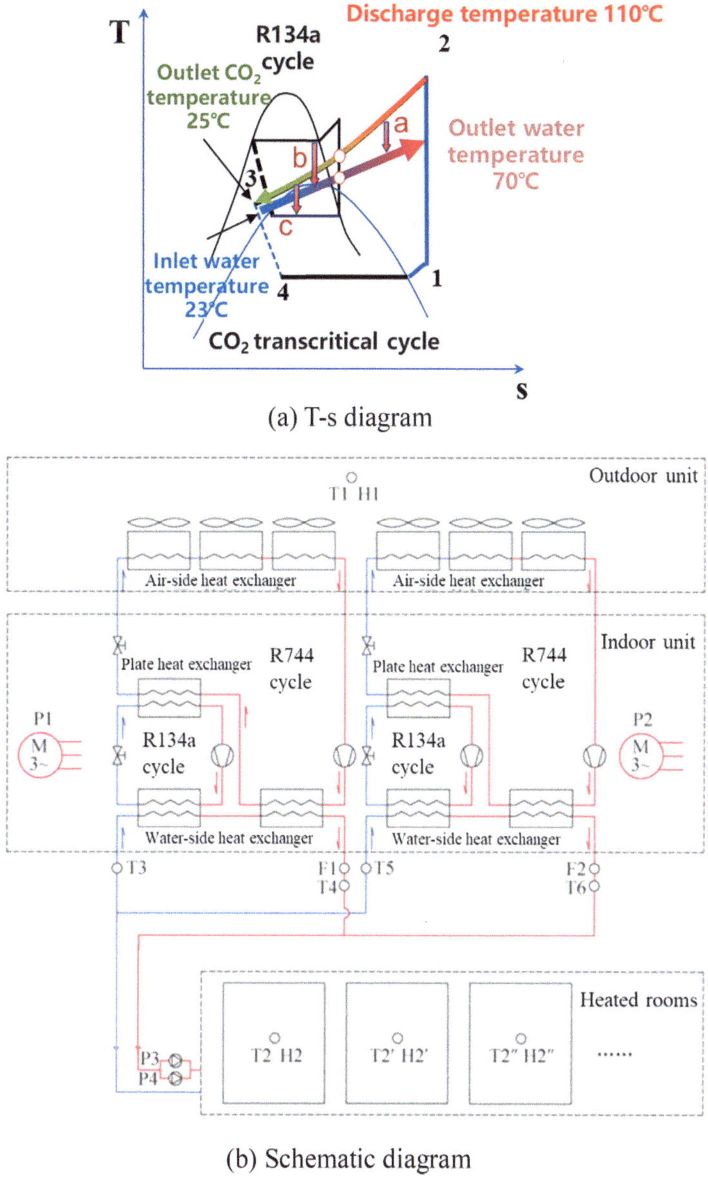

Fig. 8.13 CO2 semi-cascade air source heat pump system with a subcooler. (**a**) T-s diagram. (**b**) Schematic diagram

Unit 2 can still reach 2.19 and 2.88 respectively. Under the operating conditions with an average daily outdoor temperature of −6.6 to 12.5 °C at the Shijiazhuang project, the measured daily COP ranges for Unit 1 and Unit 2 were 2.32–2.38 and 2.21–3.06, respectively.

(a) Qinhuangdao Project (b) Shijiazhuang Project

Fig. 8.14 Field photograph of CO_2 semi-cascade air source heat pump system with subcooler. (**a**) Qinhuangdao Project. (**b**) Shijiazhuang Project

8.4.2 Heat Pump Defrosting Technology

Defrosting is another key technology that affects the efficient operation of the air source heat pump. Traditional methods, such as reverse cycle defrosting, hot gas bypass defrosting and thermal storage defrosting cannot fully achieve both high system efficiency and high comfort simultaneously. Recently, novel defrosting solutions, including direct hot gas defrosting, have been devised to optimize the defrosting process, ensuring heating continuity and reducing defrosting duration. Unlike the reverse cycle defrosting, the direct hot gas defrosting technology does not require reversal of the four-way valve during defrosting. The refrigerant flow direction and the pressure-enthalpy diagram during the defrosting process are shown in Fig. 8.15, which is the same as during normal heating operation. The high-temperature, high-pressure refrigerant discharged from the compressor first enters the indoor heat exchanger. By increasing the opening degree of the "quick-opening" electronic expansion valve, the throttling heat loss of the refrigerant is minimized during defrosting, allowing the higher-temperature refrigerant to reach the outdoor unit for defrosting. The refrigerant experiences heat transfer on both the indoor and outdoor sides, and can continue to supply partial heat to the room during the defrosting process, helping to improve indoor comfort.

The heat of the refrigerant entering the outdoor heat exchanger determines the speed of direct hot gas defrosting. Therefore, efforts must be made to minimize the heat loss in the electronic expansion valve during the $3 \rightarrow 4$ throttling process shown in Fig. 8.15. However, if the indoor heat exchanger and outdoor heat exchanger are connected, it will lead to a large amount of liquid return on the compressor suction side, causing a continuous decrease in the system's compression ratio and rapid decay of the exhaust temperature, making it difficult to maintain stable defrosting heat supply. Therefore, the direct hot gas defrosting process must ensure a certain degree of throttling to achieve continuous defrosting. Through field experiments measuring of an R32 refrigerant heat pump system, it was found that to ensure both

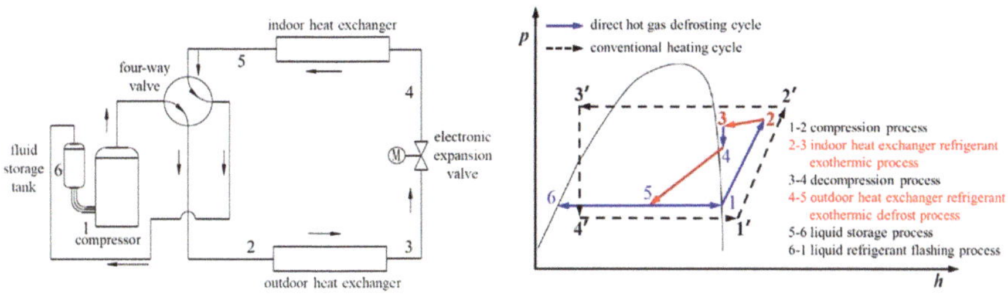

(a) Refrigerant flow diagram (b) Pressure-enthalpy diagram

Fig. 8.15 Refrigerant flow direction and process pressure-enthalpy diagram of direct hot gas defrosting technology. (**a**) Refrigerant flow diagram. (**b**) Pressure-enthalpy diagram

Table 8.3 Comparison of operation effects of different defrosting technologies under various frost accumulation levels

Initial frost accumulation level (kg)	Defrosting method	Total defrosting heat (kJ)	Defrosting duration (s)	Heating restoration duration (s)
1.4	Reverse cycle	381.4	275.5	578.5
	Direct hot gas	365.4	741.5	777.5
0.9	Reverse cycle	249.1	235.1	538.1
	Direct hot gas	260.0	486.5	522.5
0.4	Reverse cycle	128.7	175.6	478.6
	Direct hot gas	149.7	182.3	218.3

defrosting heat capacity and speed, the air flow rate corresponding to the electronic expansion valve opening during direct hot gas defrosting must be maintained in the range of 50–70 L/min.

According to the comparison between the defrosting effects of reverse cycle defrosting and direct hot gas defrosting under different frost accumulation levels, as shown in Table 8.3, it can be observed that the direct hot gas defrosting technology has a significant advantage in thin frost areas (initial frost accumulation <0.4 kg). Although the actual defrosting duration is nearly equivalent to that of reverse cycle defrosting, the heating recovery time is only 45% of reverse cycle defrosting's duration, ensuring better indoor thermal comfort.

8.4.3 High-comfort Heat Pump Terminal

Residential heat pumps are typically fitted with convective or radiant terminals. The convective terminal is capable of rapid response but offers poor comfort, particularly in winter when there are the issues such as drafts and significant vertical temperature differences indoors. On the other hand, radiant heating terminals offer excellent thermal comfort, but they suffer from drawbacks such as slow startup and low heating power. Therefore, improving existing heat pump terminals is crucial for promoting the application of heat pumps in residential buildings.

(1) Improvement of traditional terminals

Directly improve the structural characteristics of convective terminals such as heat pump air conditioners to ensure comfort in winter. For example, use the air flow attachment effect (Coanda effect) to design a cold and hot air diversion for the air outlet duct, achieving shower-like cooling and carpet-like heating. In summer, guide the cold air to rise through a large-curvature air deflector to prevent direct blowing across the whole area. In winter, guide the hot air downward, which spreads forward in the room after reaching the ground. The hot air covers the ground and then naturally floats upward. A solution for cross-flow bilateral zoned air supply has been devised, and a high-efficiency bilateral air supply technology and a two-zone three-dimensional air supply adaptive control method are designed, aiming to optimize the draft feeling and the temperature regulation in the working area.

Air source heat pump air heater is a floor-mounted direct-expansion air source heat pump unit. Its upper air outlet rapidly directs warmth to human activity areas through adjustable angles, while the lower outlet ensures ground-hugging airflow diffusion, achieving the comfort similar to underfloor heating. In recent years, it has become a primary solution for heating demands in northern regions, particularly in rural areas (Fig. 8.16).

(2) **Dual-supply system**

Refrigerant cooling & water heating system is a dual-supply system designed to create thermal comfort in both winter and summer, and has been well received by the market. In winter, hot water is circulated through floor for radiant heating, to create a "cool head and warm feet" feeling, achieving optimal thermal comfort. In summer, the system is equipped with direct expansion terminal units to efficiently produce cooling air, and overhead air supply is used to ensure rapid cooling and uniform temperature distribution indoors. In addition, there are also water cooling & water heating system and refrigerant cooling & refrigerant heating system available on the market. Currently, the market shares of water

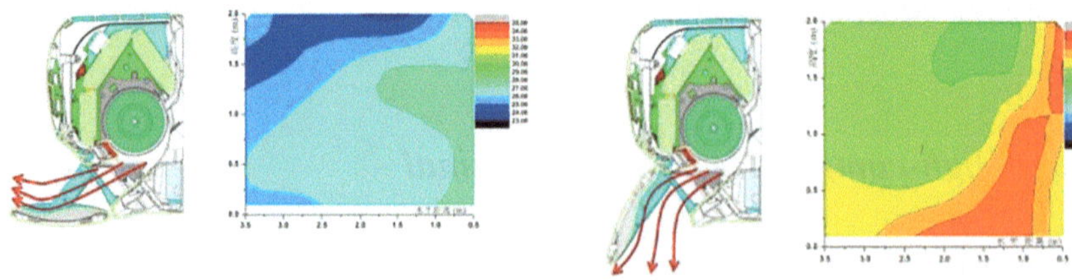

(a) Shower-like cooling (b) Carpet-like heating

Fig. 8.16 Improving the structural features of the air conditioner to achieve comfortable air supply. (**a**) Shower-like cooling. (**b**) Carpet-like heating cooling & water heating system, refrigerant cooling & water heating system, and refrigerant cooling & refrigerant heating system in dual-supply systems are 60.7%, 38.2%, and 1.1%, respectively. However, if the aforementioned terminals operate in continuous radiant heating mode in winter, the energy efficiency ratio of the unit improves, but the prolonged operation time due to continuous running may result in energy consumption reaching 4–5 times the current level, making this mode unsuitable for creating a "part-time, part-space" winter indoor built environment in hot summer and cold winter zone (Fig. 8.17).

Therefore, some experts have proposed improving the intermittency of radiant terminals to make them suitable for intermittent heating, including enhancing the heat transfer efficiency of terminals and reducing heat storage capacity, e.g. creating an air gap beneath a raised floor to increase the surface area for heat exchange, and installing fans under the raised floor to circulate indoor air through the heating floor and back into the room. However, there is currently a lack of commercialized applications for these methods.

(3) **Convective/radiant coupled terminal**

The dual-supply system integrates convective terminal and radiant terminal at the system level, where the two terminals operate in winter and summer separately. Since convective terminal and radiant terminal each have their own advantages in terms of intermittency and comfort, they can be used together during the same season. To this end, the professionals have proposed a convective/radiant coupled terminal system. It utilizes primarily convective heat transfer during the startup state and predominantly radiant heat exchange during the steady-state to provide heating and cooling for rooms. This approach achieves efficient intermittent creation of a comfortable indoor environment, enabling intelligent and adjustable innovations in environmental conditioning while ensuring human-centered comfort. The previous research has rendered the design method for convective/radiant coupled terminal, where the capacity ratio between convective and radiant units shall be greater than 4.0, ensuring rapid building startup and steady-state comfortable heating. The coupled terminals developed in recent years such as split direct expansion, packaged direct expansion and

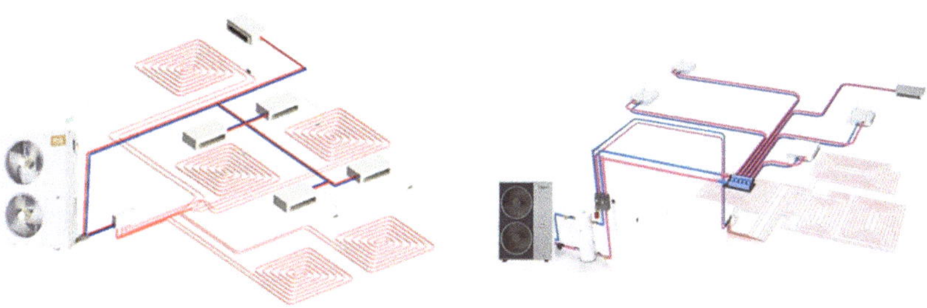

(a) Refrigerant cooling & water heating system; (b) water cooling & water heating system

Fig. 8.17 Typical dual-supply system. (**a**) Refrigerant cooling & water heating system; (**b**) water cooling & water heating system heat pipe terminals have reduced the indoor temperature rise process to within 30 min (initial temperature 10 °C), significantly improving the effectiveness of intermittent heating.

The split direct expansion terminal is the most basic convective/radiant coupled terminal, consisting of multiple sets of direct-expansion radiant panels and convective units. The direct-expansion radiant panel is fitted with microchannel flat tubes to achieve direct heat exchange between the refrigerant and air as well as building envelopes; small-diameter flat tubes are used to avoid excessive refrigerant charge; backplates and intermediate bent fins are installed to enhance the heat transfer capacity of the middle flow channels where thermal expansion causes upward movement. The split terminal system offers high flexibility, allowing free arrangement based on personnel location and needs, but it has a low level of productization. Based on this, a packaged direct expansion terminal has been developed, which combines the convective unit and radiant unit into one system. The pipeline flow path is optimized. During heating, the refrigerant sequentially passes through the convective finned tube heat exchanger (HX) A, the radiant heat exchanger and the finned tube convective HX B, ensuring uniform heat exchange on the radiant surface while avoiding system stability issues caused by overheating or insufficient supercooling. By now, split direct expansion terminal and packaged direct expansion terminal have been applied practically, with a starting time of <20 min and a 20% improvement in comfort compared to traditional convective terminals (Figs. 8.18 and 8.19).

Heat pipes have the advantages of simple and aesthetically pleasing structure, as well as rapid and uniform heat exchange. The novel integrated convective-radiant flat heat pipe terminal is built and integrated with an air source heat pump. With the phase-change heat transfer of flat heat pipes being used as the radiant surface, optimization designs have been implemented, including the addition of fins, incorporation of cross-flow fans, the heat exchanger structure composed of upper and lower channels of flat heat pipes, and perforated refrigerant channel within the heat exchanger. These enhancements increased the maximum heating capacity (with a heat source temperature of 70 °C) to 1194.6 W. On the other hand, a packaged terminal based on a three-fluid heat exchanger and a gravity heat pipe has been conceived. The three-fluid heat exchanger enables pairwise heat exchange among three media: heat source, heat pipe, and air. The results indicate that the

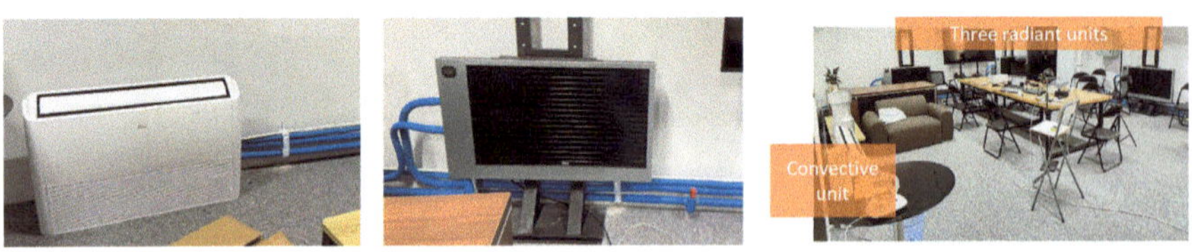

(a) Convective unit, (b) Radiant unit, (c) Practical application

Fig. 8.18 Split direct expansion terminal and its practical application. (**a**) Convective unit, (**b**) radiant unit, (**c**) practical application

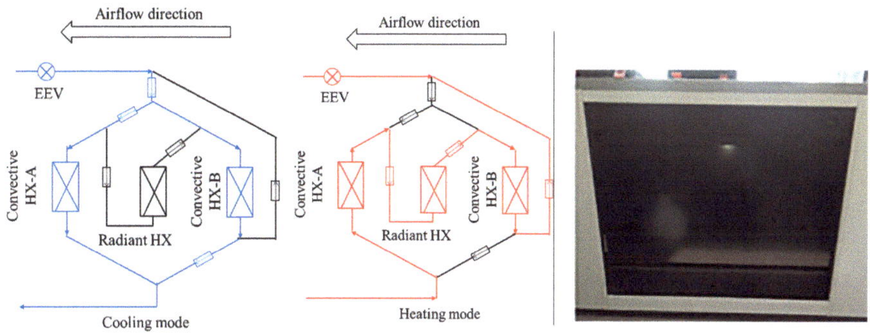

(a) Schematic diagram of cooling/heating flow path, (b) Terminal prototype

Fig. 8.19 Structure and prototype of packaged direct expansion terminal. (**a**) Schematic diagram of cooling/heating flow path, (**b**) terminal prototype

overall heat capacity of the terminal can be adjusted within the range of 673–5753 W, with the radiant heat transfer proportion varying between 5.5% and 38.7%. Additionally, the temperature difference in the vertical direction can be kept within 1 °C/m.

In the aforementioned studies on different operation strategies for direct-expansion and heat pipe terminals, variable frequency starting proves to be the most energy-efficient and optimal solution, albeit with a longer starting time. The operative temperature of the control unit can enable dual control of comfort and energy efficiency at different temperatures. In addition, adjusting convective and radiant parameters, such as increasing terminal temperature and reducing air speed, can improve the indoor comfort. Furthermore, by coupling energy simulation with computational fluid dynamics, a dynamic environment conditioning model for "heat source-terminal-indoor environment" of convective-radiant terminal has been accurately established, and the variations in dynamic indoor environment under different operating modes have been explored. The results demonstrate that the optimal control strategy for convective-radiant coupled environment conditioning under "part-time, part-space" demand conditions is the operation mode combining maximum convective capacity during startup state with equal-load radiation during steady-state (Fig. 8.20).

8.4.4 Refrigerant for Residential Heat Pump

Currently, the main refrigerants used in the residential heat pumps in China include HFC-32 and R-410A (a mixture of HFC-32 and HFC-125). However, in compliance with the *Kigali Amendment to the Montreal Protocol*, HFC-32 and R-410A must be gradually replaced by low-GWP refrigerants. China officially joined the Protocol in 2019, and is required to start gradual phase-out of the HFCs from 2024.

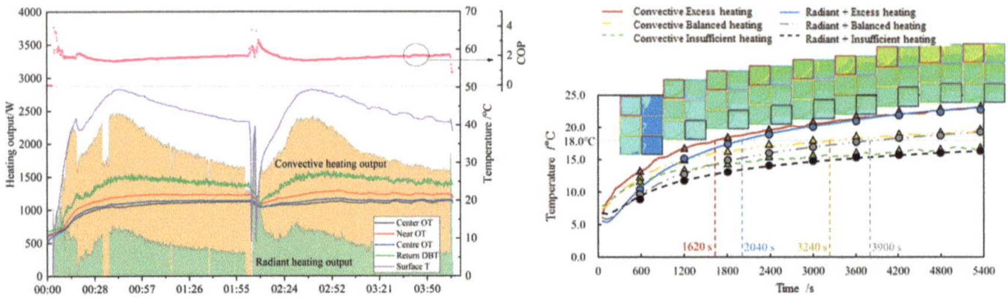

(a) Environmental laboratory terminal performance measurement results

(b) Terminal performance measurement results

Fig. 8.20 Convective-radiant coupled terminal laboratory and simulation results. (**a**) Environmental laboratory terminal performance measurement results. (**b**) Terminal performance measurement results

To address this challenge, several international organizations and major economies, particularly the European Union, the United States, and Japan, have taken concrete actions by formulating and implementing relevant regulations and policies to guide and standardize the future use of refrigerants in the heat pumps. On June 12, 2023, China published the *Recommended Catalog of Substitutes for Ozone-Depleting Substances in China*. The introduction of this catalog aims to fulfill the obligations under *The Montreal Protocol on Substances that Deplete the Ozone Layer*, promoting the gradual phase-out of ozone-depleting and phase-down high GWP substances in China. The catalog specifically highlights HC-290 (propane) and R744 (carbon dioxide) as recommended environmentally friendly refrigerant alternatives for residential heat pumps, as detailed in Table 8.4.

HC-290, as a hydrocarbon refrigerant, exhibits excellent cooling performance and environment-friendly characteristics, and is believed to be a refrigerant with significant application potential for heat pumps in residential buildings both domestically and globally. However, it is important to note that HC-290 is a Class A3 refrigerant, which means it is highly flammable, undoubtedly posing certain challenges and limitations to its application. On one hand, to ensure the safety of refrigeration systems using HC-290, it is essential to strictly control the charge amount. Recently, the technology for refrigerant charge reduction has been continuously advancing. For example, microchannel heat exchanger has emerged as a notable highly efficient component for energy conservation and carbon reduction. On the other hand, the flammability of HC-290 also imposes stricter requirements on the safety control measures for the equipment using this refrigerant. Therefore, in-depth exploration of the leakage characteristics of refrigerants, along with the development of innovative safety-enhancing measures have become an indispensable and critical issue in advancing the application of flammable refrigerants. In addition, another natural refrigerant, carbon dioxide (CO_2), has become an ideal working fluid for heat pump water heaters and heating pumps in severe cold areas due to its excellent heating characteristics. However, at the current stage, carbon dioxide (CO_2) must withstand high pressure during operation, which poses technical and

Table 8.4 Heat pump refrigerant alternatives specified by recommended catalog of substitutes for ozone-depleting substances in China

Alternative	ODP	GWP	Main applications	HCFCs to be replaced
HC-290	0	<1	Room air conditioner, **household heat pump water heater**, commercial standalone refrigeration system, industrial refrigeration system	HCFC-22
R-744	0	1	**Household heat pump water heater**, industrial or commercial heat pump water heater, industrial or commercial refrigeration system, cold storage	HCFC-22

economic challenges for the design and manufacturing of system components. Developing cost-effective key components has become a critical factor for the large-scale application of carbon dioxide in residential heat pumps. In summary, HC-290 is the primary alternative refrigerant for heat pumps in residential buildings in China, while CO_2 is expected to be widely applied in heat pump water heaters and heating pumps for severe cold areas.

8.5 Indoor Air Quality of Urban Residential Buildings

8.5.1 Current Status of Indoor Air Quality in Residential Environments

(1) Main Types of Pollutants in Residential Environments

With the acceleration of urbanization and the advancement of industrialization, indoor air quality has become a significant factor affecting human health. Especially in China, indoor pollution has become an increasingly serious health issue. The primary pollutants in the indoor environment include particulate matters (PM), formaldehyde (HCHO), volatile organic compounds (VOCs), semivolatile organic compounds (SVOCs), and microorganisms, which pose significant threats to human health.

The pollution of particulate matters in daily living environments is becoming increasingly severe. Particle is a significant pollutant affecting indoor air quality. According to the *Standards for Indoor Air Quality* (GB/T 18883-2022), particles can be classified into different types based on particle size, including inhalable particulate matter (PM10) and fine particulate matter ($PM_{2.5}$) (GB/T 18883-2022 Indoor Air Quality Standards n.d.). Inhalable particulate matter (PM10) refers to particles in the air with a diameter of 10 μm or less. These particles are widely dispersed in the air and can easily be inhaled into the human respiratory system, reaching the trachea and bronchi. Long-term inhalation may lead to respiratory diseases such as asthma, chronic bronchitis, and emphysema. PM10 primarily originates from outdoor air pollution, construction activities, and traffic emissions, and is also influenced by indoor activities such as combustion and cooking. Fine particulate matter ($PM_{2.5}$) refers to airborne particles with a diameter of 2.5 μm or less, which possess stronger penetration capabilities, enabling them to reach deep into the alveoli and even enter the bloodstream. Due to their small size and large surface area, $PM_{2.5}$ particles readily carry harmful chemical substances such as heavy metals and volatile organic compounds (VOCs), resulting in more severe health hazards. $PM_{2.5}$ is closely associated with diseases of the respiratory system and cardiovascular system, representing a significant global public health concern. Its primary sources include smoke from incomplete combustion, industrial emissions, vehicle exhaust, as well as indoor smoke from combustion processes and cooking fumes.

Gaseous pollutants include inorganic and organic categories. Among them, inorganic gaseous pollutants mainly include sulfur dioxide (SO_2), nitrogen dioxide (NO_2), carbon monoxide (CO), and ammonia (NH_3). These pollutants primarily originate from indoor fuel combustion, infiltration of external air pollutants, emissions from building materials, and other sources. Inorganic gaseous pollutants pose a significant health hazard to the human body, particularly to the respiratory system and cardiovascular system. CO can bind with hemoglobin, leading to carbon monoxide poisoning; NO_2 and SO_2 may trigger respiratory inflammation, asthma, and other diseases. NH_3 poses significant harm to the human body, strongly irritating the oral and nasal mucosa as well as the upper respiratory tract. Short-term exposure to high concentration of ammonia can lead to the symptoms such as tearing, sore throat, dizziness, nausea, vomiting and fatigue. In worse cases, it may cause pulmonary edema or adult respiratory distress syndrome.

Organic pollutants include volatile organic compounds (VOCs) and semivolatile organic compounds (SVOCs).

VOCs are significant components of indoor air pollution, commonly including formaldehyde, benzene, methylbenzene, xylene, acetaldehyde, etc. (GB/T 18883-2022 Indoor Air Quality Standards n.d.; World Health Organization (WHO) 2014). VOCs originate from furniture, decoration materials, cleaning agents, etc. Long-term exposure to these harmful gases may trigger headaches, rapid breathing and allergic reactions, and even increase cancer risks. Formaldehyde, as the most typical pollutant among VOCs, is a known carcinogen and poses particularly severe harm to human health.

SVOCs are a class of organic compounds with large molecular weights and low volatility. They come in many types, but the primary categories widely present indoors are phthalate esters (PAEs), polycyclic aromatic hydrocarbons (PAHs), polychlorinated biphenyls (PCBs), and polybrominated diphenyl ethers (PBDEs) (World Health Organization (WHO) 2014; Wang et al. 2010). Phthalate esters are known as PAEs for short.

Six types of PAEs are restricted in many standards, namely di(2-ethylhexyl) phthalate (DEHP), butyl benzyl phthalate (BBP), dibutyl phthalate (DBP), diisononyl phthalate (DINP), diisodecyl phthalate (DIDP), and di-n-octyl phthalate (DNOP). In addition to the aforementioned six categories of substances, phthalate esters detected indoors may also include compounds such as diisobutyl phthalate (DIBP). Polycyclic aromatic hydrocarbons (PAHs) primarily originate from incomplete combustion products. The Environmental Protection Agency has identified seven types of pollutants as suspected carcinogens, including benz[a]anthracene, benzo[a]pyrene, benzo[b]fluoranthene, benzo[k]fluoranthene, chrysene, dibenz[a,h]anthracene, and indeno[1,2,3-cd]pyrene. In addition, the polychlorinated biphenyls (PCBs) in the air are mainly low-chlorinated biphenyls such as dichlorobiphenyl, trichlorobiphenyl, and tetrachlorobiphenyl. These substances were previously primarily used as flame retardants in electrical equipment such as transformers. Due to their carcinogenicity, they have been banned in many countries. However, because of their long-term persistence and resistance to degradation, many of these substances still remain present indoors today. Polybrominated diphenyl ether (PBDEs) is primarily used as flame retardant in electronic devices and other applications.

In addition to particle and chemical pollution, microbial contamination in indoor air is not to be overlooked either. Microbial contamination primarily includes bacteria, fungi, and viruses, with diverse sources such as indoor moisture accumulation and improper use of air-conditioning systems (GB/T 18883-2022 Indoor Air Quality Standards n.d.; World Health Organization (WHO) 2014). These microorganisms often adhere to the surface of particles or spread through droplets generated by human coughing and sneezing, thereby further exacerbating the complexity of indoor air pollution. When these pollutants are suspended in the air, they can not only trigger allergic reactions but also lead to respiratory tract infections and other illnesses, posing multifaceted threats to human health.

In addition, radioactive contamination is also one of the significant threats to the indoor environment. Radon (^{222}Rn) is a naturally occurring colorless, odorless, non-volatile radioactive inert gas, identified by the World Health Organization (WHO) as one of the primary environmental carcinogens (World Health Organization (WHO) 2014). In residential buildings, radon comes from soil gas, and some building materials, particularly stone, also emit radon.

(2) Sources of Main Pollutants

The significant presence of VOCs in the indoor environment has a major impact on human health and is one of the primary causes of sick building syndrome. Building materials and furniture, as the primary emission sources of volatile organic compounds indoors, have a significant impact on indoor air quality. Artificial boards and coatings are fundamental decorative materials in home renovation, as well as one of the primary sources of indoor air pollutants. Over 80% of the artificial boards in China contain high content of urea-formaldehyde resin as an adhesive, ensuring a certain surplus of formaldehyde, which is the reason why these boards often emit formaldehyde during indoor use. The free formaldehyde and other hazardous substances in unqualified decoration materials are released over a long period in residential environments, causing air pollution. In addition to formaldehyde, various paints, coatings, adhesives, thinners, etc. used during home decoration can also lead to the release of VOCs such as benzene and xylene. In China, multiple building material product standards have been successively issued and continuously revised, which also set limit requirements for volatile organic compound content, such as *Technical Requirement for Low-Volatile-Organic-Compound-Content Coatings Product* (GB/T 38597-2020).

In recent years, SVOCs have gained increasing attention as a new type of pollutant. Human exposure to SVOCs may have adverse health effects and even potentially lead to tumors. The primary sources of indoor SVOCs include flame retardants added to various household products, combustion by-products from cooking and cigarettes, plasticizers, adhesives, insecticide sprays, as well as personal protective equipment such as cosmetics and deodorants (Krol et al. 2011). In terms of relevance to current environmental issues, phthalate ester (PAE) used as the main component of plasticizers and organophosphate ester used as primary constituent of flame retardants hold significant importance for the indoor environment. Regarding the SVOC content in building materials, China has established the national standard GB/T 42898-2023, which describes the test methods for the release of semivolatile organic compounds (SVOCs) from building material products, but there is currently no mandatory standard directly limiting the SVOC content in building materials.

Second, pollutants generated from indoor combustion activities such as cooking should not be overlooked either. Chinese kitchens involve extensive cooking methods such as stir-frying, pan-frying, and deep-frying, during which the emission of oil fume particles, specifically PM2.5 particles, is the primary cause of deteriorating residential air quality. In addition, the combustion of liquefied petroleum gas, coal gas, and coal also generates inorganic pollutants such as CO and NO_x, as well as SVOCs such as polycyclic aromatic hydrocarbon (PAH), which pose serious threats to human health. Reducing the use of solid cooking fuels can effectively decrease emissions of particles, black charcoal, and organic carbon, thereby lowering residents' exposure to air pollutants and yielding positive impacts on environment and health.

There are several other factors that can contribute to pollution to the residential environment. For instance, microorganisms such as dust mites, mold, bacteria, and viruses thrive in the indoor environment and can spread through the air, leading to respiratory tract infections and other illnesses. Tobacco smoke contains multiple carcinogens, and prolonged inhalation increases the risk of diseases such as lung cancer. Concrete may release trace amounts of radon (^{222}Rn), causing radioactive contamination in building interiors. Additionally, electrostatic precipitators generate corona discharges that release secondary pollutants such as O_3, thereby compromising indoor air quality.

In summary, the main sources of pollutants in residential environments in China are diverse, including various emission sources such as decoration materials, furniture, soft furnishings, and daily necessities.

Table 8.5 Standards for Indoor air quality (GB/T 18883-2022)

No.	Indicator	Unit of measurement	Requirement	Remarks
1	Ozone (O_3)	mg/m^3	≤0.16	1-hour average
2	Nitrogen dioxide (NO_2)	mg/m^3	≤0.20	1-hour average
3	Sulfur dioxide (SO_2)	mg/m^3	≤0.50	1-hour average
4	Carbon dioxide (CO_2)	%[a]	≤0.10	1-hour average
5	Carbon monoxide (CO)	mg/m^3	≤10	1-hour average
6	Ammonia (NH_3)	mg/m^3	≤0.20	1-hour average
7	Formaldehyde (HCHO)	mg/m^3	≤0.08	1-hour average
8	Benzene (C_6H_6)	mg/m^3	≤0.03	1-hour average
9	Methylbenzene (C_7H_8)	mg/m^3	≤0.20	1-hour average
10	Xylene (C_8H_{10})	mg/m^3	≤0.20	1-hour average
11	Total volatile organic compounds (TVOC)	mg/m^3	≤0.60	8-hour average
12	Trichloroethylene (C_2HCl_3)	mg/m^3	≤0.006	8-hour average
13	Tetrachloroethylene (C_2Cl_4)	mg/m^3	≤0.12	8-hour average
14	benzo[a]pyrene (BaP)[b]	ng/m^3	≤1.0	24-hour average
15	Inhalable particulate matter (PM_{10})	mg/m^3	≤0.10	24-hour average
16	Fine particulate matter ($PM_{2.5}$)	mg/m^3	≤0.05	24-hour average
17	Radon (^{222}Rn)	Bq/m^3	≤ 150	Annual average[c] (reference level[d])

[a]Volume fraction

[b]Refers to benzo[a]pyrene in inhalable particulate matter

[c]At least 3 months of sampling

[d]Represents the maximum acceptable annual average radon concentration indoors, not a strict boundary between safety and danger. When the indoor radon concentration exceeds the reference level, measures should be taken to reduce the indoor radon concentration. When the indoor radon concentration is below the reference level, protective measures can also be taken to reduce it, reflecting the principle of optimized radiation protection

(3) Standards for Major Pollutants Concentration in Residential Environment

China's current *Standards for Indoor Air Quality* (GB/T 18883-2022) covers nineteen indicators, including physical, chemical, biological, and radioactive parameters. This standard requires that indoor air must be non-toxic, harmless, and free of unusual odors. The concentration limits for major pollutants are specified in Table 8.5.

In addition, to control the indoor environment pollution caused by building materials and decorative materials, the *Standard for Indoor Environmental Pollution Control of Civil Building Engineering* (GB 50325-2020) (see Table 8.6), issued by the Ministry of Housing and Urban-Rural Development, also plays a crucial role during the completion acceptance of construction projects.

Table 8.6 Standard for Indoor environmental pollution control of civil building engineering (GB 50325-2020)

Pollutant	Class I civil building project	Class II civil building project
Radon (Bq/m^3)	≤150	≤150
Formaldehyde (mg/m^3)	≤0.07	≤0.08
Ammonia (mg/m^3)	≤0.15	≤0.20
Benzene (mg/m^3)	≤0.06	≤0.09
Methylbenzene (mg/m^3)	≤0.15	≤0.20
Xylene (mg/m^3)	≤0.20	≤0.20
TVOC (mg/m^3)	≤0.45	≤0.50

Note: Class I civil buildings include residences, hospitals, elderly care facilities, kindergartens, and schools; Class II civil buildings include office buildings, shops, hotels, cultural and entertainment venues, bookstores, libraries, exhibition halls, gymnasiums, public transportation waiting rooms, restaurants, and barbershops

8.5.2 Pollutant Concentration Levels and Health Hazards in Residential Environment

(1) **Pollutant Testing and Analysis Methods**

Sampling and testing are crucial for identifying, understanding, and quantifying pollutants to assess the health risk posed by exposure. Traditional pollutant sampling methods are primarily divided into two categories: passive sampling and active sampling. Passive sampling refers to the process where pollutants are adsorbed onto an adsorbent at a controlled rate through physical diffusion without the use of an air pump. Active sampling, on the other hand, involves using a sampling pump to draw a specific volume of air through an adsorbent, thereby capturing the target substances on the adsorbent.

Passive air samplers, due to their low cost, ease of deployment, suitability for large-scale or hazardous environment sampling, and suitability for long-term monitoring of average personal exposure levels, have been widely applied. Passive sampler can be designed as single-phase or dual-phase passive sampler. A single-phase passive sampler refers to a single adsorption medium directly exposed to the sampling matrix, such as polyurethane foam (PUF) sheets, solid-phase microextraction, or thin-film microextraction. A dual-phase passive sampler is more common, consisting of an adsorbent barrier and an adsorbent. The adsorbent barrier can eliminate or minimize the impact of environmental factors, helping to ensure a relatively stable sampling rate, which is crucial for quantitative analysis. Common dual-phase passive samplers can be categorized into tube sampler, badge-type sampler and radial sampler, depending on their morphology. Among them, the tube sampler, specifically the TENAX tube, is now a commercially available analytical equipment, making analysis convenient. It is widely used as a passive sampling device. The greatest challenge in the use of passive sampling devices is how to quantify the gaseous concentration of pollutants, and many studies have focused on this area. The most common method is to calibrate its sampling rate in the laboratory, which involves establishing a test chamber with a known concentration, placing the passive sampler inside for a fixed duration, and determining the sampled amount. The ratio of the sampled amount to the sampling duration and concentration is then defined as the sampling rate.

Traditional active sampling methods, such as TENAX tube, PUF tube, and styrene-divinylbenzene copolymer XAD® adsorption tube, have been widely used to measure the total concentration of pollutants in the indoor environment. The active method, due to its accurate and controllable sampling flow, can obtain precise gas concentration. In addition, since the sampling pump can provide a high sampling flow, active sampling can quickly complete the sampling process, thereby determining instantaneous concentration. In recent years, the emerging active sampling device, known as the needle trap device (NTD), has gained prominence. The NTD features an internal diameter of merely 0.4 mm, with its specially designed stainless steel needle internally coated or packed with adsorbent. In the sampling process, when air passes through the micro-needle, the analytes are captured by the adsorbent. The needle is then directly inserted into the chromatography inlet, and the carrier gas flow is used to desorb the captured analytes. Compared to traditional active sampling devices, which often require additional instruments (such as Tenax tube needing thermal desorption unit) and extra analytical processes (e.g. extraction and concentration), NTD directly desorbs at the chromatography inlet, simplifying pretreatment, enhancing sensitivity, and reducing analysis costs. Due to its advantages of sensitivity and convenience, NTD has been widely applied in the sampling of volatile organic compounds in indoor environments and other environments. In recent years, needle trap devices have also been introduced into the field of semivolatile organic compound analysis, improving the sampling and analysis efficiency of semivolatile organic compounds.

(2) **Pollutant Concentration Levels**

Currently, the concentration levels of common pollutants (such as formaldehyde, benzene, etc.) indoors are as follows:

1. Volatile organic pollutants

   (a) Formaldehyde

   The main sources of formaldehyde in residential environments include: artificial boards such as plywood, particleboard, and fiberboard used for interior decoration; furniture made from these artificial boards; various wall materials such as paint, coatings, and wallpaper; other construction and decoration materials such as synthetic carpets and foam plastics. Additionally, high concentrations of formaldehyde are also produced during the combustion of tobacco and fuels.

   Currently, the applicable standards for formaldehyde in residential air are the *Standards for Indoor Air Quality* (GB/T 18883-2022) (1-hour average $\leq$0.08 mg/m^3) and the *Code for Indoor Environmental Pollution Control of Civil Building Engineering* (GB 50325-2020) (1-hour average $\leq$0.08 mg/m^3).

(b) Benzene series

The main benzene-series pollutants indoors include benzene, methylbenzene and xylene, which are primary components of adhesives, solvents, and additives in interior architectural finishing materials, artificial board furniture, sofas, etc. Additionally, indoor combustion can also generate certain benzene series compounds, such as from fuels, smoking, cooking, mosquito coils, and incense burning. The current *Standards for Indoor Air Quality* (GB/T 18883-2022) stipulates that the 1-hour average concentrations of benzene, toluene, and xylene indoors must be less than 0.11, 0.20 and 0.20 mg/m^3, respectively. Xu et al. (2007) selected 1241 residences in cities such as Beijing, Tianjin, Shanghai, Chongqing, and Changchun, where the decoration had been completed within 1 year. They measured the average indoor concentrations of xylene and ethylbenzene, which were the highest among all reported data, at 0.189 and 0.107 mg/m^3, respectively. However, some residences, especially newly renovated ones, still exhibited excessive levels. Yang et al. (2007) reported that the average concentrations of benzene in kitchen air and methylbenzene in bedroom air from 11 residential houses in Jiangxi Province were the highest among all data, measuring 0.134 and 0.645 mg/m^3.

(c) TVOC

TVOC is an indicator for assessing the overall pollution status of volatile organic compounds (VOCs) indoors. The sources of volatile organic compounds in residential indoor environments are extremely complex, including: organic solvents, building materials, decoration materials, etc. Currently, GB/T 18883-2022 stipulates that the 8-h average of TVOC must be below 0.6 mg/m^3.

The pollution status of TVOC in indoor air of residential buildings in China is relatively severe, especially in newly decorated homes. The concentration range of TVOC is 0.037–1.808 mg/m^3. Xu et al. (2007) selected 1241 residences in cities including Beijing, Tianjin, Shanghai, Chongqing, Changchun, and Shizuishan, Ningxia, with interior decoration completed within 1 year, to measure indoor TVOC concentrations. Among the nine cities surveyed, except for Pingliang, Gansu, and Zhuhai, Guangdong, the average concentrations in all other cities exceeded the standard. The highest concentration was recorded in Shanghai residences at 1.808 ± 2.338 mg/m^3, surpassing the standard by more than 3 times. The newly completed and decorated residential buildings generally have TVOC levels 2–3 times higher than the national standard, with the highest exceeding the standard by 20 to 30 times. Compared with other pollutants, TVOC concentrations decrease more rapidly over time after project completion. Wang et al. (2005) found that the TVOC concentration reached 1.17 mg/m^3 2 weeks after completion, but dropped to 0.64 mg/m^3 after 1 month, and was far below the national standard after 3 months. Zhang et al. (2007) also found that the TVOC concentration in indoor air reached the national standard approximately 5 months after decoration. Moreover, the more extensive the decoration, the longer it took for the TVOC concentration to meet the national standard.

2. Semivolatile organic compounds

Based on 2,762 indoor dust samples from 18 provinces/municipalities/autonomous regions in China, Li et al. (2024) found that the concentrations of DMP, DEP, DBP, DIBP, BBP, and DEHP are 0.023–2361 μg/g, with average concentrations of 6.22, 5.95, 220, 79.0, 2.57, and 657 μg/g, respectively. They also noted that the levels of DMP, DEP, and DBP in Chinese indoor dust were higher than those in other countries (e.g., the United States, Canada, France, etc.). In a systematic review, Liu et al. (2022a, b) summarized the concentrations of phthalate ester in various indoor environments. In residential settings, the average concentrations of DMP, DEP, DiBP, DnBP, BBzP, DEHP, and DnOP were 2.1, 8.1, 182.2, 150.1, 2.6, 711.85, and 3.0 μg/g, respectively. In office environments, the average concentrations of these compounds were 2.8, 2.5, 216.3, 96.3, 12.1, 763.6, and 8.0 μg/g. In school settings, the average concentrations were 3.9, 7.5, 45.0, 137.5, 502, 2673.8, and 10.1 μg/g. Liu et al. noted that indoor exposure to phthalate ester generally exhibits an increasing trend, and in the future, indoor phthalate ester concentration levels may continue to rise, warranting significant attention.

(3) **Residential Pollutants and Human Health**

Pollutants can enter the human body through three pathways: skin contact, oral ingestion, and dermal exposure, posing certain health risks. Numerous studies have proven that the impact of indoor air pollution on human health cannot be ignored.

Volatile organic compounds (VOCs) are widely present in building materials, furniture, and cleaning products. Exposure to VOCs may cause respiratory tract irritation and nervous system damage, with potential carcinogenic risks. Studies have shown that VOC exposure is significantly associated with aggravated asthma symptoms and allergic rhinitis. In addition, VOCs such as benzene and methylbenzene may also cause symptoms like headache, fatigue, and dizziness. Formaldehyde primarily originates from wood-based compressed products, thermal insulation materials, and cleaning agents. Formaldehyde exposure can cause eye and respiratory tract irritation, and may also increase the risk of nasopharyngeal cancer. Research indicates that formaldehyde exposure also has adverse effects on the nervous system, potentially leading to attention deficits and cognitive dysfunction.

Indoor particles primarily originate from combustion, smoking, and dust. Long-term inhalation of particles can trigger respiratory diseases, such as chronic obstructive pulmonary disease (COPD), and increase the risk of lung cancer (Pope and Dockery 2006). Studies have proven that exposure to PM2.5 can increase the risk of cardiovascular disease, such as myocardial infarction and stroke. As a susceptible population, children's exposure to particles may impact their lung function development and immune system health.

Radon is a colorless, odorless radioactive gas, primarily originating from the soil and rocks beneath buildings. Exposure to radon gas increases the incidence rate of lung cancer, and is the second leading risk factor for lung cancer after smoking, next only to cigarette smoking (Lecomte et al. 2014; World Health Organization (WHO) 2009).

In addition to volatile organic compounds (VOCs), building materials and decoration materials also release semivolatile organic compounds (SVOCs). These compounds have low vapor pressure, are easily adsorbed onto indoor surfaces and dust, and are difficult to remove effectively through ventilation. Long-term exposure to SVOCs may cause serious harm to human health. Phthalate esters (PAE) are a common type of SVOCs, widely used in plastic products. This substance exhibits an endocrine disrupting effect, which may adversely affect the health of human endocrine system and reproductive system. Studies indicate that children and pregnant women are high-risk groups, and prolonged exposure may lead to reproductive toxicity, neurodevelopmental disorders, and behavioral abnormalities. Epidemiological studies have found associations between phthalate ester exposure and decreased male fertility as well as conditions such as endometriosis in females. Polybrominated diphenyl ethers (PBDEs) are a typical type of semivolatile flame retardant widely used in furniture and electronic products. Animal and in vitro studies indicate that PBDEs may interfere with thyroid function and disrupt neurotransmitter balance, thereby affecting nervous system development and leading to adverse outcomes such as attention deficits and learning disabilities. In addition, epidemiological studies suggest that exposure to PBDEs may increase the risk of thyroid dysfunction. In addition, other types of SVOCs such as polycyclic aromatic hydrocarbons (PAHs), brominated flame retardants (BFRs), and polychlorinated biphenyls (PCBs) may also be present in the indoor environment. These substances readily adhere to dust and indoor surfaces, posing potential threats to human metabolic function, immune systems, and nervous systems.

The microbial communities in the indoor environment have a significant impact on health. A damp and poorly ventilated indoor environment is prone to mold growth. Inhaling these biological pollutants can trigger allergic rhinitis, asthma attacks, and respiratory tract infections. For immunocompromised individuals, exposure may also lead to severe pulmonary infections. Fabrics, carpets, and bedding are the primary habitats for dust mites. Feces and proteins of dust mite can trigger asthma, rhinitis, and skin allergic reactions. At the same time, dust mite allergy is also one of the common inducing factors of childhood asthma. Bacteria and viruses can spread through contaminated water sources or ventilation systems. Inhalation of or contact with these pathogenic microorganisms may lead to respiratory tract infections, gastrointestinal diseases, and skin disorders. For instance, the COVID-19 occurred in 2019. Multiple studies have provided substantial evidence for its transmission in indoor air, highlighting the critical importance of ventilating the indoor environment to reduce viral load.

8.5.3 *Indoor Air Purification for Residential Environment*

(1) **Current Status of Air Cleaner Usage in Residential Settings**

$PM_{2.5}$ is a global hazardous substance that has significant adverse effects on human health. Long-term exposure to $PM_{2.5}$ is closely associated with the incidence of lung cancer, mortality from cardiovascular disease, and respiratory diseases. According to statistics, exposure to $PM_{2.5}$ caused as many as 6.5 million premature deaths worldwide in 2019. According to the 2019 Global Burden of Disease Study, outdoor $PM_{2.5}$ pollution and indoor solid fuel pollution ranked as the 7th and 10th leading risk factors for global disease burden, respectively.

Although the Government of China has implemented stringent air pollution control measures in recent years, leading to a significant reduction in outdoor $PM_{2.5}$ concentrations, the levels remain substantially higher than the 5 μg/m^3 guideline recommended by the World Health Organization (WHO) in 2021. In 2021, the average annual $PM_{2.5}$ concentration in cities at and above prefectural level in China was 30 μg/m^3, indicating that outdoor $PM_{2.5}$ pollution still poses a significant threat to health. With the acceleration of population aging and urbanization, the health impacts of $PM_{2.5}$ pollution continue to rise. Furthermore, there are significant disparities in $PM_{2.5}$ concentrations among Chinese cities, leading to an uneven distribution of health burdens.

Indoor $PM_{2.5}$ primarily originates from two major sources: outdoor pollutants entering through ventilation or infiltration, and indoor activities (such as cooking, smoking) that generate pollutants. To mitigate the health risk associated with indoor $PM_{2.5}$ exposure, China issued its first indoor $PM_{2.5}$ standard in 2022, setting a 24-hour average concentration limit of 50 μg/m^3. However, this standard is only equivalent to the second-stage transitional target of the WHO, still far exceeding the health-impact-free level recommended by the WHO.

Since people spend approximately 90% of their time indoors, indoor air purification can rapidly reduce $PM_{2.5}$ exposure and effectively improve health, including benefits such as enhanced cardiopulmonary health and reduced short-term stress hormones.

From a national perspective, the ownership rate of air cleaners in Chinese households reached 5.0% in 2021, with significant regional variations. Among the regions, Beijing has the highest ownership rate, reaching 7.0%; while Guizhou has the lowest, at just 2.0%. This regional disparity may primarily be attributed to more severe air pollution or higher economic development levels, where local residents are more strongly encouraged to purchase and use air cleaners. Correlation analysis reveals a significant positive relation between the usage rate of air cleaners and both atmospheric $PM_{2.5}$ concentration and per capita disposable income. In comparison, the penetration rate of air cleaners in South Korea reaches 70%, while it is 42% in Europe, 27% in the United States, and 17% in Japan. Although air pollution is more severe in China, the penetration rate and actual usage level of air cleaners clearly lag behind.

In addition, the actual performance of the air cleaner is significantly influenced by user behavior. Researchers conducted field surveys showing that in households equipped with air cleaners, 80% did not activate the devices, while only 20% used them intermittently, with an average daily usage duration of merely 1–4 h. Zhang et al. (2024) investigated the primary motivations for activating household air cleaners, revealing that 90% of households activated the devices in response to outdoor air pollution conditions, while 70% to 80% reported usage linked to indoor cigarette smoking. In contrast, although Chinese cooking generates a significant amount of $PM_{2.5}$, the proportion of households that consider cooking as the primary motivation is relatively low, accounting for sixty percent. In addition, some participants mentioned other motivations, including allergies and other health issues, as well as environmental elements such as temperature, humidity, odors, overcrowding, and poor ventilation. Among them, "allergy" was the most frequently mentioned motivation. In addition, there were mentions of using an air cleaner due to other health concerns, such as a family member falling ill, flu season or COVID-19 pandemic. Although the air cleaner does not have the function of regulating temperature and humidity, a small number of people still consider excessively high or low temperature and humidity as motivations to activate the air cleaner. This fact also reflects some misunderstandings among residents regarding the functions of an air cleaner.

In terms of the seasonal selection for air cleaner usage, the surveyed households showed significant differences, reflecting the varying motivations influenced by seasonal changes. The survey shows that in winter, due to prominent atmospheric $PM_{2.5}$ pollution, 60% of households most frequently activate their air cleaners. During spring and autumn, allergens such as willow catkins and pollen are the primary factors. Nearly 60% of households use an air cleaner in spring, while close to 50% opt to activate it in autumn. In comparison, fewer households use air cleaners in summer.

In addition, air cleaner users often opt to operate it in "low speed" or "automatic mode," while the "high-speed mode" is underutilized due to concerns over noise and energy consumption. Comparative studies have revealed that even in areas with severe $PM_{2.5}$ pollution, the utilization of air cleaners remains insufficient. Especially between large northern cities and middle-income regions, this disparity is even more pronounced. Field investigations revealed that with the air cleaner in operation, the indoor-to-outdoor ratio decreased from 0.36–1.31 to 0.23–0.77, demonstrating that under proper usage conditions, the air cleaner exhibits significant effectiveness in reducing $PM_{2.5}$ concentrations.

Some air cleaner not only remove indoor $PM_{2.5}$, but also eliminate chemical pollutants indoors, such as formaldehyde, methylbenzene, and other VOCs. Currently, depending on their purification principles, the more mature or popular chemical substance purification technologies can be categorized into adsorption filtration type, catalytic reaction type, ionization type, and plant purification type. Various methods exhibit significant differences in technical performance regarding the removal of indoor chemical pollutants. Adsorption filtration technology is currently the most mature and widely used commercial technology, demonstrating high removal efficiency for medium- and low-volatility compounds. However, its effectiveness is limited for highly volatile compounds such as formaldehyde and acetaldehyde, and the materials have a short lifespan requiring regular replacement. Ultraviolet photocatalytic oxidation (UV-PCO) technology has shown performance comparable to adsorption filtration in experiments, capable of decomposing VOCs into carbon dioxide and water. Nevertheless, challenges include the generation of by-products, performance limitations due to environmental conditions, and the technology's immaturity. Ozonation and air ionization technologies exhibit limited effectiveness in VOC removal and carry the risk of excessive ozone levels. Plant-based air purification shows some efficacy against specific compounds such as formaldehyde and acetaldehyde, but its overall performance remains weak. Overall, existing technologies each have their own advantages and disadvantages in terms of efficiency, stability, and safety. The appropriate purification technology should be selected based on the specific characteristics of pollutants and practical application requirements.

Overall, the penetration rate and usage level of air cleaners in China are still in the early stages, characterized by low penetration rate, insufficient utilization, and suboptimal performance. To enhance the actual effectiveness of air cleaners, it is essential to focus on family education and awareness-raising, promote correct practices of air cleaner usage, extend operation duration, and select high-efficiency modes. Besides, it is necessary to advance technological progress while reducing costs, and develop highly efficient equipment with low noise and energy consumption, to minimize operation costs. Policies can also provide corresponding support and promotion, strengthen policy support in highly polluted areas, offer subsidy strategies, and promote wider application of air cleaners.

(2) **Operation Performance of Residential Air Cleaners**

Using an air cleaner is one of the important ways to reduce indoor pollutant concentrations. Regarding traditional particle pollution, numerous studies have demonstrated that air cleaners can effectively reduce indoor $PM_{2.5}$ concentrations and are efficient in removing $PM_{2.5}$ generated from specific emission sources. During daily cooking, the air cleaner can work in synergy with the range hood to effectively remove particles generated during the cooking process: When both the air cleaner and range hood are turned on during cooking and for a period afterward, compared to using only the range hood, the amount concentration of $PM_{0.3-2.5}$ and $PM_{2.5-10}$ can be reduced by an additional 84% and 88%, respectively. Studies also indicate that the cleaning efficiency of filtration-based air cleaners for $PM_{2.5}$ produced by smoking, incense burning, and cooking is in the range of 61.79%–86.41%, 43.48%–83.98%, and 51.10%–61.62%, respectively. In research where wood burning was the primary pollution source, using an air cleaner reduced indoor $PM_{2.5}$ concentrations by 33%–60%. Similarly, studies focusing on outdoor traffic emissions as the main pollution source showed that air purification could lower indoor $PM_{2.5}$ concentrations by 36%–60% (Zhu et al. 2021).

In the purification of gaseous chemical pollutants, a study investigated the indoor VOC conditions in 20 Chinese households, and indicated that using an air cleaner can improve indoor air quality, with a reduction in TVOC concentration. Testing on commercial air cleaner equipment revealed that under indoor conditions with VOC concentrations ranging from 100 ppb to over 1 ppm, the use of an air cleaner could reduce VOC levels by approximately 40%. Additionally, for sudden spikes in VOC emissions, the air cleaner could restore indoor VOC concentrations to initial levels within 2 h. Another study examined 13 different brands and models of air cleaners available in the Chinese market and found that those employing activated carbon adsorption technology or hybrid technologies such as plasma and photocatalysis demonstrated significant effectiveness in formaldehyde removal, achieving a maximum cleaning efficiency of 90.4% (Wang et al. 2015).

Meanwhile, with the continuous improvement of China's atmospheric environment and the outbreak of COVID-19, public concern over environmental quality has shifted from particle ($PM_{2.5}$) to gaseous pollutants (TVOC, formaldehyde, etc.) and other contaminants such as microorganisms. This has raised newer and higher demands for residential indoor environmental quality and the use of air cleaners.

Against this backdrop, China revised the standard *Air Cleaner* (GB/T 18801-2015) in 2022, which was officially implemented in May 2023. The revised standard *Air Cleaner* (GB/T 18801-2022) has made the following improvements in normative aspects: enhanced technical requirements for the removal of particles and gaseous pollutants, established the correlation between CADR and CCM values for particles and gaseous pollutants. Here, CADR (clean air delivery rate) represents the volume of clean air an air cleaner can purify per unit time, while CCM (cumulative clean mass) indicates the cumulative purification capacity of the air cleaner for target pollutants. The standard also refined the evaluation method for single-component high-concentration gaseous pollutant removal from the 2015 version, introducing mixed loading test methods and dynamic equilibrium test methods for gaseous pollutants. Additionally, it incorporated performance evaluation methods for virus removal. In terms of informative content, the standard added evaluation methods for odor removal/simulated secondary odor, allergen removal methods, and dynamic equilibrium test methods for gaseous pollutant removal (using ozone as an example) (Zhao et al. 2022). The revision and implementation of the standard meet the current market conditions and guide the technological development of the industry.

In addition, experts have conducted in-depth research on the economic benefits of using air cleaners: air cleaners can significantly reduce indoor pollutant concentrations, thereby effectively lowering people's exposure to pollutants and subsequently reducing the associated disease burden. By applying the cost-benefit evaluation model of air cleaner usage to assess the current status of cleaner utilization in China, corresponding optimization strategies can be proposed: To enhance benefits, residents should increase both operational duration and power settings during practical use while keeping windows and doors closed when operating the air cleaner. For cost reduction, it is essential to reasonably determine the CADR of air cleaners based on indoor purification targets and the degree of air pollution, as well as optimize filter replacement frequency according to operational conditions, thereby improving the net benefit (Zhang 2024).

8.6 Natural Ventilation Design Optimization for Urban Residential Buildings

When the wind blows towards a building, due to the obstruction of the building, the air flow bypasses the roof, sides, and back of the building, and an air flow eddy zone will be generated on the leeward side of the building, as shown in Fig. 8.21. The length of the eddy current zone increases with the increase of the height and the windward width of the building, and decreases with the increase of the depth of the building. The greater the building height, the smaller the depth and the larger the length, the larger the negative pressure area on the back, which is beneficial to its own ventilation. However, for the buildings behind it, the ventilation is very unfavorable.

Good natural ventilation is the key for a residential building to achieve a good indoor environment with low energy consumption. Achieving good natural ventilation depends on building planning and layout as well as single building design. Different planning layouts of buildings within the residential area, as well as the varying shapes and internal divisions of single buildings, can lead to significantly different effects of natural ventilation.

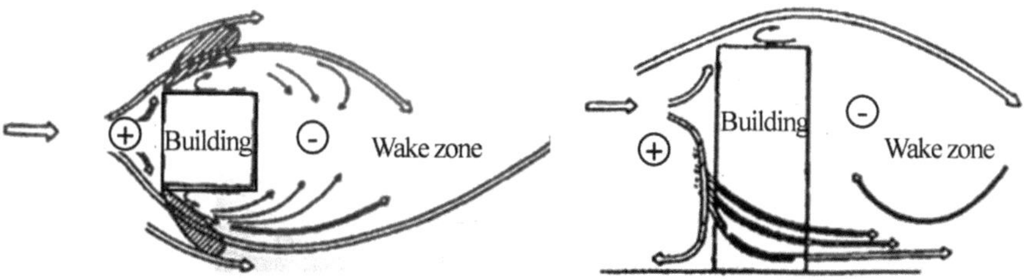

Fig. 8.21 Air flow distribution around the building

8.6.1 Selection of Residential Building Groups Layout

8.6.1.1 Form of Residential Building Groups Layout

The planning layout of a residential building is influenced by various factors such as the shape of the plot, daylighting, and design concepts. The residential building layouts in China primarily include three fundamental patterns: "row-style," "cluster-style," and "enclosure-style". Current residential area building layouts mainly combine these three basic patterns.

"Row-style residential area," also known as "plate-type building residential area," refers to a residential zone primarily composed of strip-shaped slab-type buildings as the main housing form. The width of a slab-type building is approximately 18–20 m, while the length varies significantly, generally in the range of 40–120 m. Traditional slab-type residential buildings are typically 6-storey structures, but in recent years, densely populated cities have seen an increase in high-rise slab-type buildings exceeding 12 storeys. The building layout follows a neat row-column arrangement based on the orientation of the strip-shaped buildings, as shown in Fig. 8.22. This layout of residential buildings offers superior sunlight exposure and ventilation conditions, facilitating pipeline installation and industrialized construction, making it the primary form of early residential areas in China. Currently, this type of layout still accounts for a high proportion of residential compounds in China's urban areas.

"Cluster-style residential area", also known as a "tower residential area," is a type of residential community where the primary housing form consists of point-style towers with length-to-width ratios close to 1:1. The length and width of the tower varies within the range of 30–40 m, while the height typically varies between 12–24 floors. The towers are mostly arranged in a staggered layout, as shown in Fig. 8.23, to facilitate daylighting for different buildings and ventilation in the residential area. A tower residential area has become a popular architectural layout due to its advantages such as small building footprint, high degree of spatial division freedom, and the feasibility of implementing greenery in open spaces.

Fig. 8.22 Row-style residential area

Fig. 8.23 **Cluster-style residential area**

Fig. 8.24 **Enclosure-style residential area**

Enclosure-style residential areas are closed or semi-closed residential communities formed through a perimeter-style layout structure, as shown in Fig. 8.24. Enclosed buildings are mostly slab-type buildings, right-angled buildings, or U-shaped buildings. In this type of residential area, the enclosed central zone is the core activity space, and it is connected with the outside through several passageways. Due to the poor ventilation of enclosure-style layouts and the generally suboptimal lighting conditions in buildings formed by such arrangements, this type of residential area currently accounts for a relatively small proportion in design practices.

8.6.1.2 Impact of Group Layout on Building Ventilation

From the perspective of natural ventilation in buildings, structures within residential areas may create two effects on incoming airflow: obstruction and guidance.

For row-style residential areas, when the orientation of strip-shaped buildings is perpendicular to the incoming wind direction, the buildings obstruct airflow. The airflow is blocked by the buildings and detours around them from both sides and above, creating corner effect on either side of the buildings where velocity increases. A three-dimensional vortex forms behind the buildings, hindering air circulation, as shown in Fig. 8.25. In this case, both the height and width of the strip-shaped buildings influence the airflow direction behind them. When the orientation of strip-shaped buildings aligns with the incoming wind direction, the buildings guide the airflow. Air can smoothly pass through the spaces between the buildings, carrying away heat and pollutants, as illustrated in Fig. 8.26. When the incoming wind direction forms an angle of less than 90° with the building orientation, the airflow exhibits a mixed pattern of guidance and obstruction, as depicted in Fig. 8.27. Therefore, the row-style layout generally has a greater guiding effect on ventilation than an obstructive one, making this type of layout suitable for all climate zones in China.

For cluster-style residential areas, the impact of buildings on airflow primarily manifests as a mixed effect of obstruction and guidance. Due to the relatively narrow windward width of cluster-style buildings, the alteration of airflow caused by the buildings' obstruction is minimal, resulting in weaker corner effects. The air continues to flow along the corridor between the two rows of buildings, creating a guiding effect. Thus, cluster-style buildings exhibit a mixed effect of both obstructing and guiding airflow, as shown in Fig. 8.28.

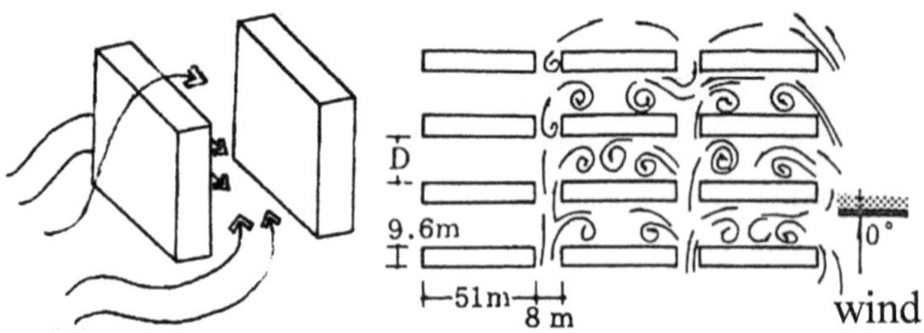

Fig. 8.25 Obstructing effect of row-style layout on airflow

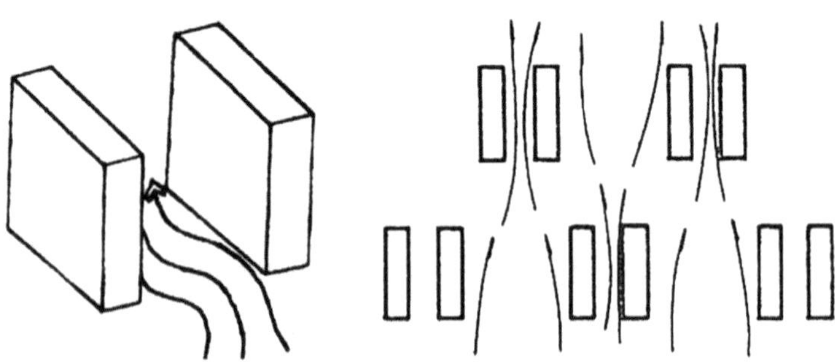

Fig. 8.26 Guiding effect of row-style layout on airflow

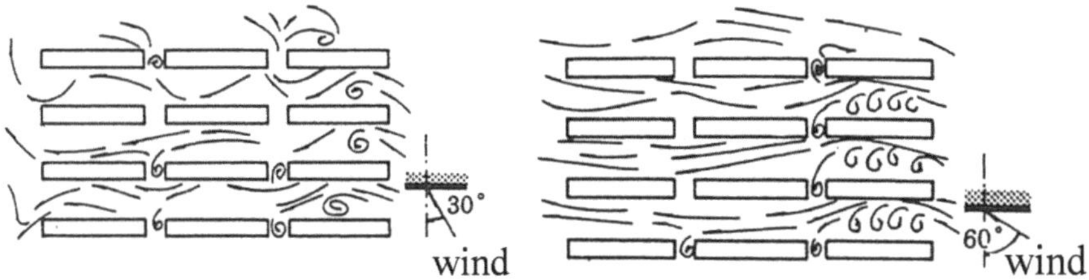

Fig. 8.27 Obstructing effect and guiding effect of row-style layout on airflow

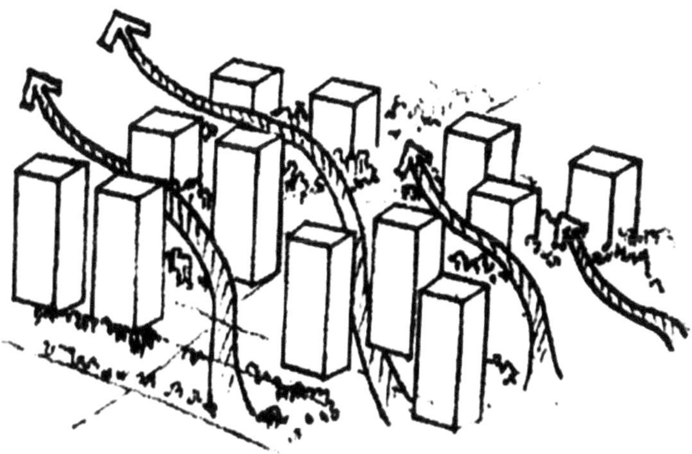

Fig. 8.28 Obstructing effect and guiding effect of cluster-style layout on airflow

The staggered arrangement of buildings can indirectly increase the spacing between them. Meanwhile, the buildings within the cluster should not be entirely aligned with the prevailing wind direction in summer, so as to facilitate natural ventilation. When there is a large green space within the building area, it can be arranged in the center and perpendicular to the prevailing wind direction, dividing the building area into two or more smaller zones and increasing the spacing between them to mitigate the attenuation of airflow velocity in the downwind zone. In the regions with relatively cold winters, it is necessary to comprehensively consider the comfort in both winter and summer seasons. This means ensuring good ventilation in summer while also blocking cold winds from invading the building cluster in winter, as shown in Fig. 8.29. In cluster-style layout, residential buildings can be flexibly arranged based on needs, terrain, topography, and climate conditions. Proper planning can maximize the buildings' role in guiding or obstructing airflow, thereby achieving the goal of enhancing ventilation or wind protection for the building complex. As a result, this type of layout is suitable for all climate zones in China.

For enclosure-style residential areas, buildings obstruct incoming airflow from all directions, as shown in Fig. 8.30. The effect is similar to the obstruction caused

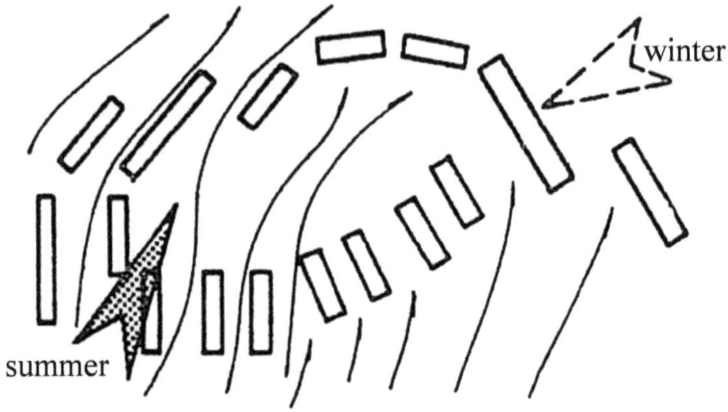

Fig. 8.29 Cluster-style layout considering summer ventilation and winter wind protection

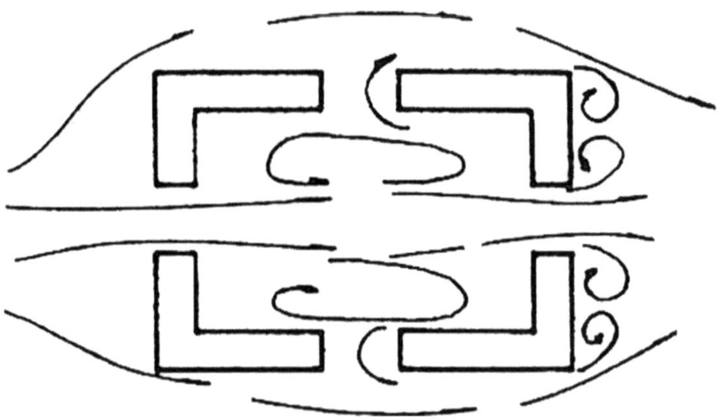

Fig. 8.30 Obstructing effect of enclosure-style layout on airflow

by row-style buildings, resulting in poor air circulation within the residential area. Particularly, buildings at the center of the enclosure have limited pathways for interaction with the external environment, leading to extremely poor natural ventilation. Among all layout types, this layout is the least favorable for natural ventilation during summer and transitional seasons. It is only suitable for cold northern regions with wind protection needs in winter but is not suitable for hot and humid climate zones.

In the relation between building wind environment and humans, ventilation and wind protection significantly impact human comfort. For residential buildings, effective arrangement of natural ventilation is required during summer to meet human comfort needs without using air conditioning, thereby reducing energy consumption; in winter, appropriate wind protection measures should be taken to minimize building energy consumption while ensuring adequate air exchange.

In regions with hot and humid climates, although natural ventilation introduces outdoor heat, the effects of perspiration and dehumidification it brings are indispensable means for humans to combat extreme heat. All-day natural ventilation is crucial for meeting human comfort needs. In hot and dry climate zones, the prerequisite for creating a comfortable indoor environment during summer is to reduce ventilation and solar radiation during high-temperature periods, avoiding additional heat from airflow. Meanwhile, ample natural ventilation should be introduced at night when outdoor air temperatures are lower, utilizing the cool nighttime air to achieve building cooling. Therefore, enhancing natural ventilation is a crucial requirement for residential buildings in high-temperature regions. Residential compounds should be built in row-style or cluster-style layouts to facilitate natural ventilation in buildings, and should avoid enclosed or semi-enclosed residential area layouts. Buildings within a cluster should not all face the prevailing wind direction in summer. A staggered arrangement with a regular "high-low alternating" pattern is recommended, which effectively increases the building spacing, mitigates wind speed attenuation, and facilitates air circulation throughout the cluster. When there is a large green space within the residential area, it can be arranged in the center and perpendicular to the prevailing wind direction, dividing the residential area into two or more smaller zones. This increases the spacing between the zones and helps mitigate the attenuation of air velocity in the downwind area.

In the regions with cold winters or strong winter winds, it is essential to balance comfort in both winter and summer. This means ensuring adequate ventilation in summer while blocking cold winds from penetrating the building complex in winter, thereby preventing excessive heat loss of both occupants and structures due to ventilation. Residential areas can be arranged in semi-enclosed layouts, cluster-style layouts, or row-style layouts, with slab-type buildings, right-angle-shaped buildings, or U-shaped buildings arranged on the windward side of the prevailing wind direction in winter. Through enclosed or semi-enclosed layout structures, closed or semi-closed residential areas are formed, with the central enclosed area serving as the core activity space of the residential area. Buildings are used to block cold winds, reducing wind speed within the residential area. The opening direction of the enclosed or semi-enclosed layout avoids the direction of cold currents, and the building orientation also stays clear of the cold current direction. Additionally, buildings can be arranged in a way that positions the high-rise buildings to face away from the prevailing winter wind direction, thereby reducing the impact of cold winds on low-rise buildings, mid-rise structures, and courtyards. Additionally, measures such as installing windbreaks, barriers or windproof belts can be implemented to block cold winds from invading the residential compound. For instance, the wind-resistant vegetation growing in the primary windward direction during winter can help shield the residential area while also reducing heat loss of buildings. Common windbreak tree species are listed in Table 8.7.

For enclosed building layouts, the wind exit is small, and the flow velocity weakens, creating large vortices within the courtyard. This results in extensive low-pressure zones around the buildings, making it the least favorable layout for natural ventilation during summer and transitional seasons. It is only suitable for regions

Table 8.7 Common windbreak tree species

Wind-resistant performance	Tree species
Strongest	Juniper, ginkgo, quince, and willow
Strong	Oriental arborvitae, peach laurel, dracaena marginata, trachycarpus fortunei, fig tree, elm, ligustrum lucidum, hibiscus syriacus, zelkova schneideriana, albizzia, bamboo, Chinese scholar tree, ternstroemia gymnanthera, myrica rubra, loquat tree, ficus microcarpa, and tulip tree
Slightly strong	Dragon juniper, black pine, oleander, coral tree, pittosporum, walnut tree, cherry tree, and bodhi tree

with cold, windy winters and minimal demand for summer ventilation, cooling, and dehumidification.

8.6.2 Selecting the Form of Single Residential Building

8.6.2.1 Form of Single Residential Building

According to the type of building structure/planar shape characteristics, residential buildings can be divided into two main categories: slab-type buildings and towers. A slab-type building refers to a residential structure whose length is significantly greater than its width. A typical slab-type building features a two-units-per-floor design with a single staircase. Low-rise residential buildings are predominantly designed as strip-shaped slab-type buildings, while slab-type buildings also account for the highest proportion among high-rise residential buildings, as shown in Fig. 8.31a. A tower refers to a high-rise residential building that resembles a tower in appearance, with roughly equal length and width. It is designed with multiple housing units centered around shared staircases and elevators. Typically, each floor of a high-rise tower accommodates more than six households. Common forms of towers include V-shaped, cross-shaped, square, and butterfly-shaped forms, as shown in Fig. 8.31.

8.6.2.2 Relation Between the Opening Position and the Airflow Path

For Buildings with natural ventilation, when wind blows towards the structure, the direction of the airflow jet entering through the inlet is determined by the surrounding lateral pressure outside the entrance. When the surrounding pressures are equal, the jet direction remains unchanged and continues along its original path, as illustrated in Figs. 8.32 and 8.33. When the air inlet is offset to one side, upwards, or downwards, the pressure on either side or above and below becomes unequal. As the airflow passes through the inlet, the jet stream deflects toward the side with lower pressure, as shown in Figs. 8.34 and 8.35.

Fig. 8.31 Common forms of residential buildings. (**a**) Plate-type high-rise residence. (**b**) V-shaped tower residence. (**c**) Cross-shaped tower residence. (**d**) Square tower residence. (**e**) Butterfly-shaped tower residence

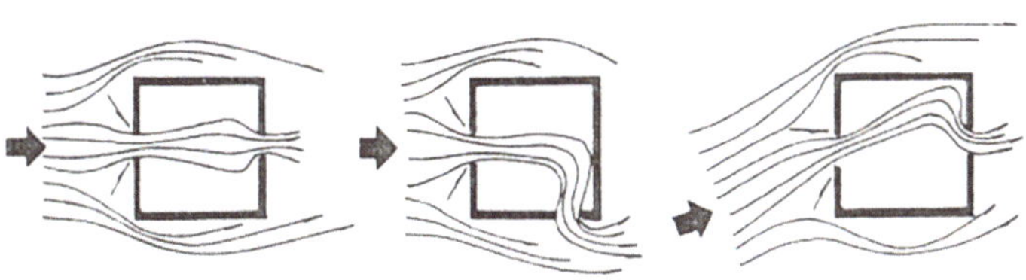

Fig. 8.32 The air inlet is centered on the plane, with equal pressure on both sides, and the jet direction of the airflow remains unchanged

For buildings with only an air inlet and no air outlet in natural ventilation, or when the positions of the air inlet and air outlet are parallel to the outdoor airflow direction, the disturbance to indoor air is minimal, as shown in Figs. 8.36 and 8.37.

When the openings are on both sides, ventilation is unfavorable. However, by rationally organizing the positive and negative pressure zones, such as arranging them in an alternating pattern, where one opening is located in the positive pressure zone and the other in the negative pressure zone, it is possible to allow airflow to pass through the interior, forming a natural ventilation path. See Fig. 8.38.

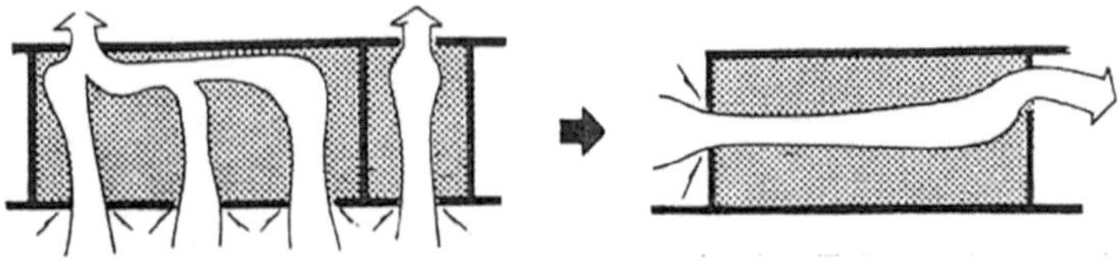

Fig. 8.33 Windows uniformly arranged, centered in cross-section, with unchanged jet direction

Fig. 8.34 The air inlet is offset to one side, causing the jet to deflect toward that side; uneven window spacing results in the jet direction deflecting toward denser side

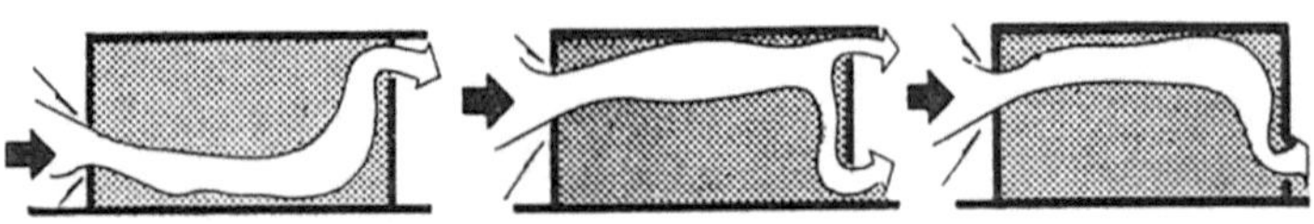

Fig. 8.35 When the air inlet is positioned lower or higher on the profile, the direction of the air-flow jet also deflects downward or upward accordingly

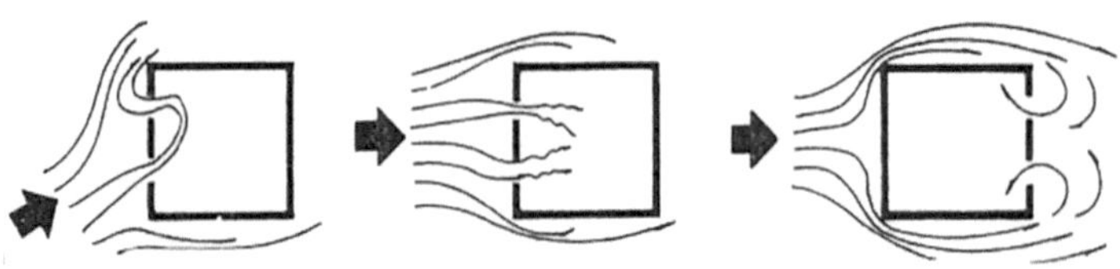

Fig. 8.36 The opening is only on the windward or leeward side, causing minimal disturbance to indoor air without enabling ventilation

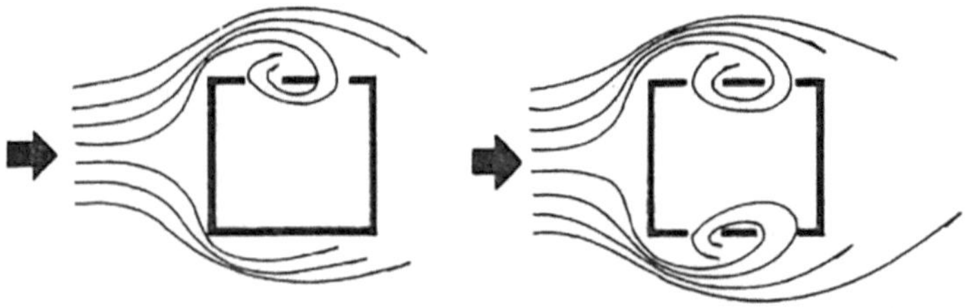

Fig. 8.37 The opening is on the side, with equal pressure on both sides, causing only a slight disturbance to the indoor air

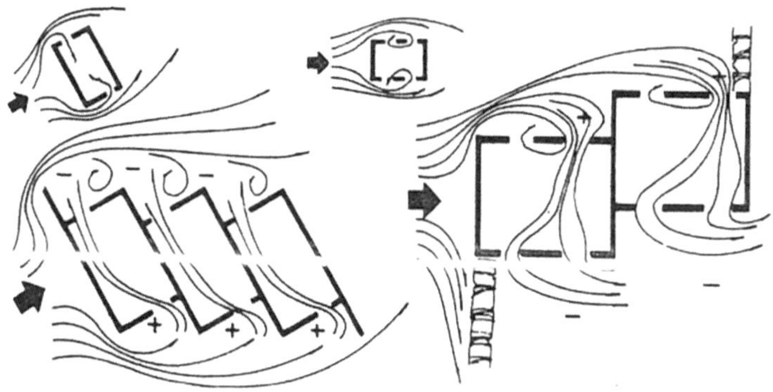

Fig. 8.38 One opening is in the positive pressure zone, while the other is in the negative pressure zone, with airflow passing through the interior

8.6.2.3 Impact of Single Building Form on Building Ventilation

Among the residential building forms, slab-type buildings and towers each have their own advantages and disadvantages. The large facade width of the slab-type building ensures excellent lighting effects, while the north-south cross-ventilation layout guarantees optimal natural ventilation. The residential density of the tower is higher than that of slab-type buildings, with most unit layouts featuring single-sided windows, resulting in poorer lighting and ventilation effects. However, compared to slab-type buildings, towers offer advantages such as lower construction costs, excellent seismic performance, and greater flexibility in space division, making them a common residential type in densely populated major cities.

The layout of a building directly influences its internal structure, spatial division, and the placement of ventilation openings, thereby having a significant impact on the effectiveness of natural ventilation. This study selects four common residential building layouts in China, namely the four-staircase slab-type building, V-shaped tower, cross-shaped tower, and square tower. Taking the climatic conditions of Beijing as an example, the natural ventilation performance of these different residential building layouts is analyzed. The selected residential building layouts are shown in Figs. 8.39, 8.40, 8.41 and 8.42, while the number of units per typical floor and floor area are presented in Table 8.8.

Figure 8.43 presents the average air change rate of natural ventilation for the isolated buildings of the aforementioned four layouts under a meteorological wind speed of 1 m/s (scale 1): slab-type residential building, V-shaped tower, cross-shaped tower, and square tower. It can be seen that under the same climatic conditions and natural ventilation strategy, the air change rate of natural ventilation varies significantly among residential buildings with different layouts. In slab-type residential buildings, the alignment of doors and windows creates short, straight, and smooth air passages, reducing windward surfaces and effectively realizing cross ventilation. This ensures unimpeded airflow into indoor spaces, guaranteeing smooth ventilation during summer and transitional seasons. Residential tower

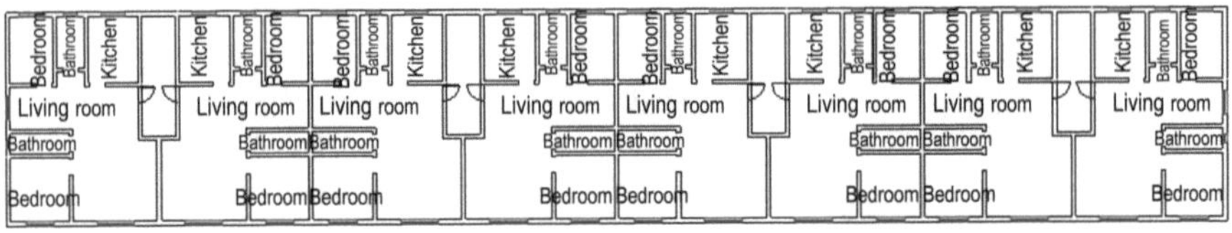

Fig. 8.39 Layout of four-staircase slab-type residence

Fig. 8.40 Layout of V-shaped tower residence

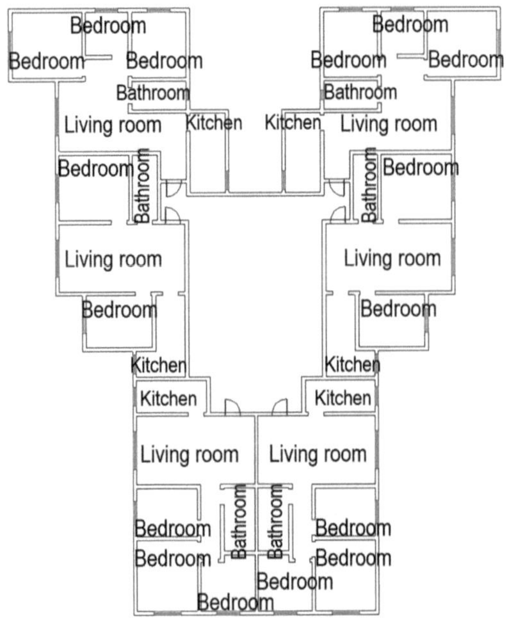

buildings, mostly relying on one-side ventilation, cannot achieve cross ventilation, resulting in significantly lower ventilation rates compared to slab-type residential buildings with north-south cross-ventilation layout. The natural ventilation effect of slab-type residential buildings is superior to all tower residences, while the layouts of all forms of residential tower buildings are unfavorable for natural ventilation during summer and transitional seasons.

8.6.2.4 Impact of Building Density on Building Ventilation

The layout of building clusters influences the wind environment around the buildings and alters the wind pressure distribution on the building surfaces, thereby affecting the effectiveness of natural ventilation. All other conditions being equal, the more densely buildings are arranged and the smaller the spacing between front and rear buildings, the less conducive it is to natural ventilation, resulting in a lower

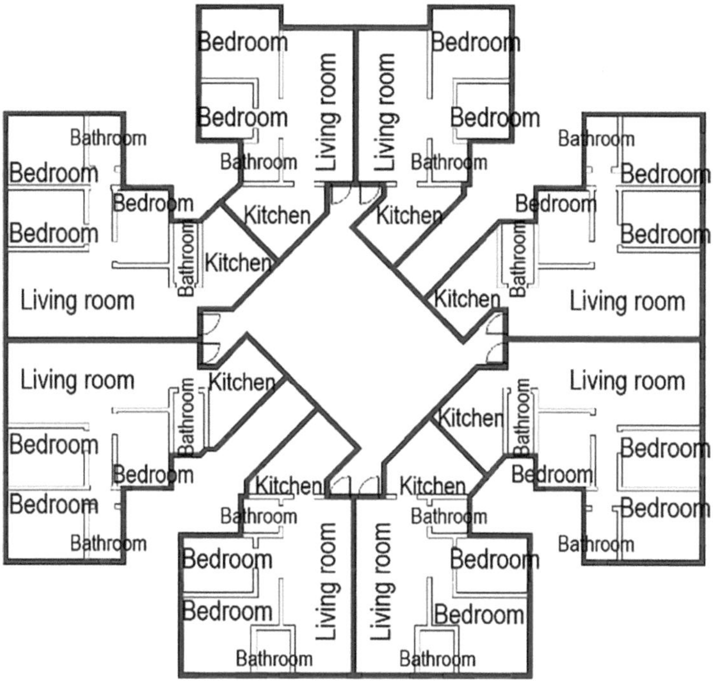

Fig. 8.41 Layout of cross-shaped tower residence

natural ventilation rate. The density of the surrounding layout of buildings can be characterized by the building density PAD (as shown in Fig. 8.44). According to the *Code of Urban Residential Areas Planning & Design*, the building density of high-rise residential buildings shall not exceed 20%.

Figure 8.45 presents the average natural ventilation air change rate of slab-type residential buildings (with layouts shown in Fig. 8.46) under meteorological wind speeds of 1 m/s (scale 1), comparing isolated buildings with four different building density scenarios: 5%, 10%, 15%, and 20%. Compared to the ideally ventilated isolated building, as can be observed, when the building density increases to 20%, which represents the densest high-rise slab-type residential buildings, the natural ventilation air change rate decreases by 36%.

Figure 8.46 shows two residential compounds in Beijing with different building densities. Among them, (a) depicts a compound with relatively sparse building layout, while (b) illustrates a compound with higher building density. Under the same natural ventilation strategy, the annual average natural ventilation rate of residences in Compound B will be 21% less than that in Compound A, meaning the fresh air rate in Compound B is only 79% of that in Compound A.

For residential towers, most unit types have windows on only one side, which cannot achieve cross ventilation, resulting in poorer ventilation effectiveness compared to slab-type residential buildings. Figure 8.47 presents the average natural

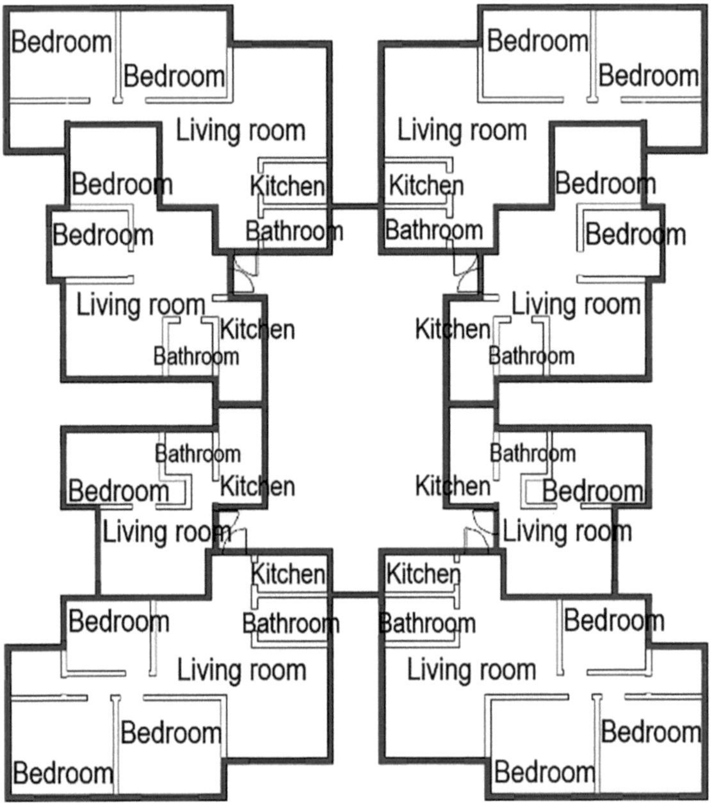

Fig. 8.42 Layout of square tower residence

Table 8.8 Statistical table of basic information on selected residential building forms

Building type	Four-staircase plate-type building	V-shaped tower	Cross-shaped tower	Square tower
Household per floor (households)	8	6	8	8
Floor area per floor (m^2)	1008	776	818	769
Number of ventilation windows per floor (units)	34	22	28	24
Number of ventilation doors per floor (units)	24	16	20	16
Number of floors (floor)	18	18	18	18

ventilation air change rate of V-shaped tower residences (with layout as shown in Fig. 8.20) under a meteorological wind speed of 1 m/s (scale 1), comparing isolated buildings with four different building densities: 5%, 10%, 15%, and 20%. It can be seen that compared to the ideal isolated building for ventilation, when the building density increases to 20%, which represents the most densely arranged V-shaped tower residence, the natural ventilation air change rate decreases by 28%.

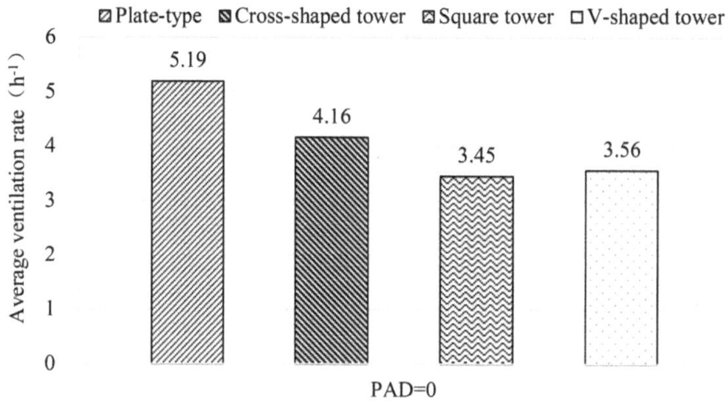

Fig. 8.43 Comparison of natural ventilation rates in isolated residential buildings with different layouts under identical climatic conditions and natural ventilation strategies

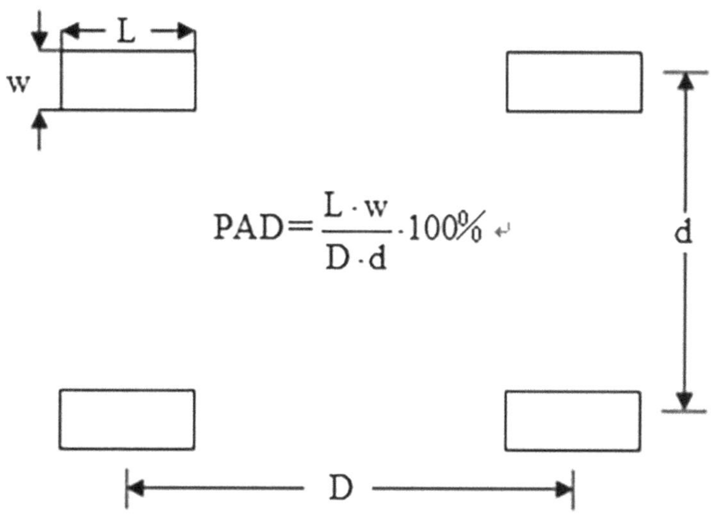

Fig. 8.44 Definition of building density PAD

Figure 8.48 presents the average natural ventilation air change rate of cross-shaped tower residences (with layout as shown in Fig. 8.22) under a meteorological wind speed of 1 m/s (scale 1), comparing isolated buildings with four different building densities: 5%, 10%, 15%, and 20%. It can be seen that compared to the ideal isolated building for ventilation, when the building density increases to 20%, which represents the most densely arranged cross-shaped tower residence, the natural ventilation air change rate decreases by 17%.

Figure 8.49 presents the average natural ventilation air change rate of square tower residences (with layout as shown in Fig. 8.23) under a meteorological wind speed of 1 m/s (scale 1), comparing isolated buildings with four different building

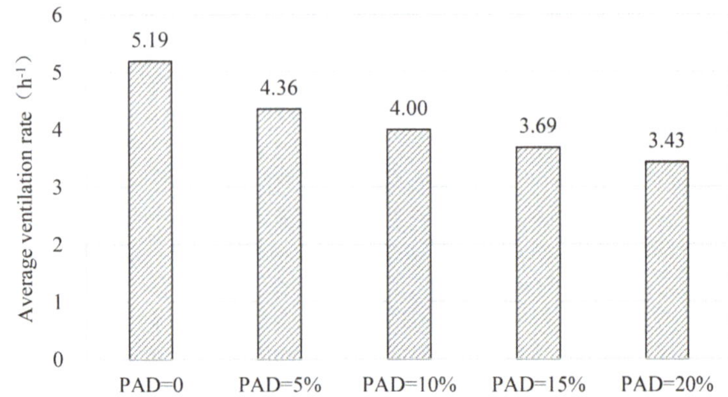

Fig. 8.45 Natural ventilation air exchange volume of four-slab residential buildings under different building density

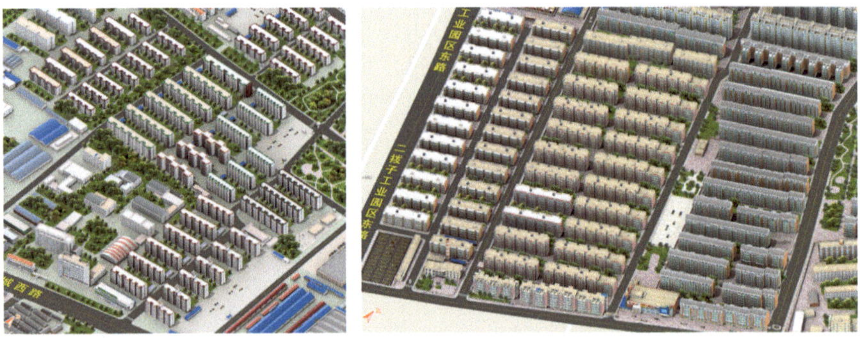

(a) Residential compound with a building density of 5% **(b) Residential compound with a building density of 20%**

Fig. 8.46 Two residential compounds with different building densities in Beijing. (**a**) Residential compound with a building density of 5%. (**b**) Residential compound with a building density of 20%

densities: 5%, 10%, 15%, and 20%. It can be seen that compared to the ideal isolated building for ventilation, when the building density increases to 20%, which represents the most densely arranged square tower residence, the natural ventilation air change rate decreases by 22%.

Figure 8.50 shows a comparison of the average natural ventilation air change rates for the aforementioned slab-type residential buildings, V-shaped towers, cross-shaped towers, and square towers under different building densities at a meteorological wind speed of 1 m/s (scale 1). It can be observed that under the same climatic conditions and natural ventilation strategy, the air change rate of natural ventilation varies significantly among residential buildings with different layouts. The layouts of all forms of residential towers are unfavorable for natural ventilation during

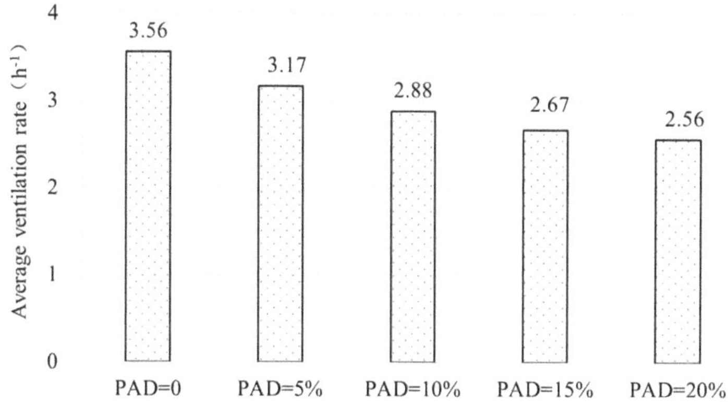

Fig. 8.47 Natural ventilation air exchange volume of V-shaped tower residence under different building densities

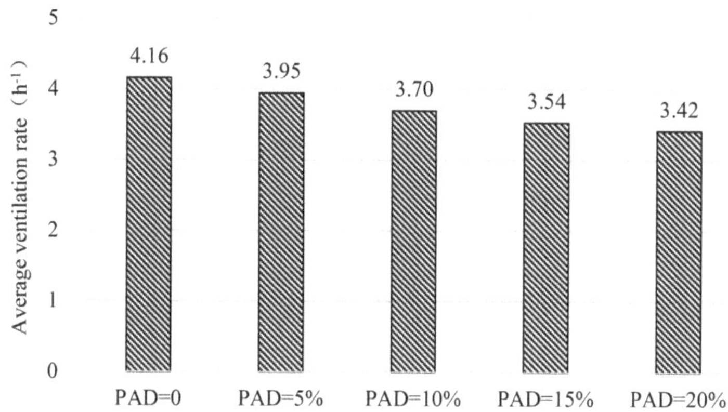

Fig. 8.48 Natural ventilation air exchange volume of cross-shaped tower residence under different building densities

summer and transitional seasons. Even in sparsely arranged residential towers, the effectiveness of natural ventilation may not necessarily surpass that of more densely arranged slab-type residential buildings.

In summary, to enhance the effectiveness of natural ventilation in residential buildings, slab-type buildings are the preferred choice for facilitating natural ventilation during summer and transitional seasons. However, when the outdoor temperature drops in winter, measures such as reducing the opening of doors and windows or sealing open halls with doors and windows can be implemented to block the infiltration and circulation of cold air, thereby improving the building's adaptability to outdoor climatic conditions.

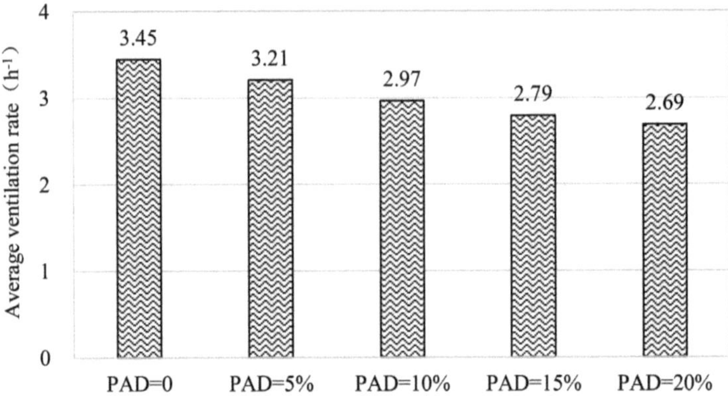

Fig. 8.49 Natural ventilation air exchange volume of square tower residence under different building densities

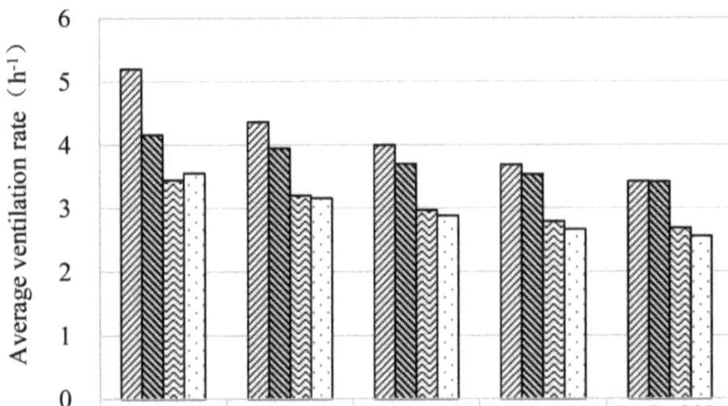

	PAD=0	PAD=5%	PAD=10%	PAD=15%	PAD=20%
Plate-type	5.19	4.36	4.00	3.69	3.43
Cross-shaped tower	4.16	3.95	3.70	3.54	3.42
Square tower	3.45	3.21	2.97	2.79	2.69
V-shaped tower	3.56	3.17	2.88	2.67	2.56

Fig. 8.50 Comparison of natural ventilation rates between slab-type buildings and V-shaped towers under different building densities

8.6.3 Selection of Ventilation Methods for Residential Buildings

The ventilation method of residential buildings plays a crucial role in the health and comfort of occupants, as well as the energy consumption level of the building. When the outdoor temperature and humidity are suitable, ventilation can effectively remove the waste heat and excess moisture generated indoors, thereby reducing the operating time of air conditioning and lowering energy consumption. However, in high outdoor temperature and humidity or extremely cold outdoor environment, ventilation can lead to indoor heat and moisture loads or cooling loads, resulting in

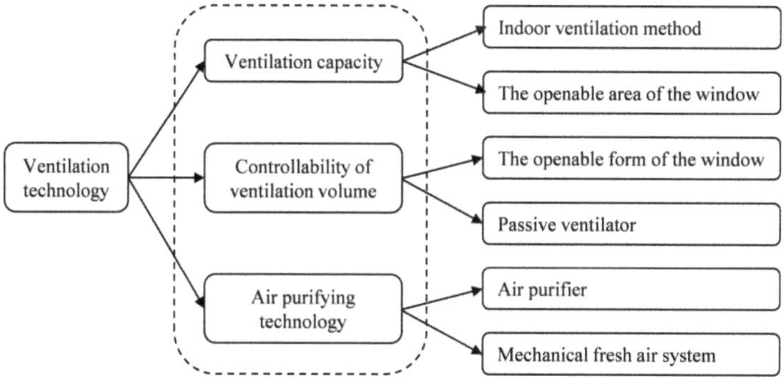

Fig. 8.51 Three requirements for ventilation technology

increased energy consumption of air conditioning. At the same time, ventilation is also a crucial means of maintaining good indoor air quality by introducing outdoor air and removing various indoor pollutants. However, when there is severe outdoor pollution, ventilation can introduce more inhalable particulate matter from outside into the indoor environment. Therefore, appropriate ventilation technologies need to be employed to create a favorable indoor environment (including thermal and humid environment and health requirements) while minimizing building energy consumption to meet environmental demands.

Through effective technical means, the three requirements for ventilation technologies can be met: (1) When ventilation is needed, a high air exchange rate can be achieved; (2) When ventilation is undesired, excellent airtightness can be maintained; when an appropriate ventilation rate is required, the desired air exchange can be achieved through adjustment; (3) The indoor air can be purified, as shown in Fig. 8.51. Next, from the aspect of achieving ventilation, we will analyze the effectiveness and suitability of six ventilation technologies.

8.6.3.1 Indoor Ventilation Methods for Building

The effectiveness of natural ventilation in a building is closely related to its morphological characteristics, layout, window and door positions, and other factors. Good design can enhance ventilation, while poor design may obstruct natural ventilation.

In the selection of building forms, the north-south cross-ventilation layout of slab-type residential buildings facilitates the formation of cross ventilation, which can quickly dissipate indoor heat; whereas residential tower units often feature windows on only one side, resulting in relatively poorer ventilation efficiency. For residential towers, it is advisable to enhance the building's external contours by creating protrusions and recesses, which can facilitate windows on different orientations, thereby optimizing indoor ventilation and air distribution.

In terms of layout, the arrangement of partition walls should follow the direction of airflow to ensure clear ventilation paths and avoid obstruction, thereby guaranteeing effective ventilation. For house types with a complex planar layout, methods such as low walls and hollowed partitions can be considered. As shown in Fig. 8.52. This will minimize the impact on the ventilation air flow.

In selecting the positions of doors and windows, consideration should be given to both room functionality and the prevailing wind direction. Openings should ideally face the prevailing wind direction, to allow outdoor air to smoothly enter through these openings, circulate through the rooms, and then exit.

Additionally, the chimney effect of high-rise spaces can be utilized by installing air outlets at the top of the building roof and vertical sections (such as the staircase, atrium, etc.), allowing heated indoor air to be expelled from the roof, thereby removing heat.

In summary, to create a favorable indoor ventilation environment, slab-type residential buildings should be prioritized in architectural form; in terms of interior layout, the impact of partition walls on airflow should be minimized; for door and window placement, openings should ideally face the prevailing wind direction; additionally, consideration can be given to installing vents at the roof and the upper part of vertical section.

8.6.3.2 Openable Area of the Window

The air volume of natural ventilation in a building is directly related to the openable area of windows. Assuming a wind speed of 0.2 m/s, the fresh air rate passing through a 1 m^2 opening area per hour is 720 m^3/h. This air volume far exceeds the air volume of common household fresh air fans currently available in the market.

Fig. 8.52 Low wall partition (left figure) and hollowed partition (right figure)

The *Residential Building Code* (GB 50386-2005) stipulates that "the opening area for ventilation in each residential building shall not be <5% of the floor area." Generally, the area of exterior windows is sufficient, but to achieve good natural ventilation, it is necessary to ensure as high an openable proportion as possible. Taking a residence with a usable area of 100 m^2 as an example, the required opening area for ventilation must not be <5 m^2. Assuming the exterior wall area of this residence is 50 m^2 with a window-to-wall ratio of 30%, the window area can be calculated as 15 m^2. That is, in the case of using windows for natural ventilation, at least 1/3 of the window area must be guaranteed to open 100%.

In summary, it is evident that opening windows for ventilation is an economical and efficient method of building ventilation. When designing and constructing residential buildings, it is necessary to ensure the window opening area complies with national standards, and to reasonably arrange the openable window area based on the total window area, in order to guarantee indoor natural ventilation.

8.6.3.3 Window Opening Method

In terms of opening method, common windows include side-hung window, sliding window, top-hung window, and bottom-hung window, as shown in Fig. 8.53. With technological advancements, the windows with composite opening methods such as bottom-hung side-hung windows have emerged in recent years. The opening method of windows primarily affects natural ventilation in three aspects: the openable area, adjustability, and ventilation experience (whether there is a sense of airflow).

In terms of openable area, the side-hung window can achieve full opening, with an operable area ratio of nearly 100%; the sliding window can usually only open up to half, with an operable area ratio of about 50%; the awning window has the lowest operable ratio, typically only around 30%. All other conditions being the same, the larger the opening area, the greater the ventilation volume and the better the ventilation effect.

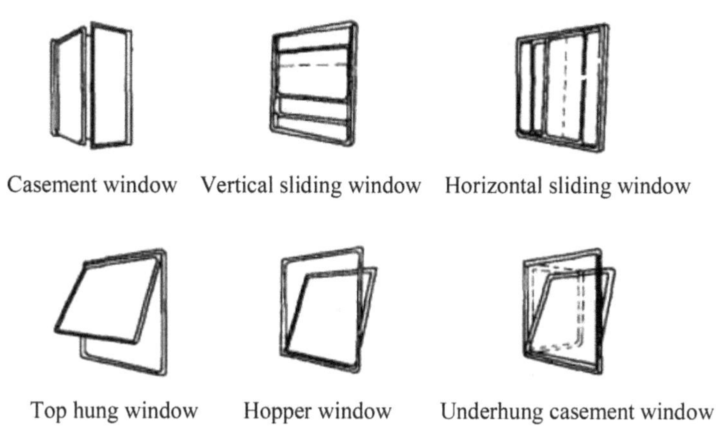

Fig. 8.53 Common opening methods of windows

In terms of adjustability, both side-hung windows and sliding windows can be opened to varying degrees as needed, allowing flexible control of airflow; whereas the opening angle of an awning window is typically preset and cannot be adjusted flexibly according to requirements.

In terms of ventilation experience, outdoor air entering through side-hung windows or sliding windows creates direct drafts, resulting in noticeable wind sensation; whereas airflow through awning windows enters at an angle, avoiding significant draft discomfort and offering a more comfortable subjective experience.

In addition, in selecting the window opening methods, safety and airtightness should also be considered. In high-rise buildings, the use of side-hung windows should be avoided due to the risk of objects falling from heights. In buildings with high air-tightness requirements, sliding windows should be used with caution as it is difficult to achieve high air-tightness.

In summary, it can be seen that different window opening methods each have their own advantages and disadvantages, and the choice should be made based on specific circumstances. For residential buildings, side-hung windows or sliding windows can be prioritized to achieve better ventilation; however, for safety considerations, high-rise buildings should avoid using side-hung windows. For residential buildings with higher air-tightness requirements, sliding windows should be used with caution.

8.6.3.4 Room Natural Ventilator

The natural ventilator is a method of air exchange that operates based on local pressure differences caused by the natural environment and the principle of gas diffusion. Since this device does not require mechanical power and relies entirely on natural ventilation to complete indoor and outdoor air exchange, it has the advantage of energy conservation.

The natural ventilator allows for ventilation without the need to open windows, making it usable regardless of time or season. It effectively prevents rain, snow, insects, noise, and sand particles from entering the room. Even when the room is unoccupied or during sleep, it can be used with confidence, as it eliminates concerns about burglars or cross ventilation intrusion, thereby ensuring the safety of doors and windows.

The ventilation efficiency of a natural ventilator primarily depends on the following three factors: (1) The difference between indoor and outdoor temperatures: the temperature disparity between indoor and outdoor air creates localized pressure differences, driving air movement. (2) The magnitude of outdoor natural wind: the wind pressure generated by outdoor natural wind creates localized air pressure differences between indoor and outdoor environments, driving air flow. (3) Air temperature: when the air temperature is higher, gas molecules are more active, and their molecular diffusion capacity is stronger, promoting mutual diffusion and infiltration between indoor and outdoor air. Therefore, natural ventilator is mainly suitable for areas or seasons with a significant difference between indoor and outdoor temperatures, or regions with abundant wind resources.

The types of natural ventilators include window-type natural ventilator and roof-type natural ventilator. Among them, the roof-type natural ventilator is a ventilation and air exchange equipment installed on the roof of buildings, suitable for industrial plants, warehouses, and large spaces in civil buildings; while the window-type natural ventilator is designed for residential buildings.

The window-type natural ventilator is installed on doors, windows, or curtain walls, utilizing the indoor-outdoor air pressure difference to achieve natural ventilation. Its function is to replace traditional window-opening ventilation methods, enabling indoor-outdoor air exchange, and it is suitable for high-rise airtight buildings. Figure 8.54 shows the external appearance and schematic diagram of a horizontal window-type natural ventilator with sound absorbing material, while Fig. 8.55 illustrates the external appearance and schematic diagram of a vertical window-type natural ventilator.

In summary, window-type natural ventilators are suitable for residential buildings, primarily in regions or seasons with a significant difference between indoor and outdoor temperatures, or areas with abundant wind resources. They are also suitable for the high-rise buildings that are airtight yet require ventilation.

8.6.3.5 Air Cleaner

An air cleaner refers to a device capable of adsorbing, decomposing, or transforming various air pollutants to circulate and purify indoor air, with its principle illustrated in Fig. 8.56.

There are two types of air cleaners on the market: those that remove indoor inhalable particulate matter and those that target chemical pollutants.

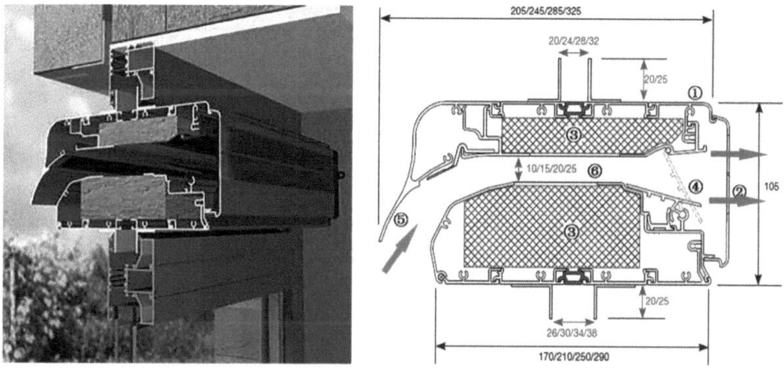

① Sound-absorbing ventilator frame, ② Removable internal air outlet, ③ Sound-absorbing component, ④ Internal air volume control valve, ⑤ Outdoor air inlet, ⑥ Airflow channel

Fig. 8.54 Schematic diagram of the appearance and structure of a sound-absorbing window-type natural ventilator (horizontal type)

① Indoor air supply outlet, ② Air volume adjustment grille near the indoor side, ③ Air volume adjustment grille near the outdoor side, ④ Outdoor air inlet, ⑤ Airflow channel

Fig. 8.55 Schematic diagram for appearance and structure of vertical window-type natural ventilator

Fig. 8.56 Schematic diagram of air cleaner principle

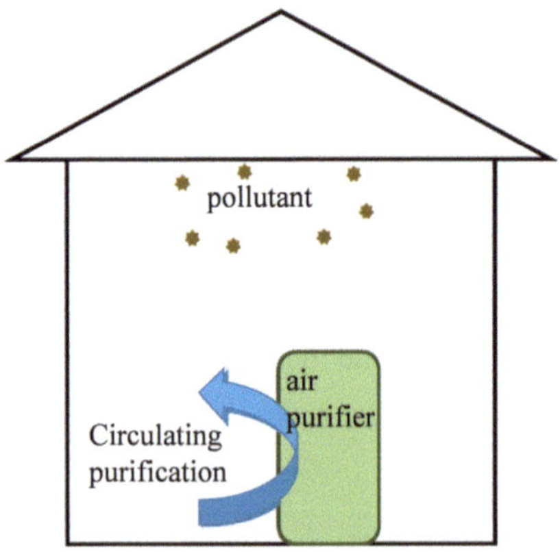

Air cleaners for removing inhalable particulate matter mainly come in two types: filtration type and electrostatic type. Filtration-type air cleaners are typically equipped with high-efficiency filtration screens, effectively removing solid, liquid, and biological particles through interception, inertial collision, diffusion, sedimentation, and sieving. They currently dominate the indoor air cleaner market. The electrostatic air cleaner, based on the principle of electrostatic adsorption, is used to remove particulate pollutants from the air. However, some electrostatic air cleaners may produce byproducts such as ozone due to improper design, and prolonged operation could lead to excessive ozone levels. In using an air cleaner, the cleaning and maintenance measures are also simple and effective: the strainer is cleaned regularly, while the medium or high-efficiency filter cores only need to be replaced periodically. These air cleaners that remove inhalable particulate matter can achieve a cleaning efficiency of 80%–95% for inhalable particulate matter. In fact, we do not necessarily pursue its dust capture efficiency, but rather the "clean air delivery rate" it provides. The indoor air cleaning effect is determined by the clean air delivery rate provided by the air cleaner, where the clean air delivery rate is the product of the circulating air rate and the filtration efficiency.

An air cleaner for removing indoor chemical pollutants can be in two types: physical adsorption type and catalytic oxidation type. Air cleaners based on the principle of physical adsorption capture chemical pollutants like VOCs in materials such as activated carbon. However, for activated carbon, periodic regeneration is essential; otherwise, its adsorption capacity will gradually diminish. An air cleaner based on the principle of catalytic oxidation typically employs photocatalysis technology using materials such as titanium nano-particles to convert harmful gases such as VOC entering the cleaner into CO_2 and water. Experiments have shown that while current photocatalysis technology can effectively eliminate VOCs, its reaction products may contain substances with toxicity far exceeding that of VOCs. Therefore, it does not currently appear to be an effective solution for VOC pollution. Therefore, installing an air cleaner alone to remove indoor chemical pollutants currently appears insufficient to fully meet the requirements.

In summary, it can be seen that if the goal is to remove physical pollutants indoors, installing an air cleaner is an effective method. The key metric for evaluating the cleaning efficiency of an air cleaner is its clean air delivery rate. However, if the goal is to eliminate chemical pollutants indoors, relying solely on an air cleaner currently cannot fully meet the requirements.

8.6.3.6 Mechanical Fresh Air System

The mechanical fresh air system uses a fan to deliver outdoor air into the room, providing the fresh air rate through the ventilation system. To ensure air quality, the fresh air system is generally equipped with filtration devices to purify the fresh air.

Currently, household mechanical fresh air systems can be divided into unidirectional airflow fresh air systems and bidirectional airflow fresh air systems, as shown in Figs. 8.57 and 8.58. A unidirectional airflow fresh air system utilizes natural air intake and forced exhaust; a bidirectional airflow fresh air system employs forced intake and exhaust. However, when both forced exhaust and indoor exhaust fans are activated simultaneously, it may increase room air infiltration.

A bidirectional airflow fresh air system can be equipped with a heat recovery device, the purpose of which is energy saving. However, in reality, residential buildings do not always need to recover heat or cold from the exhaust air. It makes sense to recover the heat or cold from the exhaust air through heat exchange only when the outdoor temperature is higher than the indoor temperature or when the outdoor temperature is significantly lower than the indoor temperature. Additionally, precautions should be taken to prevent leakage between exhaust air and fresh air during the heat recovery process, which could lead to secondary pollution.

The mechanical fresh air system also has its drawbacks and issues in practical use.

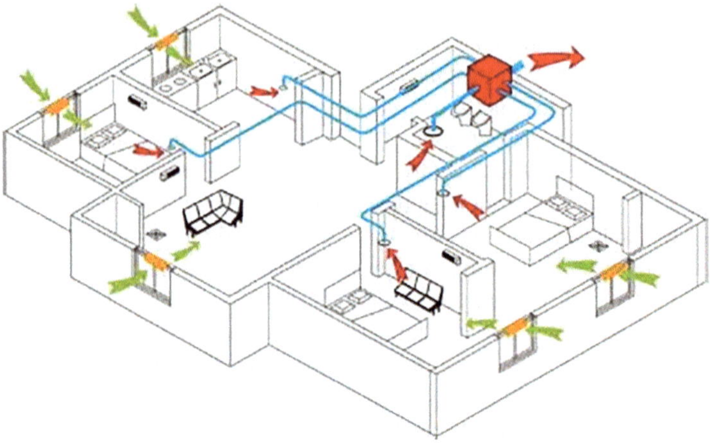

Fig. 8.57 Schematic diagram for the principle of unidirectional airflow fresh air system

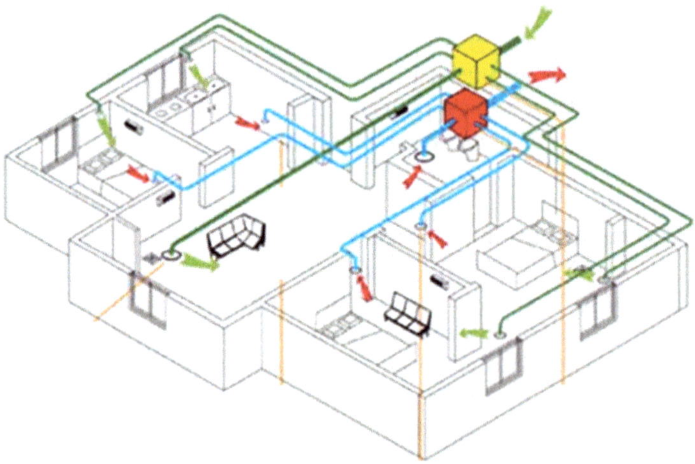

Fig. 8.58 Schematic diagram for the principle of bidirectional airflow fresh air system

Due to considerations of noise, architectural space utilization, and construction costs, the air volume of fresh air systems is generally not set too large. The air volume of a mechanical fresh air system is generally determined based on the minimum fresh air rate required for the room, providing an outdoor air volume of only 0.5–1 times per hour, thus typically operating continuously throughout the year. Within this range, the significance of adjusting indoor and outdoor ventilation rates becomes minimal.

Meanwhile, the mechanical fresh air system still cannot prevent outdoor air from infiltrating through gaps in exterior windows. This air can also carry outdoor pollutants into the room, and the mechanical fresh air system is unable to remove these pollutants. Therefore, when outdoor air quality is poor, the mechanical fresh air system struggles to effectively purify indoor air, and an air cleaner is still required to address this issue.

Due to the rapid accumulation of dust in the fresh air system, filters should be cleaned and replaced regularly to prevent them from becoming a secondary pollution source, ensuring the fresh air system can filter effectively.

In summary, the mechanical fresh air system is generally equipped with a filtration device, which can provide clean fresh air for rooms. However, this system faces the issue of limited airflow capacity, making it difficult to fully utilize outside air for room cooling. Consequently, its application in residential buildings has certain limitations, and further improvements are still needed in areas such as airflow volume, noise control, and maintenance.

8.7 Analysis of Airflow Short-Circuiting Issues in Urban Residential Air Conditioning Outdoor Units

In actual operation, the two main external factors affecting the energy efficiency and electricity consumption of an air conditioner are the indoor and outdoor operating conditions, specifically the air intake volume and the dry-bulb/wet-bulb temperature of the incoming air at both the indoor and outdoor units. Among these, the setting of indoor temperature and airflow depends on the user's comfort requirements. Guided by the concept of behavior-based energy conservation, during cooling operation, the indoor temperature is generally set within the range of 26–28 °C, which is a controllable factor. On the other hand, the ambient temperature of the outdoor unit's working environment is primarily determined by the meteorological conditions of the outdoor environment and the installation method of the outdoor unit. In practical use, even with exactly the same meteorological conditions in the outdoor environment, the air conditioner can hardly achieve the test performance under laboratory conditions.

In the past, since air conditioners were not essential equipment in buildings, the installation location of outdoor units for air conditioners was not considered during residential design. Therefore, when installing air conditioners in such buildings, it is common to mount triangular brackets on the exterior wall to bear the outdoor unit, as shown in Fig. 8.59. For ordinary low- to mid-rise buildings, the triangular bracket installation platform is the primary type of platform for outdoor unit installation due to its flexibility and convenience. With this installation method, the issue of exhaust air recirculation is not prominent. Currently, most residential buildings are high-rise buildings with exterior wall insulation materials, making the triangular bracket installation method unsuitable for the large-scale installation of outdoor units for air conditioning in high-rise buildings. In current residential design, architects often install air conditioner outdoor units in recessed

Fig. 8.59 Actual installation position of air conditioning outdoor units in residential buildings

Fig. 8.60 Actual installation position of air conditioning outdoor units in residential buildings

slots to ensure aesthetic appeal and save space, concealing them with specially designed louvers, as shown in Fig. 8.60. In addition, there are some other issues with the installation platforms for outdoor units in existing residential buildings. The main problems include unreasonable placement of outdoor units, obstruction by surrounding objects, and even excessively dense concentration of outdoor units, as shown in Fig. 8.61. Improper installation of the outdoor unit can lead to severe exhaust air recirculation, both within a single outdoor unit and between multiple units. This results in a widespread phenomenon where the intake air temperature deviates from the actual ambient conditions, adversely affecting system performance. Such issues significantly reduce the equipment's energy efficiency and cause unnecessary energy waste.

It is evident that unreasonable installation platforms or methods for outdoor units are significant factors leading to the decline in the actual operational performance of air conditioners, thereby becoming a key factor affecting building energy efficiency and consumption reduction in residential buildings. How to reasonably arrange the outdoor unit and optimize its surrounding thermal environment is a question worthy of in-depth discussion.

Fig. 8.61 Actual installation position of air conditioning outdoor units in residential buildings

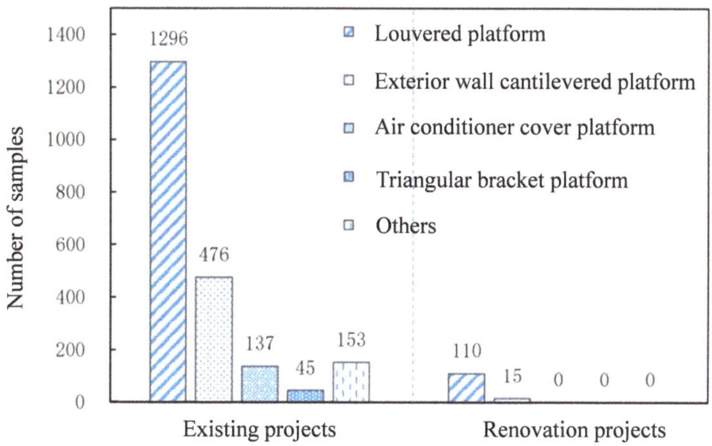

Fig. 8.62 Distribution count of outdoor unit installation platforms in the sample

8.7.1 Type of Air Conditioning Outdoor Unit Installation Platform in Residential Building

Currently, there are many installation methods for outdoor units in residential buildings in China. However, by categorizing them based on the characteristics of their installation platforms, there are 4 types of outdoor unit installation platforms in residential buildings, including triangular bracket platform, louvered platform, exterior wall cantilevered platform with railings (abbreviated as exterior wall cantilevered platform), and air conditioner cover platform.

Figure 8.62 presents the survey results on the quantity distribution of various installation platforms across 2232 commodity housing projects (including 2107 existing projects and 125 planned projects) in major cities of 15 provinces and municipalities in China.

As can be seen, the louvered platform has been widely applied in existing projects and is also the primary choice for installation platforms in upcoming projects. It stands as the mainstream type of installation platform, accounting for 63% of all samples. The usage of exterior wall cantilevered platform in existing projects is second only to louvered platform, accounting for 22% of the total sample. The air conditioner cover platform is primarily used in facade aesthetic renovation projects, accounting for 6% of all samples. It should be noted that "Others" refers to properties in the survey sample where air conditioner placement was not uniformly planned. According to the survey results, the vast majority of completed and upcoming projects have uniformly placed outdoor unit installation platforms, indicating that standardized planning for installation platforms is becoming a trend. Additionally, although triangular bracket platforms are uncommon in existing buildings and will not be applied in upcoming projects, they remain a primary installation type in older structures, accounting for 2% of all samples.

(1) Louvered platform

The louvered platform is currently and will continue to be the primary form of installation platform for air conditioner outdoor units. In recent years, real estate developers in China have gradually incorporated the installation of outdoor units into residential development phase, reserving space for outdoor units in building layouts. Currently, there are various engineering practices, such as reserving an air conditioner installation space in the middle of two balconies and using grilles to conceal the outdoor unit to ensure the aesthetic appeal of the facade, as shown in Fig. 8.63. There is a significant issue with heat dissipation from the outdoor unit on the louvered platform.

(2) Exterior wall cantilevered platform

The exterior wall cantilevered platforms are a series of uniformly arranged, vertically aligned support platforms protruding perpendicularly from the building's exterior wall, designed for placing air conditioner outdoor units. The application of exterior wall cantilevered platforms is also quite common nowadays. These platforms are typically located in less visible areas such as the rear of buildings, corners, or courtyards. To prevent the outdoor unit from falling,

Fig. 8.63 Schematic diagram of louvered platform

metal railings are usually installed around the platform, which to some extent maintains the aesthetic appeal of the building facade. However, the main drawbacks of this method are that it still poses certain risks during installation and can easily affect the thermal operating environment of the outdoor unit. When using an exterior wall cantilevered platform, there is a significant vertical temperature gradient in the local outdoor air temperature, resulting in the intake air temperature of the upper-level air conditioner outdoor unit being notably higher than that of the lower-level outdoor unit. Additionally, as developers continue to enhance the quality of their real estate products, the exterior wall cantilevered platform is increasingly integrated with the balcony for installation convenience, becoming a part of the balcony. This represents the future mainstream approach for exterior wall cantilevered platforms. Figure 8.64 illustrates its typical structure and implementation.

(3) Air conditioner cover platform

In recent years, for aesthetic considerations of the exterior wall, some outdoor units are enclosed and concealed by air conditioner covers after installation. The outdoor unit is exposed to sun and rain all year round, making it highly susceptible to rust and dust blockage, which can affect the normal operation of the air conditioner. The air conditioner cover platform not only serves to protect the outdoor unit of the air conditioner, but also shields the front and sides of the outdoor unit, enhancing the aesthetic appearance of the residential area.

Currently, there is a wide variety of outdoor unit covers available on the market, which lack manufacturing standards. Most of them are in the forms such as perforated panels or wrought iron patterned railings, as shown in Fig. 8.65, aiming to provide both front-side shielding and decoration for the outdoor unit. From the perspective of practical application, the air conditioner cover platform, similar to the triangular bracket platform during the installation and maintenance of air conditioners, also poses significant safety risks. Therefore, this method is only suitable for renovation projects, and the extent of its impact on the air conditioner's operational performance requires further study.

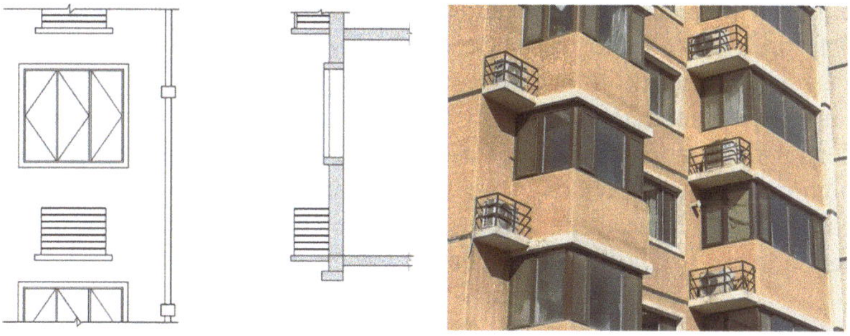

Fig. 8.64 Schematic diagram of exterior wall cantilevered platform

Fig. 8.65 Schematic diagram of perforated panel and louvered air conditioner cover platform

Fig. 8.66 Schematic diagram of triangular bracket platform

(4) Triangular bracket platform

The triangular bracket platform is widely applied in existing residential buildings, commonly found on the exterior walls of older residential buildings with a long service life. Its structure is as shown in Fig. 8.66. First, the triangular bracket provided with the air conditioner is secured onto the exterior wall, and then the bolt module is used to fix the outdoor unit onto the bracket. This platform is easy and convenient to install, requiring no design, and can be mounted anytime, anywhere, making it widely applicable. However, as the number of residential floors continues to increase, the use of such installation platforms in high-rise residential buildings presents safety hazards such as falling risks and compromises the aesthetic integrity of the structures. Consequently, their application in newly constructed buildings has been gradually declining.

8.7.2 *Impact of Installation Platform on Air Conditioner Performance*

The heat exchange capacity of the outdoor heat exchanger is primarily influenced by the intake air volume and intake air temperature. For outdoor units installed on outdoor unit platforms, their intake air temperature is affected by the exhaust air.

When the outdoor heat exchanger has a heat dissipation load, the exhaust air recirculation will cause the inlet air temperature of the heat exchanger to be higher than the ambient temperature. Taking a 1.5 hp outdoor unit as an example, the impact of several typical installation platforms on the operational performance of the air conditioner was studied using the CFD simulation method. The characteristic parameter settings of each installation platform are shown in Table 8.9 and Fig. 8.67.

Figure 8.68 illustrates the average air intake temperature rise of the heat exchanger for triangular brackets, exterior wall cantilevered platforms, Air conditioner cover platforms, and regular and optimized louver platforms under different heat dissipation loads. It can be observed that regular louvered platforms not only cause airflow attenuation in the outdoor unit but also lead to exhaust air recirculation, resulting in an average intake air temperature rise Δt 3–5 °C relative to the outdoor temperature. In contrast, the Δt for the other three types of platforms ranges between 0.5–2 °C, with the triangular bracket exhibiting the smallest value. The average Δt of the optimized louver platform ranges between 1.5–2 °C, which is basically on par with the exterior wall cantilevered platform.

The higher the temperature of the air entering the outdoor unit, the more work the compressor must perform to achieve the same cooling effect. As the compressor works more, the outdoor unit releases more heat into the surrounding air, further raising the ambient temperature near the unit. This creates a vicious cycle, significantly reducing the efficiency of the air conditioning system. When the intake air temperature of the outdoor unit rises by 1 °C, the air conditioner's coefficient of performance (COP) decreases by 3%. When the inlet temperature exceeds 43–46 °C, it may trigger the safety protection of the compressor, causing the air conditioning equipment to stop operating.

Table 8.9 Characteristic parameters of installation platform

	Distance (mm)					Louver parameters		
Installation platform type	L1	L2	L3	L4	L5	Aperture ratio R	Distance δ	Angle θ
Triangular bracket	–	150	–	–	–	–	–	–
Exterior wall cantilever	100	100	100	100	–	0.3	–	–
Air conditioner cover	100	100	100	100	-	0.3	-	-
Regular louver	100	100	100	150	50	0.3	20 mm	45°
Optimized louver	200	200	200	100	100	0.5	50 mm	30°

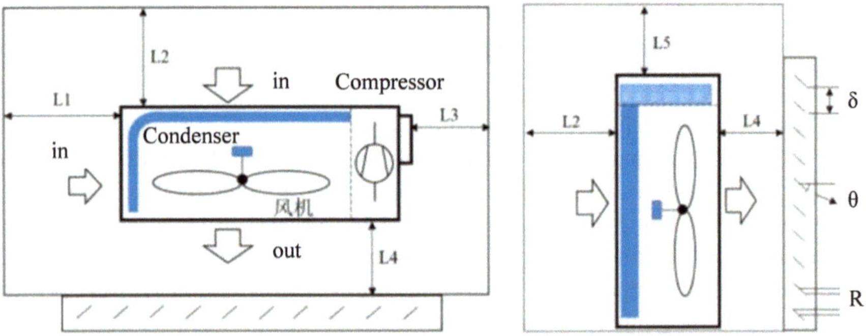

Fig. 8.67 Characteristic dimensions of the installation platform

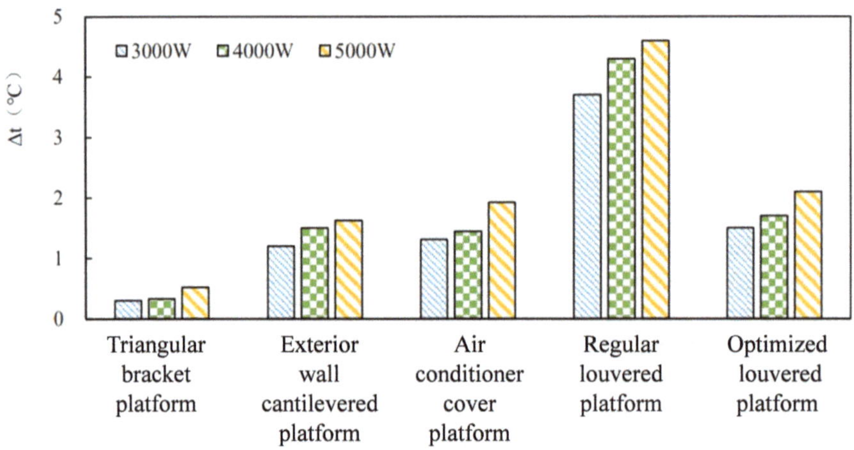

Fig. 8.68 Average inlet air temperature rise of outdoor heat exchanger under different heat exhaust capacities

8.7.3 Optimization Measures for Outdoor Unit Installation Platform

For a typical split air conditioner outdoor unit, the surrounding obstructions mainly include the front grille or louvers, the rear wall, and the side walls. These obstructions can cause some of the hot air discharged by the air conditioner's outdoor unit to flow back into the intake air of the outdoor unit, leading to a reduction in the refrigerating capacity of the outdoor unit and consequently a decline in the performance of the unit. The proportion of this part of hot air return to the total intake air of the outdoor unit is defined as the hot air recirculation ratio, which can be used to evaluate the performance changes of the unit. Specifically, a higher hot air recirculation ratio leads to a greater decline in the cooling performance of the unit, while a lower hot air recirculation ratio indicates better operational performance of the unit.

8.7.3.1 Spacing Between Outdoor Unit and Obstruction

Here, the impact of obstructions in the horizontal direction of outdoor unit installation platform on the unit's performance is considered, including side walls, rear walls, and front louvered barriers. The positions are illustrated in Fig. 8.69.

(1) Rear wall

Take an air conditioning unit with a refrigerating capacity of 1hp and an air outlet speed of 2.6 m/s at an outdoor dry-bulb temperature of 35 °C as an example. For ease of installation, the unit should not be placed too far from the rear wall, otherwise the return air of the unit will be affected by the rear wall, leading to poor return air flow. During the experiment, only the rear wall served as an obstruction. The distance between the outdoor unit and the rear wall was adjusted to gradually increase the space for air return. As the distance between the outdoor unit and the rear wall increases, the hot air recirculation ratio decreases, and the performance of the unit improves. When the distance is 10 cm, the hot air recirculation ratio is 12%, as shown in Fig. 8.70. When the distance increases to 40 cm, the hot air recirculation ratio drops to 6%, and the operational performance of the unit is significantly enhanced.

(2) Side wall

Similarly, take an air conditioning unit with a refrigerating capacity of 1hp and an air outlet speed of 2.6 m/s at an outdoor dry-bulb temperature of 35 °C as an example. When the distance between the outdoor unit and the side wall increases from 10 cm to 40 cm, the hot air recirculation ratio of the unit is as shown in Fig. 8.71. As can be seen from the figure, when the distance between the outdoor unit and the side wall reaches 40 cm, the hot air recirculation ratio of the unit has dropped below 5%. It can be considered that the performance degradation of the unit is negligible, indicating this distance as the optimal installation spacing.

(3) Louver

From the perspective of architectural aesthetics, louvers are often added at the air outlet of the outdoor unit. The obstruction of the louvers can cause reflection of

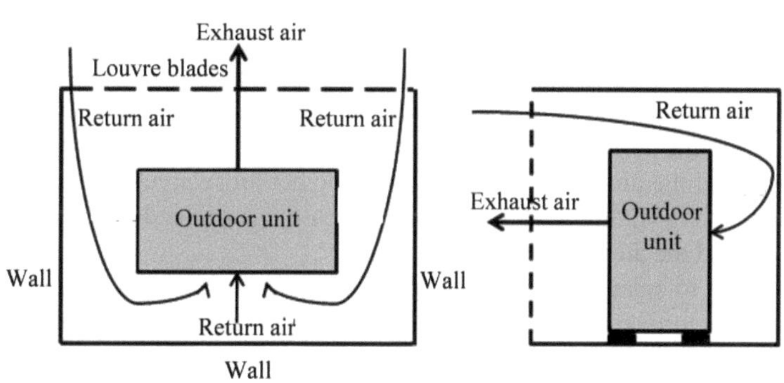

Fig. 8.69 Schematic diagram of air return and exhaust path for outdoor unit

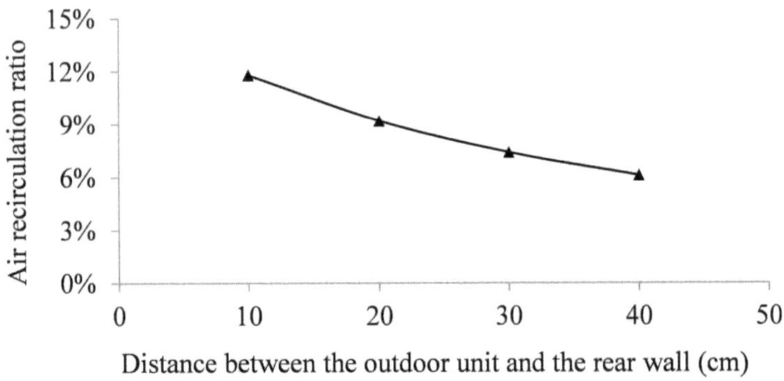

Fig. 8.70 Relation between the distance of the outdoor unit from the rear wall and the hot air recirculation ratio

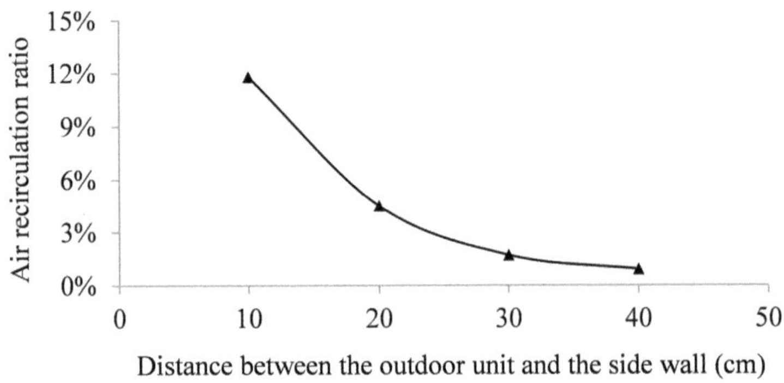

Fig. 8.71 Relation between the distance of outdoor unit from side wall and hot air recirculation ratio

the airflow, leading to an increase in local ambient temperature and a reduction in the condensation efficiency of the air-conditioning system. Similarly, take an air conditioning unit with a refrigerating capacity of 1hp and an air outlet speed of 2.6 m/s at an outdoor dry-bulb temperature of 35 °C as an example. When the distance between the front louvers and the outdoor unit increases from 10 cm to 40 cm, the hot air recirculation ratio of the unit is as shown in Fig. 8.72. As can be seen from the figure, as the distance between the louver and the outdoor unit increases, the performance of the unit improves. When the distance between the louver and the air outlet reaches 40 cm, the hot air recirculation ratio is about 6%, which is significantly better than the case where the distance between the louver and the air outlet is 10 cm.

In summary, to ensure the efficient operation of the air conditioning system, the primary improvement methods and measures are as follows: maintain a minimum clearance of 30 cm between the outdoor unit and surrounding obstructions; depending on the project's specific conditions, where feasible, it is

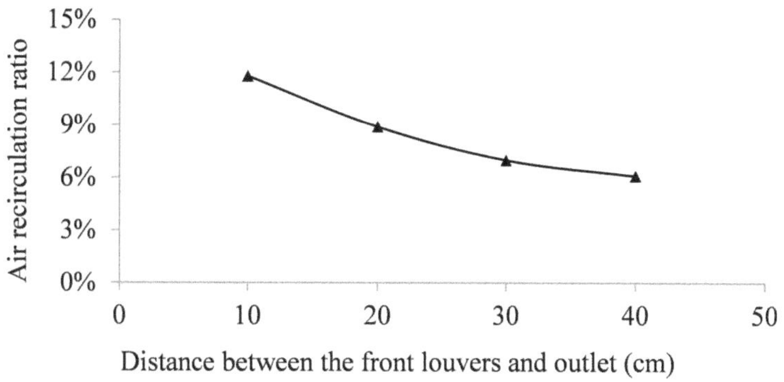

Fig. 8.72 Relation between the distance of front louvers to the air outlet and the hot air recirculation ratio

recommended to replace solid walls on both sides of the outdoor unit installation area with decorative permeable components such as open louvers or perforated railings, to enhance heat dissipation pathways from the recessed space.

8.7.3.2 Louver Opening Ratio and Tilt Angle

(1) Louver opening ratio

To meet the aesthetic requirements of the building, louvers are typically used to enclose the installation platform of the air conditioner outdoor unit. However, the louver transmittance will directly affect the ventilation and heat exchange performance of the unit. As the opening ratio gradually increases, the air flow rate of the outdoor fan first decreases and then increases, while the β value transitions from small to large before gradually decreasing, and the α value progressively diminishes, as shown in Fig. 8.73. In other words, when the louver is completely sealed ($R = 0$) or nearly sealed ($R = 0.1, 0.2$), the attenuation of the outdoor unit's air flow rate is minimal, but almost all the air flow recirculates back to the heat exchanger's air inlet. Conversely, when the louver opening ratio is large ($R = 1$, i.e., the front louver is removed), the outdoor unit's air flow rate approaches the fan's rated air flow rate, and the short-circuiting of exhaust air is also minimal. This trend aligns with the measurement results obtained from the experimental installation platform, thereby confirming that the aforementioned model can effectively describe the impact of the installation platform's structural parameters on the actual air flow rate of the outdoor unit. To meet the operational requirements of the unit, the louver's transmittance should be no <0.7.

(2) Louver tilt angle

Figure 8.74 illustrates the impact trend of grille tilt angle α (ranging from $-60°$ to $60°$) on air conditioner performance under different cooling conditions, when the surrounding clearance of the outdoor unit matches the grille spacing.

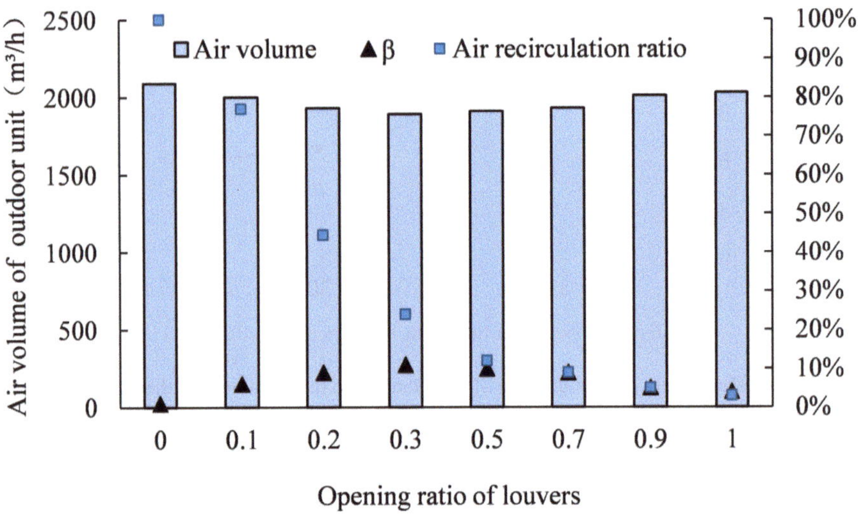

Fig. 8.73 Relation between louver opening ratio, air volume attenuation coefficient, and recirculation ratio

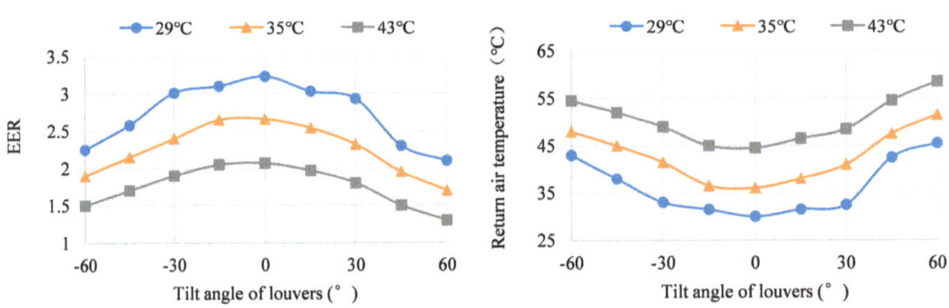

Fig. 8.74 Influence of louver tilt angle on air conditioner performance

The influence of the louver grille tilt angle α on performance increases as α grows. When the louver grille is arranged horizontally (i.e., α = 0°), the performance is optimal. At α = 15°, 30°, 45°, and 60°, the EER decreases by approximately 5%, 12%, 28%, and 36%, respectively, compared to when α = 0°. Conversely, at α = −15°, −30°, −45°, and −60°, the EER decreases by about 1%, 8%, 18%, and 27%, respectively. Clearly, the larger α is, the more severe the performance degradation becomes. When the grille tilt angle is the same, upward tilting (positive angle) has a greater impact on cooling performance than downward tilting (negative angle). An upward-tilted grille is more likely to cause heat exhaust recirculation in the outdoor unit, resulting in more severe airflow short-circuiting and higher intake air temperature for the outdoor unit. To meet the operational requirements of this unit, if louvers must be installed, it is preferable to set them horizontally or slightly downward. This configuration effectively guides airflow and reduces the occurrence of "short-circuiting" airflow.

8.7.3.3 Grille Spacing

Figure 8.75 presents the combined effects of grille spacing h and tilt angle α on the performance of the air conditioner when installed using louver grooves. As can be seen, when the louver grille is tilted upward, the refrigerating capacity gradually increases with the enlargement of the louver grille spacing. After reaching the optimum spacing value, further increasing the spacing leads to a decline in refrigerating capacity. When $\alpha \geq 30°$, the optimum spacing of the grille is 150 mm, while for $\alpha < 30°$, the optimum-performance spacing is 120 mm. When the grille is tilted downward, the situation is more complex than when the louver grille is tilted upward. At the small tilt angle $\alpha > -30°$, the changing trend of the refrigerating capacity is almost the same as that of tilting upward. As the spacing increases, the refrigerating capacity gradually improves. After reaching the optimum spacing, continuing to increase the spacing will instead reduce the refrigerating capacity. At the large tilt angle $\alpha \leq -30°$, the performance degradation is small when the small spacing h = 30 mm. As the grille spacing increases (in the figure, h increases from 30 mm to 60 mm), the performance degradation becomes larger. After reaching the lowest value, the degradation of the refrigerating capacity decreases. When the spacing further increases after reaching the maximum value, the refrigerating capacity instead decreases. As can be seen, there is an optimum spacing for the louver grille, and it is related to the tilt direction and angle of the grille. When $-15° \leq \alpha < 30°$, the optimum spacing is 120 mm. When $-60° \leq \alpha < -15°$ or $\alpha \geq 30°$, the optimum spacing is 150 mm.

8.7.3.4 Influence Between Air Conditioner Outdoor Units

The heat released by the air conditioner outdoor unit will raise the ambient air temperature of the surrounding environment, leading to natural upward movement of heat and air. This causes an increase in the ambient air temperature of the upper floors and a rise in the return air temperature of the outdoor unit. In addition, the heat released by the air conditioner outdoor unit will also be absorbed by other outdoor units on the same floor. Therefore, adequate attention should be paid to the rising heat and air flow, and efforts should be made to minimize their adverse effects.

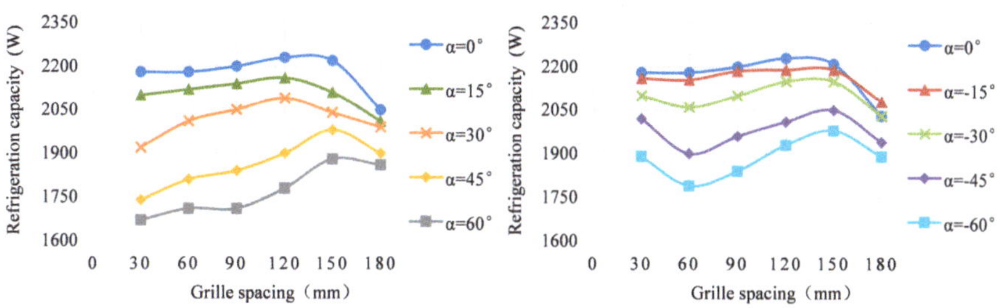

Fig. 8.75 Influence of grille spacing on the performance of the air conditioner

Taking a building with a floor height of 3m and 10 floors as an example, the CFD simulation method was used to investigate the mutual influence between units under calm outdoor wind conditions when the capacities of the air conditioning units were 1 hp, 1.5 hp, and 2 hp, corresponding to outdoor unit air outlet velocities of 2.6 m/s, 3.2 m/s, and 4.0 m/s, respectively. Among them, the dimensions of the outdoor unit are: length × width × height = 900 mm × 750 mm × 300 mm. The dimensions of the outdoor platform are: length × width × height = 1500 mm × 1000 mm × 600 mm. The distances between the outdoor unit and the louvers, rear wall, and side wall are 100 mm, 200 mm, and 300 mm respectively.

(1) The impact between the upper and lower outdoor units of air conditioners
For conditions where only one outdoor unit is arranged on each floor, the exhaust air recirculation ratio of the units located on each floor is as shown in Fig. 8.76. It can be seen in the figure that the exhaust air recirculation ratio of the outdoor unit increases continuously as the floor number increases. The greater the exhaust air velocity, the smaller the influence of lower-floor outdoor units on upper-floor outdoor units. When the floor number increases by 1, the exhaust air recirculation ratio of the outdoor unit increases by approximately 2%, resulting in a corresponding 2% decrease in the cooling performance of the unit.
(2) The influence of horizontal spacing between air conditioner outdoor units

The influence of horizontal spacing between outdoor units on the exhaust air recirculation ratio is shown in Fig. 8.77. As can be seen, the greater the spacing between outdoor units, the smaller the mutual influence between outdoor units on the same level, and the smaller the impact of lower-level outdoor units on adjacent upper-level outdoor units. It can be observed that when the horizontal spacing between outdoor units is 1 m, 2 m, 3 m, and 5 m respectively, the exhaust air recirculation ratio of the unit increases by approximately 1%–5%, 1%–4%, 1%–3.5%,

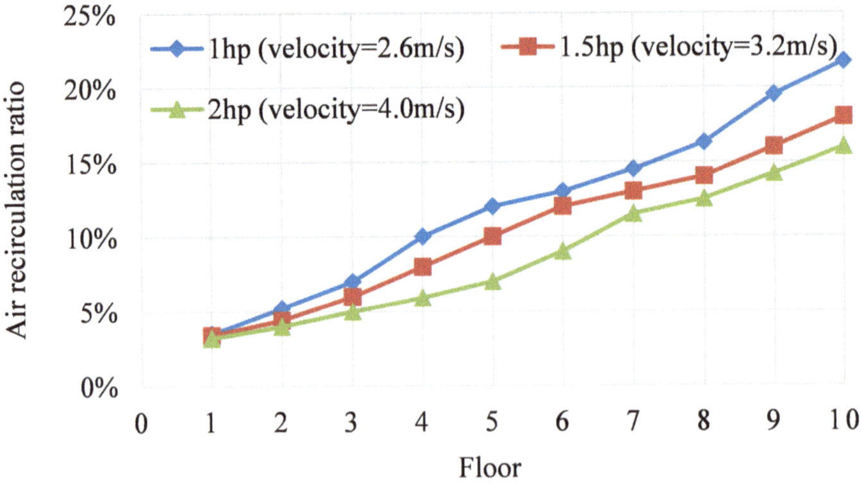

Fig. 8.76 Relation between exhaust air recirculation ratio of outdoor unit and floor number

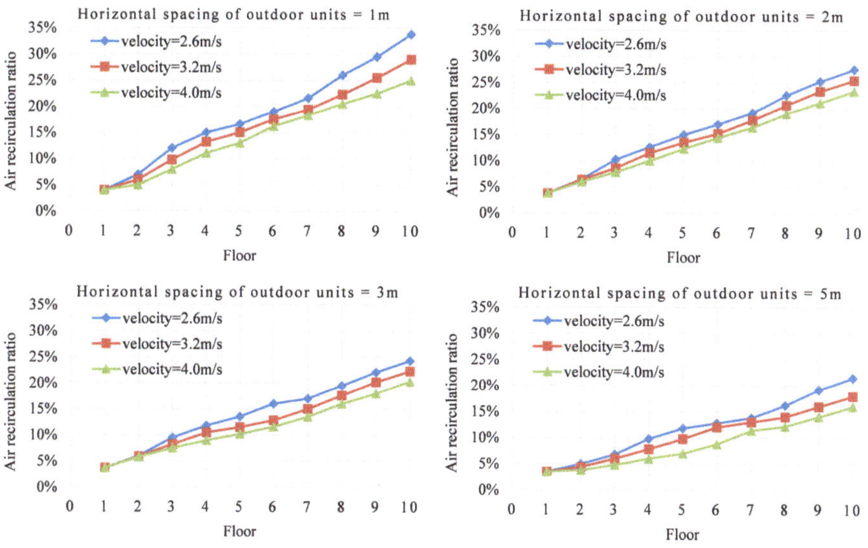

Fig. 8.77 Relation between horizontal spacing of outdoor units and exhaust air recirculation ratio

and 0%–3% when the floor number increase by 1. When the horizontal spacing between outdoor units exceeds 5m, the mutual influence between units can essentially be neglected. For every 5-floor increase in building height, the exhaust air recirculation ratio of the unit increases by approximately 10%–17%, 10%–13%, 8%–12%, and 5%–10% respectively; when the horizontal spacing between outdoor units exceeds 5 m, the mutual influence between units remains within 10%.

In summary, the heat released by the outdoor unit will increase the ambient air temperature, leading to higher return air temperatures for outdoor units on the same floor and upper floors, thereby reducing the cooling performance of the unit. Therefore, to meet the operational requirements of the unit, whenever feasible, the spacing between outdoor units on the same floor or horizontal plane should be maximized, or outdoor units on different floors should be staggered vertically to ensure sufficient surrounding space for air circulation and heat dissipation.

8.8 Analysis of Cold Island Effect in Air Source Heat Pump Heating for Urban Residential Buildings

8.8.1 Impact of Air Source Heat Pump Heating on Urban Heat Island Effect

The urban heat island (UHI) index is a crucial indicator of urban microclimate. The urban heat island effect refers to the phenomenon of urban "warming" caused by factors such as extensive artificial heat emissions, increased high-heat-capacity structures like buildings and roads, and reduced green space (Kalnay 2003; Howard 1833; Mhoff 2010 Sheng 2017; OKE 1973). The use of air source heat pumps for winter heating can affect the urban heat island, making it necessary to study the heat island index.

8.8.1.1 Variation Characteristics of the Urban Heat Island Index

Research indicates that the urban heat island index exhibits significant diurnal and seasonal variations (Yang 2013; Santamouris 2015; Rosenzweig 2005; Yang et al. 2017; Yang 2023). Based on the hourly temperature observation data from automatic weather stations during 2016–2020, this study analyzes the spatial distribution and variation characteristics of the urban heat island index in 30 typical Chinese cities (Yang 2023). Comprehensive analysis results indicate that in densely populated northern urban areas with central heating systems during winter, the heat island index ranges between 1.6–3.0 °C, as shown in Fig. 8.78, where the median UHI is 2.3 °C. In southern urban areas without a city-wide central heating system, the winter urban heat island (UHI) index ranges between 0.9–1.8 °C, with a median UHI of 1.3 °C (Yang 2023).

The average diurnal range of the urban heat island index for 30 target cities from 2016 to 2020 is shown in Fig. 8.79. The average diurnal temperature range of UHI in northern urban areas during winter is approximately 0.9 °C higher than that in

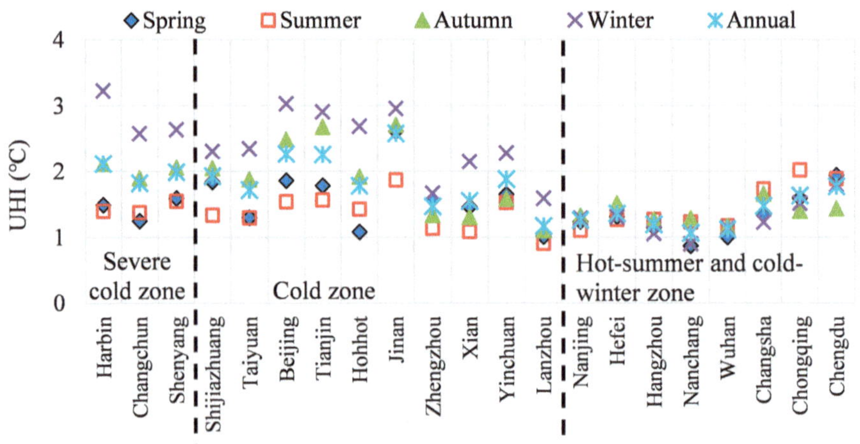

Fig. 8.78 Seasonal distribution of urban heat island index in 30 typical cities (2016–2020)

southern cities. This increase in UHI is attributed to the substantial release of heat from buildings due to winter heating.

8.8.1.2 Impact of Air Source Heat Pump Heating on Urban Heat Island Index

Based on the actual urban heat island index level mentioned above, the average urban heat island index in northern urban areas due to winter heating is approximately 0.9 °C. Currently, the heating seasonal performance factor of low-temperature air source heat pump units on the market (at a water temperature of 55 °C) is 2.30. Based on this, it can be estimated that replacing central heating with air source heat pump units would lead to an average reduction of approximately 0.5 °C in the urban heat island (UHI) effect during winter in northern urban areas. Taking Beijing as an example, the average winter UHI will decrease from 3.0 °C to approximately 2.5 °C.

The urban heat island index also exhibits diurnal variation characteristics. The diurnal variation trends of average temperatures at urban and rural observation stations in all the aforementioned typical cities are similar, namely, the urban heat island effect is stronger at night than during the day. Taking Beijing as an example, the current average daytime UHI is approximately 0.6 °C, while the average nighttime UHI is around 3.6 °C, as shown in Fig. 8.80. If air source heat pumps completely replace central heating, the average urban heat island index in Beijing during the day will not change significantly, while the average urban heat island index at night will decrease from 3.6 °C to 2.3 °C.

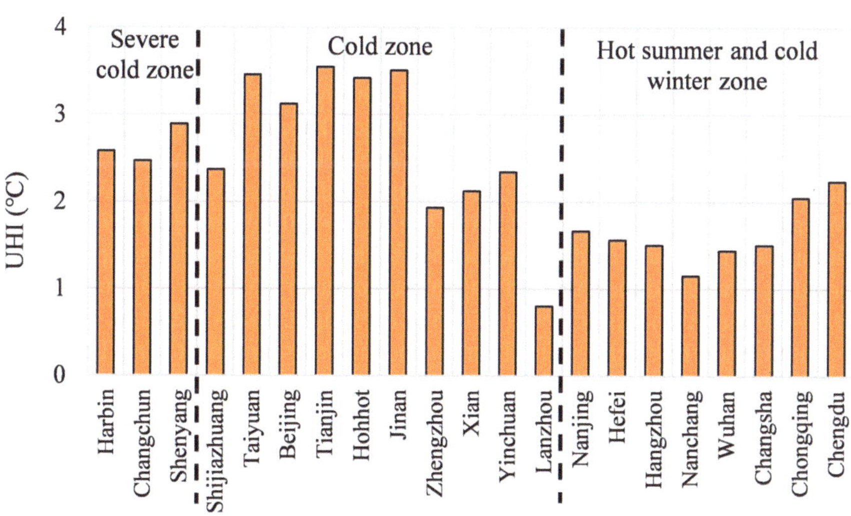

Fig. 8.79 Diurnal temperature range of average urban heat island index in 30 typical cities (2016–2020)

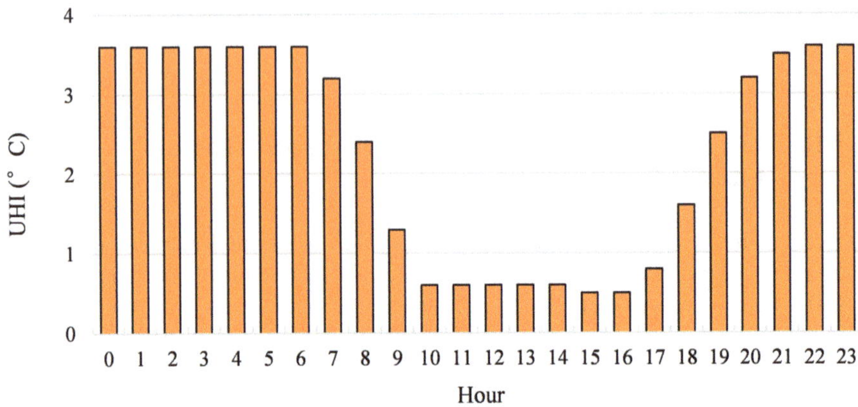

Fig. 8.80 Diurnal variation trend of average urban heat island index in Beijing (2016–2020)

8.8.2 Analysis of Local Cold Island Issue in Air Source Heat Pump Heating System

In winter, the heat pump unit absorbs thermal energy from the air and discharges cold air outward. The reflux of cold air not only affects the performance of the unit but also leads to a decrease in the ambient temperature of the surrounding area, forming a localized cold island (Fig. 8.81). The following research and analysis of this issue are conducted through CFD simulation methods.

8.8.2.1 Introduction to Study Cases and Parameters

Market research indicates that the most commonly used air source heat pump air-cooled units in the Chinese market are dual-fan system units, with single-fan and four-fan units also available. The common models have rated heating capacities of 70 kW, 85 kW, 140 kW, and 170 kW respectively. Typically, the units with 65 kW and 80 kW capacity are dual-fan system units, while those with 130 kW and 160 kW capacity are quad-fan system units. The 65 kW, 80 kW, and 130 kW units can combine up to 16 unit modules, while the 160 kW units can combine up to 8 unit modules. Overall dimensions of the most common dual-fan unit: (a) Width: 1950–2300 mm; (b) Depth: 960–1,050 mm; (c) Height: 1950–2320 mm.

Based on the research results, the most common air source heat pump unit on the market was selected as the study object. The unit's rated heating capacity is 70 kW, with an average COP of 2.3 during the heating season. Each unit is equipped with 2 fans, providing a total air volume of 28,000 m^3/h. The appearance of the unit is shown in Fig. 8.82.

The basic parameters of the CFD simulation model are shown in Table 8.10. In cold areas, the base height of the unit is 0.5 m above the ground. Based on the requirements of the Manufacturer, this model of unit can combine up to 16 unit

Fig. 8.81 On-site installation diagram of air source heat pump outdoor unit

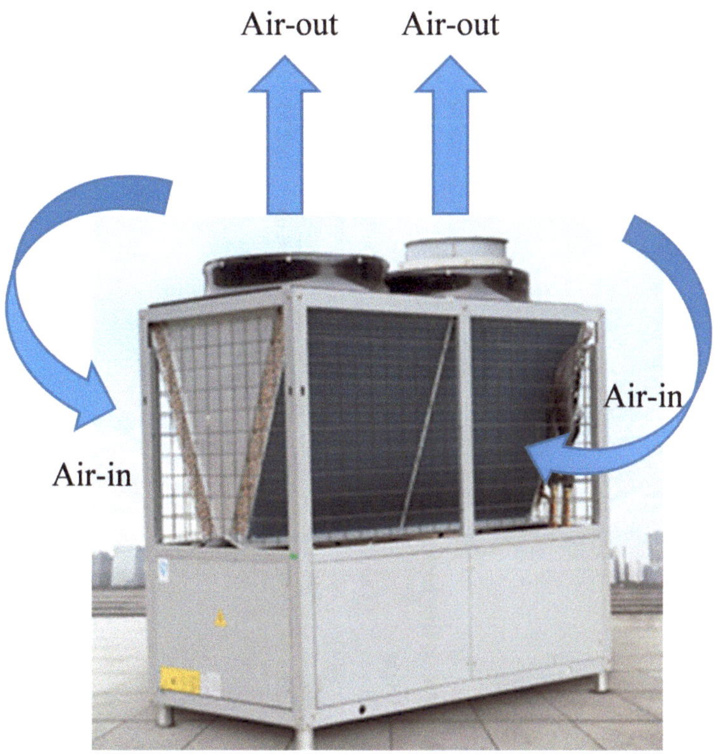

Fig. 8.82 Appearance of common air source heat pump air-cooled units

modules. The unit is applied in Beijing, and the diurnal variation of surface wind speed in the Beijing area is shown in Fig. 8.83.

Through the study of various influencing factors, it was found that the unit spacing and arrangement are the main factors affecting the surrounding thermal

Table 8.10 Basic parameters of air source heat pump unit

Unit length (m)	Unit width (m)	Unit height (m)	Height above ground (m)	Outdoor wind speed (m/s)	Air flow rate (m^3/h)	Number of fans (pc)	Outlet air speed (m/s)	Inlet air speed (m/s)	Heating capacity (kW)	COP	Inlet air temperature (°C)	Inlet and outlet air temperature difference (°C)
2.0	1.0	2.2	0.5	≤0.2	28000	2	≈6.08	≈2.70	70	2.3	10	4.26

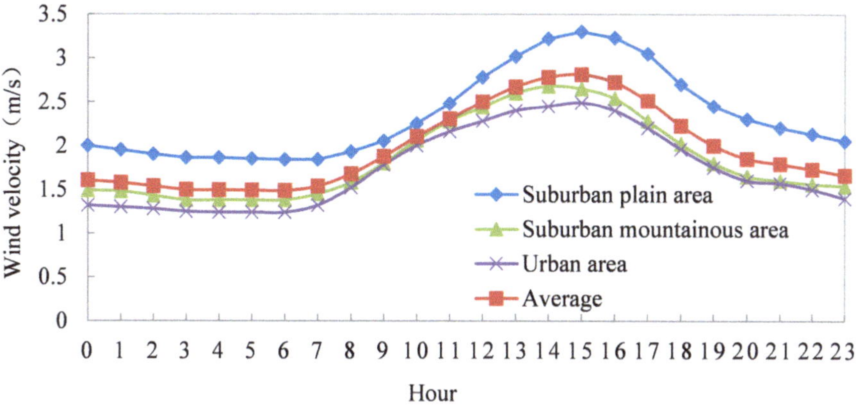

Fig. 8.83 Diurnal variation of surface wind speed in Beijing

environment. Moreover, higher ambient wind speeds are more conducive to the diffusion of local cold air, resulting in a weaker cold island effect in the surrounding air. Therefore, the main factors affecting the thermal environment around the unit and its operational performance in this study include unit scale, unit spacing, and incoming air speed.

Based on the actual conditions in Beijing, during the heating season, the high-load operation of heat pump units and low ambient temperature conditions primarily occur at night. Additionally, the prevalence of quasi-calm wind or low wind speed conditions during local nighttime is relatively common, meaning that the quasi-calm wind scenario at night represents a more unfavorable operating condition. Therefore, the simulation cases in this study primarily cover unfavorable quasi-calm wind conditions with unit cluster scales of 4, 9, 16, 25, and 36 units. The analyzed arrangements include five types: 2 × 2 array, 3 × 3 array, 4 × 4 array, 5 × 5 array, and 6 × 6 array. The study further examines the impact of wind speed on unit cluster performance, simulating ambient wind speeds of 0.2 m/s, 1 m/s, 2 m/s, 3 m/s, 4 m/s, and 5 m/s. The unit spacing varies from 1m to 4m, including five configurations: bilateral spacing = 1 m, windward spacing = 1.5 m & leeward spacing = 1 m, bilateral spacing = 2 m, bilateral spacing = 3 m, and bilateral spacing = 4 m. For details, refer to Figs. 8.84 and 8.85.

8.8.2.2 Impact of Unit Cluster Scale on Local Cold Island

The total number of unit clusters has a very significant impact on the local cold island. Figures 8.86, 8.87, 8.88, 8.89, and 8.90 take the case where the spacing between both sides of the unit is 1m as an example, and demonstrate the distribution of surrounding cold island intensity under different unit cluster scales in the least favorable conditions (quasi-calm wind conditions). It can be observed that as the scale of the unit cluster increases, the range of the local cold island effect around the

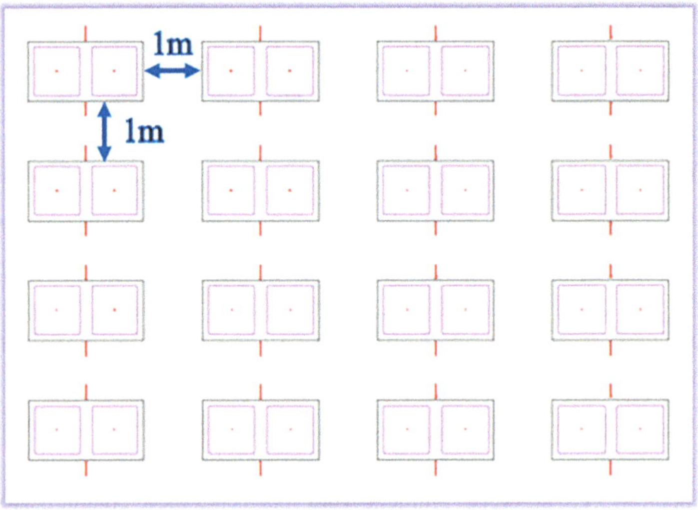

Fig. 8.84 Spacing configuration for unit layout—side spacing =1 m (demonstrated with 4 × 4 module array as example)

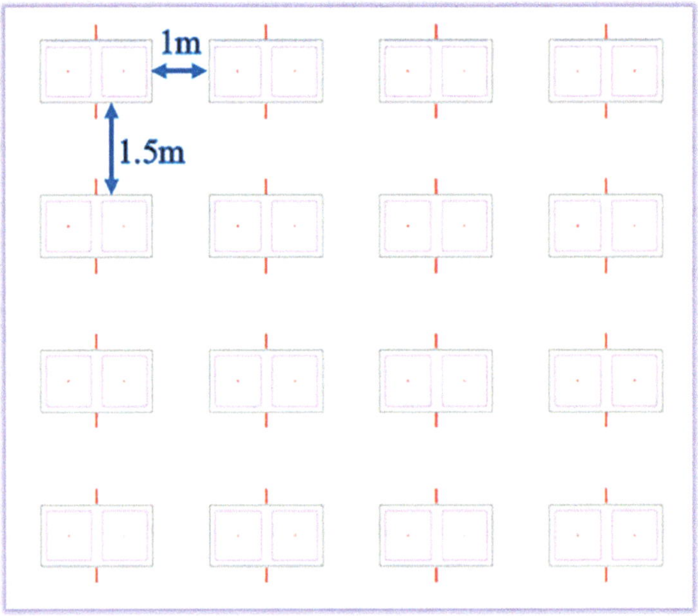

Fig. 8.85 Spacing configuration for unit layout—windward spacing = 1.5 m & other side spacing = 1 m (demonstrated with a 4 × 4 module array as example)

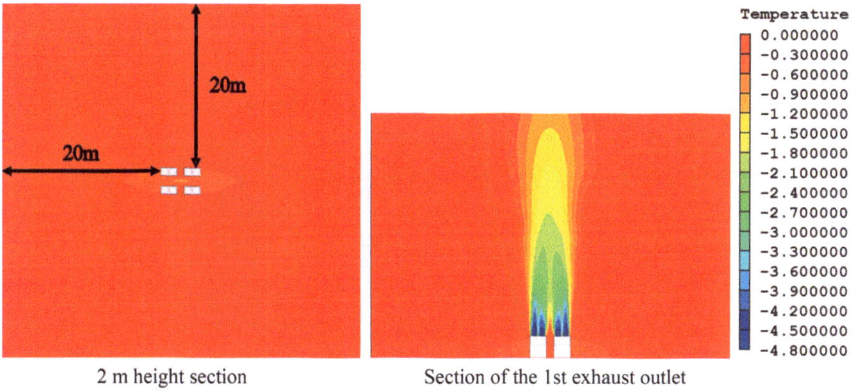

Fig. 8.86 Local cold island intensity under quasi-clam wind conditions with 2 × 2 unit layout (unit spacing = 1 m)

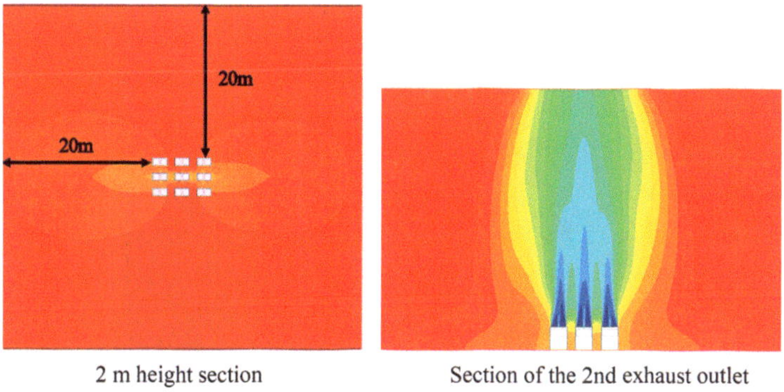

Fig. 8.87 Local cold island intensity under quasi-clam wind conditions with 3 × 3 unit layout (unit spacing = 1 m, scale same as above)

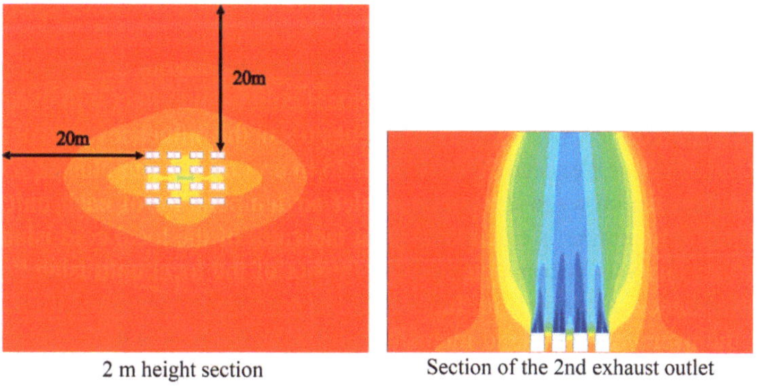

Fig. 8.88 Local cold island intensity under quasi-clam wind conditions with 4 × 4 unit layout (unit spacing = 1 m, scale same as above)

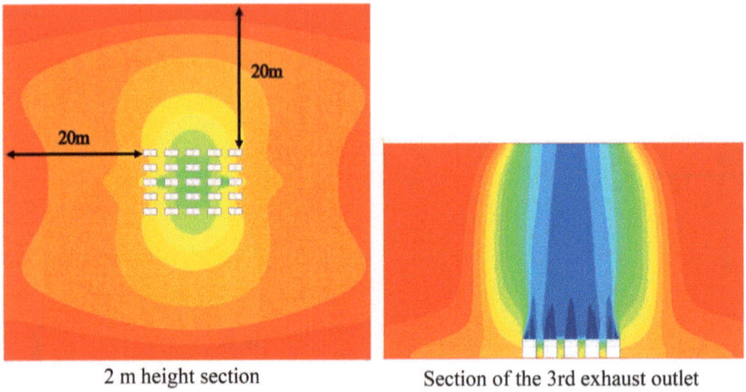

Fig. 8.89 Local cold island intensity under quasi-clam wind conditions with 5 × 5 unit layout (unit spacing = 1 m, scale same as above)

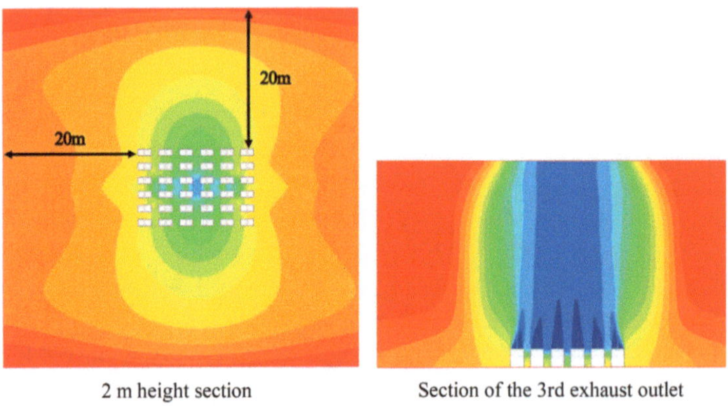

Fig. 8.90 Local cold island intensity under quasi-clam wind conditions with 6 × 6 unit layout (unit spacing = 1 m, scale same as above)

unit cluster also becomes greater. When the centralized layout of units is arranged in a 4 × 4 array, the influence range of the local cold island effect will exceed 15 m around the unit cluster; when the array reaches 6 × 6, the influence range of the local cold island effect will exceed 20 m or even 25 m around the unit cluster.

The difference between the average inlet air temperature of each unit and the ambient air temperature is calculated as an indicator of the local cold island intensity. Further analysis is conducted on the impact of the local cold island effect on unit performance, as shown in Table 8.11. According to the simulation analysis results, the average cold island intensity for 2 × 2, 3 × 3, 4 × 4, 5 × 5, and 6 × 6 arrays is approximately 0.9 °C, 1.3 °C, 2.6 °C, 4.0 °C, and 5.5 °C, respectively. Based on the unit performance curve provided by the equipment manufacturer, the corresponding average COP reduction of the units is about 3%, 5%, 9%, 14%, and 19%, respectively.

Table 8.11 Impact of local cold island intensity on unit performance under different layout scenarios

Unit spacing	The spacing on both sides of the unit is 1 m				
Layout_spacing number	2 × 2_1-1	3 × 3_1-1	4 × 4_1-1	5 × 5_1-1	6 × 6_1-1
Average recirculation ratio of unit cluster	11%	17%	26%	37%	55%
Distance from the unit cluster in area where the temperature drops below 0.5 °C	<3 m	12 m	15 m	18 m	>20 m
Local cold island intensity (°C)	0.9	1.3	2.6	4.0	5.5
Impact on unit COP	↓3%	↓5%	↓9%	↓14%	↓19%

Table 8.12 Local cold island intensity and its impact on unit performance under different unit spacing conditions

Spacing on both sides of the unit		1 m	2 m	3 m	4 m	1.5_1 m
2 × 2	Average recirculation ratio of unit cluster	11%	20%	22%	22%	11%
	Local cold island intensity (°C)	0.9	1.9	2.1	2.1	0.7
	Impact on unit COP	↓3%	↓7%	↓8%	↓8%	↓3%
4 × 4	Average recirculation ratio of unit cluster	26%	34%	35%	37%	17%
	Local cold island intensity (°C)	2.6	3.8	4.0	4.0	1.7
	Impact on unit COP	↓9%	↓13%	↓14%	↓14%	↓6%

In summary, the total number of units has a significant impact on the local cold island, leading to the following conclusions.

(1) It is recommended that the number of units in a concentrated arrangement should not exceed 9, which can ensure the average local cold island remains below 1.5 °C and maintain the basic performance of the units unchanged. When 16 units are concentrated, the cold island intensity around the unit cluster is approximately 2–3 °C, leading to a performance decrease of about 5%–10% in the units.
(2) The cold island intensity in areas more than 15 m away from the unit cluster is less than 0.5 °C. Therefore, a concentrated arrangement of 9 units requires a footprint of 1300 m^2, while 16 units require 1600 m^2.

8.8.2.3 Influence of Unit Spacing on Local Cold Island

Taking the 2 × 2 array and 4 × 4 array as study objects, this study further analyzes the impact of unit spacing on unit operational performance. The statistical analysis results are shown in Table 8.12. The simulation results indicate that a larger spacing between units does not necessarily lead to a weaker local cold island effect. On the contrary, increasing the unit spacing reduces the relative pressure in the air inlet area of the units, making it easier for the discharged air to form a reflux and be reabsorbed into the air inlet. This finding aligns with previous research conclusions

(Wang 2019). Therefore, when multiple units are concentrated, it is recommended to space the units at a distance of 1.0–1.5 m on the windward side.

8.8.2.4 Influence of Ambient Wind Speed on Local Cold Island

The ambient wind facilitates the diffusion of local cold air, especially around the units on the windward side, thereby relatively reducing the cold island intensity of the surrounding air. However, as the ambient wind speed gradually increases, the wind pressure also rises. Excessive wind pressure will adversely affect the unit's air intake, leading to a reduction in the unit's air inflow and negatively impacting its operational performance. Table 8.13 illustrates the impact of incoming wind speed on the cold island intensity around the units and their operational performance under a 4 × 4 array layout. Table 8.13 indicates that when the ambient wind speed reaches 4 m/s, the average recirculation ratio of the unit cluster increases to a level similar to that under quasi-calm wind conditions, meaning it also approximately equals the least favorable conditions at this point.

Figures 8.91, 8.92, 8.93, 8.94, 8.95 and 8.96 illustrate the localized cold island distribution around the unit cluster and the wake diffusion under incoming wind speeds of 0.2, 1, 2, 3, 4, and 5 m/s respectively. As can be seen from the figure, the influence of ambient wind speed on the local cold island effect is not only seen within the unit cluster, but also more significant on the thermal environment in the downstream area of the unit cluster. The greater the ambient wind speed, the larger the downstream area affected by the local cold island effect.

8.8.3 Analysis of Maximum Heating Capacity of Air Source Heat Pump

Based on the above conclusions, for an urban ground area of 10,000 m^2, the maximum winter heating capacity provided by centralized air source heat pumps and the heated floor area are as follows:

(1) Analysis of usable area for equipment installation: The building density is 25%, with roads occupying 20%, and green space accounting for no less than 40%.

Table 8.13 Impact of local cold island intensity on unit performance under different incoming wind speeds

Incoming wind speed (m/s)		0.2	1	2	3	4	5
4 × 4, spacing of 1 m	Average recirculation ratio of unit cluster	26%	9%	17%	22%	26%	26%
	Local cold island intensity (°C)	2.6	0.8	1.5	2.1	2.6	2.6
	Impact on unit COP	↓9%	↓3%	↓6%	↓8%	↓9%	↓9%

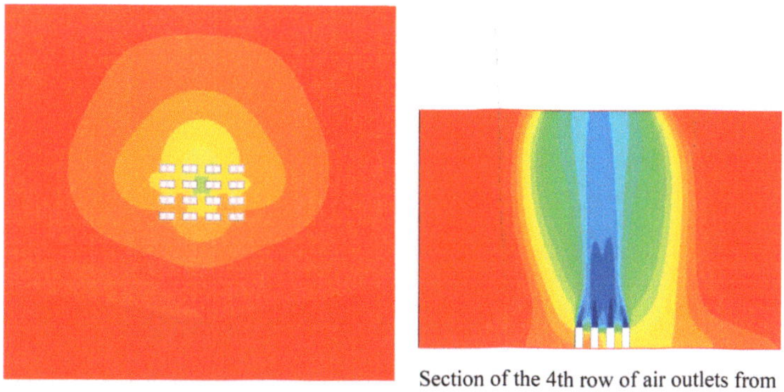

2 m height section

Section of the 4th row of air outlets from the left

Fig. 8.91 Local cold island intensity at incoming wind speed = 0.2 m/s (unit spacing = 1 m, scale same as above)

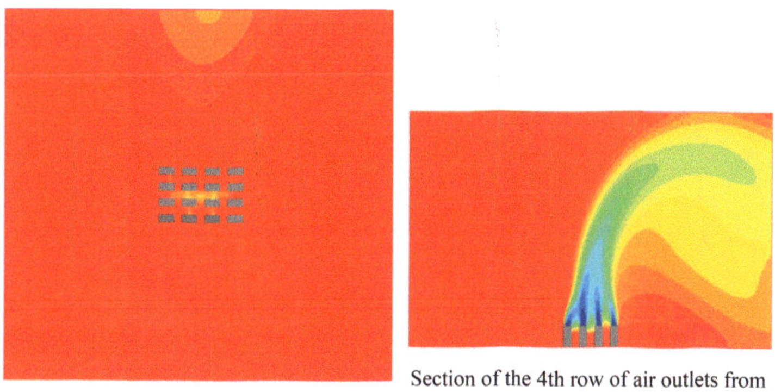

2 m height section

Section of the 4th row of air outlets from the left

Fig. 8.92 Local cold island intensity at incoming wind speed = 1 m/s (unit spacing = 1 m, scale same as above)

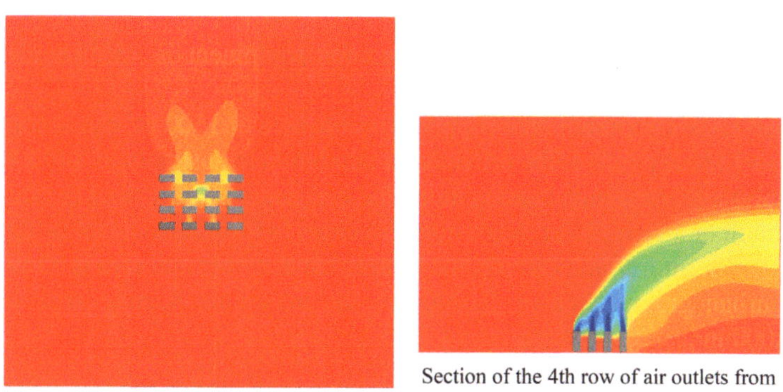

2 m height section

Section of the 4th row of air outlets from the left

Fig. 8.93 Local cold island intensity at incoming wind speed = 2 m/s (unit spacing = 1 m, scale same as above)

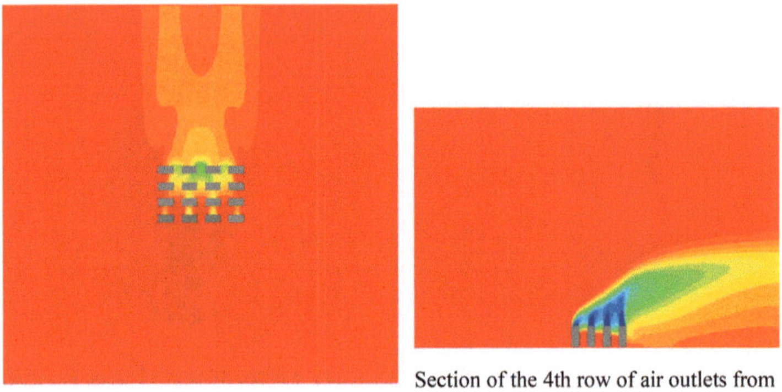

2 m height section

Section of the 4th row of air outlets from the left

Fig. 8.94 Local cold island intensity at incoming wind speed = 3 m/s (unit spacing = 1 m, scale same as above)

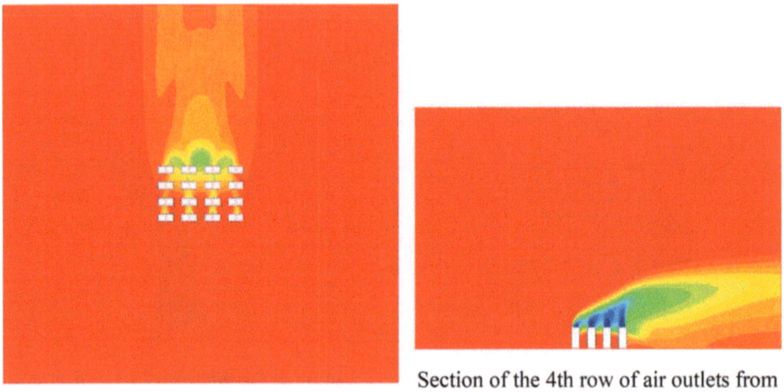

2 m height section

Section of the 4th row of air outlets from the left

Fig. 8.95 Local cold island intensity at incoming wind speed = 4 m/s (unit spacing = 1 m, scale same as above)

Therefore, the maximum usable area for equipment is 15%, equivalent to 1500 m^2.

(2) Analysis of maximum heat supply: a usable area of 1500 m^2 can accommodate up to 12 heat pump units, providing a maximum heat supply of 840 kW.

(3) Analysis of maximum heating area: The maximum heating floor area and maximum floor area ratio for major northern cities in China are shown in Table 8.14. Taking Beijing as an example, the heating capacity per unit area is 35 W/m^2. For an urban ground area of 10,000 m^2, the maximum heating area would be 24,000 m^2.

2 m height section

Section of the 4th row of air outlets from the left

Fig. 8.96 Local cold island intensity at incoming wind speed = 5 m/s (unit spacing = 1 m, scale same as above)

Table 8.14 Maximum heating floor area and maximum floor area ratio in major northern cities of China

City	Outdoor heating design temperature(°C)	Heating load per unit floor area (W/m²)	Maximum heating area (10,000 m²)	Maximum floor area ratio
Beijing	−7.5	35	2.4	2.4
Xi'an	−3.2	29	2.9	2.9
Zhengzhou	−3.8	30	2.8	2.8
Jinan	−5.2	32	2.6	2.6
Lanzhou	−8.8	37	2.3	2.3
Taiyuan	−9.9	38	2.2	2.2
Shenyang	−16.8	48	1.8	1.8
Hohhot	−16.8	48	1.8	1.8

Chapter 9
Governance Models for Future Urban Communities

9.1 Transformation of Urban Residential Pattern in China

After more than two decades of housing reform in China, the urban living pattern has transformed from the original integrated work-living and unit-affiliated compound model to a separated work-living and residential compound model. In the early stages of reform and opening up, urban housing in China was primarily allocated as welfare benefits by employers, tightly binding employment and residence. The dominant living patterns were "employer allocated compounds" or "family dormitory compounds." Under this model, the management of personnel in the compound, the operation and maintenance of housing, and public services including recreational and cultural activities were all organized by the employer. Therefore, despite the poor material conditions during this period, governance was orderly.

Since 1998, China's housing system reform abolished welfare-based housing allocation and initiated the commodity housing market nationwide. Urban habitable space began shifting from "employer-dependence" to "market choice," thereby giving rise to and gradually forming the dominant residential model in China today—the gated community. With the large-scale development of real estate, hundreds of thousands of new residential compounds have been built across the country. Some of these compounds are as small as a single building with just a few dozen households, while others are so vast that they exceed ordinary imagination. For example, Beijing's famous Tiantongyuan Residential Compound has over 600 buildings and a permanent population of one million. As of 2023, approximately 70%–85% of urban residents in China live in residential compounds, including commodity housing compounds, government-subsidized housing compounds, and employer reformed housing compounds, typically equipped with property service. In addition, a small number of urban residents live in non-community-style housing such as urban villages and self-built houses. However, with the advancement of national urban renewal, these residential forms will also undergo a comprehensive transformation into the residential compound model in the future.

Building Energy Research Center of Tsinghua University, *Decarbonizing Urban Residential Buildings*, https://doi.org/10.1007/978-981-95-8200-6_9

At present, urban areas in China have essentially established the residential compound living model, which is physically characterized by multiple high-rise collective dwellings arranged in an enclosed layout, collectively presenting a large-scale gated community pattern. Unlike the open block system commonly seen in Western cities, in China's enclosed residential areas, residential buildings are surrounded by walls or green spaces, with internal roads, green spaces, and clubhouses being shared spaces for owners. The entrances to these areas are under quasi-enclosed management. This form of residence represents a continuation of traditional employer allocated compounds in the era of the market economy. However, due to changes in property ownership and management models, it has also developed distinct characteristics that set it apart from traditional employer allocated compounds.

First, residential buildings are becoming taller and more densely packed. Urban collective housing in China is not only vast in quantity but also predominantly high-rise. The high-rise development of residential buildings improves land use efficiency, but also increases the difficulty of daily management and safety assurance, while posing higher demands on environmental health, security, and fire protection. A large number of residents are crowded into limited spaces, making daily life inevitably disruptive to each other, and minor disputes are more likely to escalate. This is reflected in various aspects of community life: disputes between neighbors are endless, such as upstairs residents' renovation noise disturbing downstairs residents, or cluttered public corridors affecting the environment. The high population density also poses challenges to safety and order, such as the occupation of fire-fighting access, high-rise residents throwing objects from heights threatening public safety, and haphazard parking obstructing roads, which may lead to severe consequences in emergencies. Neighbors are often unfamiliar with each other, lacking private communication channels when issues arise. Minor conflicts frequently require external mediation, further increasing community management costs.

Second, gated communities contain a large amount of public space and public facilities. Most residential compounds in China are enclosed by walls, which helps to enhance residents' sense of security and community cohesion. However, this enclosed layout also brings about the management of numerous shared facilities and public spaces, such as elevators, corridors, gardens, and fitness areas. The presence of shared facilities and spaces makes community management far more complex than in an open block. Under the open block model, the parts jointly owned and managed by owners are primarily limited to internal building facilities such as corridors and elevators, while public facilities fall under the centralized management of the municipal government. In large enclosed residential compounds, everything from security access control, road parking spaces to green space clubhouses requires maintenance and management by the residential compound.

In addition to the transformation of physical living spaces, the radical changes in property rights structures brought about by housing commercialization have profoundly reshaped the fundamental framework of community governance. Before the reform, urban housing was primarily provided by the state and employers, with residents only having the right to use it; after the reform, commodity housing became private property, and housing ownership shifted from public owners to private owners. Urban residents have transformed from dwellers dependent on welfare housing allocations by their employers to independent private property owners. In this model, residents' workplaces and residences are completely separated, and the governance rights over personnel, housing, and facilities begin to be distributed among diverse entities. The residents management is generally still the responsibility of their respective employers or companies, while housing is constructed by real estate developers. However, properties may be managed by property management companies or similar institutions. In most cases, community facilities are also maintained by property management companies, with residents paying fees to these companies in the form of property management fees. However, in the initial stage of property rights reform, the supporting systems were deficient, leading to many governance challenges: First, the management functions and public service functions originally borne by the state and employers were absent, and these needed to be provided by residents' self-governance or the market, but a mature operation mechanism has not yet been formed. Second, conflicts began to emerge among different stakeholders. For example, there were continuous disputes among developers, property management companies and owners regarding the income distribution of common property. Third, with the awakening of private property rights awareness, residents' attention to housing-related affairs has increased, and various property disputes have occurred more frequently. Moreover, the large number of residents, diverse demographics, and high mobility within the residential compound have made community governance structure more complex, posing a challenge to the traditional acquaintance-based society governance model.

Overall, the transformation of the physical form of residential compounds and the commercialization of housing property rights have jointly shaped the spatial foundation of urban communities with Chinese characteristics, while also presenting new challenges to governance. In 2023, the floor area of urban residential buildings in China reached 33.1 billion m^2. The per capita floor area in urban areas increased from 15.5 m^2 in 2000 to 35.5 m^2 in 2023, essentially reaching the level of developed countries in Europe and Asia. This marks a comprehensive and significant improvement in the living standards of the Chinese people. As China's urbanization has entered the second half of the process, a large number of urban residential buildings and communities have reached the stage requiring maintenance, repair, or renovation. The focus of national housing construction is shifting from large-scale new construction projects to the maintenance and renovation of existing residential buildings. This process is also accompanied by the in-depth implementation of national strategies such as building energy efficiency, carbon reduction, and low-carbon energy transition. The specific measures include electrification of building energy system, installation of distributed PV systems on residential rooftops, and creation of charging point networks for serving the electric vehicles that are quite common now. These emerging trends present entirely new challenges and higher demands for the community governance and facility management of urban residential compounds. In the face of the emerging issues in community governance, there is an urgent need to establish the community organizations and governance models that can manage the residential compounds as the primary living units. This will optimize the management of personnel, housing, and facilities within the community, further promote building energy efficiency and low-carbon retrofits, enhance public facility services, and elevate the quality of residents' cultural life. These are all critical issues in the future development of urban living space in our country.

9.2 Challenges in the Management of Residential Compounds in China

The governance of the current residential compounds primarily faces challenges in two aspects: one is the daily operation and maintenance management of hardware facilities, and the other is the organizational coordination of personnel at the governance system level. These intertwined issues manifest in every aspect of community governance.

From the perspective of hardware facilities, urban residential compounds possess extensive infrastructure in addition to the building itself. This primarily includes the following: **energy supply facilities:** covering basic energy supply systems such as electricity, gas, central heating, and tap water; **community service facilities**: including public activity areas, fitness and recreation equipment, parking lots, etc.; **environmental health facilities**: such as waste sorting collection points and disposal systems, community greening, and rainwater and wastewater drainage networks. **New Infrastructure**: with the advancement of green development strategies, many residential compounds have begun constructing distributed PV systems and adding new energy facilities such as charging points for electric vehicles.

As most urban residential buildings in China's residential compounds have been in operation for over twenty years, the hardware facilities within these compounds have begun to exhibit a series of issues, and various amenities and functionalities are gradually failing to meet current societal demands. In terms of the hardware of the residential compounds, the primary issue is the aged building structure and poor thermal performance. In China, most old multi-storey residential buildings constructed before 2000 were built as masonry-concrete structures or precast slab structures. At that time, the architectural design standards were relatively low, resulting in weaker bearing capacity and seismic performance of these building structures, which fall short of current technical specification requirements. Additionally, due to inadequate subsequent maintenance and management, issues such as differential settlement of foundations and aging of building structures are very common. Moreover, approximately 4.9 billion m^2 of urban residential buildings in China were constructed before 1990, and 8.3 billion m^2 between 1990 and 2000. The performance of the building envelope in these structures is significantly inferior to those constructed after 2010, resulting in notably higher heating demands during winter compared to buildings compliant with the 2010 energy efficiency design standards. Appropriate policy mechanisms and technical approaches should be implemented to promote energy-efficient retrofits for these buildings. Second, there is a severe shortage or aging of infrastructure, which urgently requires enhanced operation and maintenance, renovation, and even reconstruction. The trend of population aging in China has become increasingly pronounced. By the end of 2023, the population aged 65 and above accounted for 15.38% of the total population. However, in contrast, most residential buildings in China with 7 or fewer floors are not equipped with elevators, and there is a lack of various barrier-free and elderly-friendly facilities. Moreover, most old residential compounds in China face issues such as insufficient public activity spaces and inadequate public service facilities.

9.2.1 Ambiguous Property Rights of Public Space and Facilities

In the process of the transformation from the employer managed residential compound to the commercial residential compound in the country, there have been some ambiguous aspects regarding the property rights of public space and infrastructure. In a commodity housing residential compound, infrastructure such as heating pipelines, power distribution equipment, water supply and drainage pipelines, and PV generating units are generally regarded as part of the community's public facilities. The construction cost of such infrastructure has already been included in the owner's payments during the development of the residential compound, and their ownership legally belongs to all owners collectively. However, in the early stage of governance after the completion of the residential compound, the developer often appoints a preliminary property management company to control the residential compound, while the owners' committee has not yet been established. With ambiguous collective ownership entities, the developer and property management company may jointly dominate public resources. In such cases, the owner's rights and interests lack representation, making management chaos and profit conflicts more likely to occur. This often leads to situations where property management companies infringe upon the owner's rights and interests. Although the shared facilities in the residential compound are primarily intended to serve the owners, they may also generate economic benefits. For example, income generated from charging fees for residential parking spaces and charging points, advertising spaces in elevators, temporary rentals of public areas, and even subsidies earned by selling electricity from rooftop PV power generation to the power grid. According to laws and policies, the income generated from public facilities should be collectively owned by all owners and used for public expenses of the residential compound. However, in practice, the ownership and utilization of such public income often lack transparency. Property management company frequently fails to distribute the earnings to the owners, instead diverting them to cover operational costs or generate profits. In the long run, this will harm the interests of owners and sow the seeds of disputes: once owners discover that the property management company has privately profited from communal assets without benefiting the community, trust will erode, potentially triggering owner rights protection actions and contractual conflicts.

9.2.2 Management of Various Infrastructures by Multiple Entities

In a residential compound composed of commodity housing, infrastructure such as heating pipelines, power distribution equipment, water supply and drainage pipes, waste disposal, landscaping management, and PV generating units involves multiple management authorities. Water supply, electricity, and heating are managed by different state-owned enterprises, while the property management company handles environmental maintenance and interfaces with numerous owners. Management by multiple entities often leads to overlapping responsibilities and management gaps, creating a situation akin to "nine dragons governing the water." Unclear authority and lack of coordination result in either mutual buck-passing or multiple directives, ultimately leading to either neglect or increasing chaos in management. For example, a water supply pipeline leak in a residential area may involve three parties: the property management company, the water utility company, and the owner. The property management may argue that the municipal pipeline should be repaired by the water utility company, while the water utility company may claim that the area within the property boundary falls under the maintenance responsibility of the property management. Meanwhile, the owner is primarily concerned with having the issue resolved as quickly as possible. In this case, unclear authorities and responsibilities will lead to the issues being unresolved. Another example is the courtyard pipe network of the heating system in northern regions, where the ownership belongs to the owner of the residential compound, while the operation and management are handled by the heat supply company. This results in property management lacking the motivation to conduct routine maintenance and proper management of the pipe network, let alone the motivation for energy-saving renovations and operational optimization. Moreover, the multitude and dispersion of property rights holders result in a lack of clear responsibility and efficient decision-making mechanisms for public facility maintenance, leading to vulnerable infrastructure that is difficult to repair and a decline in environmental quality. Typical issues include: the elevator is shared by all residents of the building, but routine property management fees alone cannot cover the maintenance or replacement costs. Some owners are unwilling to bear the high maintenance cost, leading to prolonged unresolved elevator malfunctions; the upkeep of public spaces such as green spaces and children's playgrounds requires continuous investment. However, the "free-rider" mentality causes some owners to refuse to pay the associated fees, resulting in unresolved public issues and the emergence of the "tragedy of the commons."

9.2.3 *Unclear Mechanism of Revenue Distribution*

Due to ambiguities in property rights and management responsibilities, the revenue distribution mechanism also faces uncertainties, which may lead to infringement of owner rights. More importantly, the lack of sustainable revenue distribution and incentive mechanisms results in property management company having no motivation to actively invest in, operate, and maintain the residential compound. Consequently, the entire residential compound cannot sustain its daily operations, maintenance, and renovations through public income distribution, nor can it improve the living environment through reasonable investments and upgrades, making it difficult to achieve a positive cycle of financial returns. For instance, in the case of rooftop PV or the operation and maintenance of public spaces, if the property management company cannot derive operational benefits from optimized operations, they may only fulfill basic maintenance duties. There would be little incentive to further reduce energy consumption or enhance services, as diligent management of public facilities does not yield additional financial returns. In addition, for facilities requiring long-term maintenance investments (such as elevators and fire protection systems), if the benefit attribution is unclear and funds are insufficient, both the property management company and the owners may be reluctant to invest, leading to aging and disrepair of the facilities. Unclear and non-transparent income distribution not only dampens the enthusiasm of property management companies but also directly undermines the owners' motivation to participate in self-governance and maintain public facilities. The owners believe that the profits are taken by the property management company and they gain no benefit, thus they are unwilling to cooperate in matters such as paying for equipment replacement or introducing new technologies. Many residential compounds struggle to convene owners meeting or establish owners' committee due to unclear allocation of public revenues, leaving owners to feel their rights are undermined and lose confidence in collective management.

9.2.4 *Insufficient Sustainable Funding Sources for Operations and Maintenance*

The lack of funds is the direct cause of inadequate maintenance and facility issues in the residential compound. Property fees are the primary source of funding for daily maintenance and repairs, but many older residential compounds suffer from low or insufficient fee collection. Some residential communities, particularly those with welfare housing allocated by employers, even lack property management entirely. According to a survey conducted by Associate Professor Liu Jiayan from Tsinghua University on old residential compounds in Beijing, most property management fees are only dozens of fen (RMB 1 yuan = 100 fen) per square meter per month. In many residential compounds, the payment rate is as low as 20%–30%, and some residents even refuse to pay property fees, opting only to contribute a monthly sanitation fee of RMB 3. This results in property fees only covering basic security and cleaning services, serving as a safety net, and falling short of routine maintenance standards, making it difficult to establish daily inspections and periodic maintenance plans, let alone further planning for the optimization and improvement of the communal environment and facilities.

Residential special maintenance fund is another source of funding to ensure the normal operation of public facilities in residential compounds. According to China's *Property Management Regulations* and *Management Measures for Special Maintenance Funds for Residential Buildings*, all owners of commercial residential properties and public housing are required to contribute to the Residential Special Maintenance Fund. This fund is uniformly supervised by government departments or entrusted agencies of the owners meeting and is designated for the maintenance, renewal, and renovation of shared areas and facilities after the warranty period expires. This includes foundational components of the residence such as the bearing wall, walls, lobby, and staircase, as well as facilities like the elevator, fire protection systems, roads, and drainage pipes. However, the reality is that not all residential compounds have this fund. The system was established in 1998, and many older residential compounds built earlier did not collect maintenance funds, nor were they required to make retroactive payments afterward. Moreover, the process of utilizing this fund is extremely cumbersome, making it difficult to achieve in actual operating process. First, the property management company or the owners propose usage of the fund; then, it requires approval through discussion by two-thirds of the owners or at an owners meeting; in the third step, the property management company or relevant owners organize and implement the usage plan, submitting necessary materials to the relevant authorities to apply for fund allocation. Only after government authorities review and approve the application will the dedicated account of the management bank transfer the funds to the maintenance organization. In this process, the owner negotiation or convening of an owner's meeting often gets impeded. Data shows that by the end of 2022, the cumulative maintenance fund for commercial residential properties in Beijing had reached RMB 78.29 billion, with only RMB 7.807 billion utilized, resulting in a utilization rate of <10%. The utilization rate is even lower in many cities, with a large amount of public maintenance funds locked in bank accounts.

With the development of new energy and electric vehicles, many residential compounds have begun to introduce new infrastructure such as distributed PV power generation and electric vehicle charging points. However, issues arising from ambiguous income attribution, unclear management responsibilities, and undefined income distribution models have become particularly prominent. For instance, there are currently no clear regulations on who should invest in and construct the distributed PV systems installed in residential compounds, who should operate them, or to whom the profits should belong. In reality, some projects lack owner participation in decision-making, potentially leading to the developer or property management company directly signing contracts with energy companies. This may result in owner dissatisfaction with the income distribution later, giving rise to disputes.

Another example is the installation of charging points in residential compounds, where conflicts often arise between property management company and owners: the former worries that a large number of charging points may compromise the compound's power supply safety and increase management costs without necessarily benefiting from it, while owners fear unauthorized fees collection or obstruction from the property management company. Previously, some residential compounds experienced situations where property management company prohibited owners from installing private charging points under various pretexts or charged operators exorbitant "access fees." These issues essentially stemmed from conflicts arising from unclear income distribution, operational maintenance costs, and undefined safety responsibilities.

The unclear delineation of property rights, management rights, and income rights, coupled with the lack of funding sources, will not only directly impact the daily operational efficiency of the residential compound but also, in the long run, affect its environmental quality and property values. Clarifying and protecting the ownership of rights ensures that the economic benefits generated by the property and public assets owned by the owners belong to the owners. These benefits can then be reinvested into the further maintenance and management of the residential compound, directly contributing to the improvement of living standards and the preservation and appreciation of property values. This approach not only safeguards the rights and interests of the owners but also provides positive incentives for the enhancement of residential compound operations and services, ultimately promoting the healthy and sustainable development of residential compounds in the country.

9.3 Status Quo and Problems of Compound Governance System

Communities are the fundamental units and foundational building blocks of social governance in the country. Conceptually, a **residential compound** typically refers to a walled area composed of commodity housing, whereas a "**community**" often denotes an "**administrative community**." Thus, one community may encompass either a single residential compound or multiple compounds. Community governance encompasses not only the management of public affairs and assets but also involves providing community services (such as elderly care and childcare), handling public affairs (such as census-taking and epidemic prevention), shaping daily community culture (for instance, organizing community cultural activities), and mediating conflicts and disputes (such as resolving neighborhood disagreements), among other aspects. To reform primary communities, invigorate their vitality, and promote the energy-saving and low-carbon operation renovation of residential compounds, the most crucial driver is the reform and coordination of primary organizations and management. The recurring issues in hardware facilities also reflect the shortcomings in China's community governance, including organizational management models, property rights distribution, and fund operation.

Currently, there are multiple organization and management mechanisms in the community governance of urban residents. Government-led entities primarily include the subdistrict office and the neighborhood committee. The subdistrict office, as an outpost of the district/county government, serves all residents (including non-owners) with jurisdiction over community-wide public affairs. Its authority stems from government delegation, tasked with coordinating public matters, guiding the neighborhood committee, and mobilizing government resources to address resident needs. Both the neighborhood committee and the owners' committee are autonomous organizations. After reforms, the neighborhood committee has gradually taken on a dual role of executing government tasks while fostering resident autonomy. The subdistrict office directly guides and supervises the work of the neighborhood committee and indirectly participates in matters related to the owners' committee through collaboration between the neighborhood committee and the owners' committee. Meanwhile, the subdistrict office supervises and guides the establishment and reelection of the owners' committee in accordance with the *Property Management Regulations*, and intervenes in mediation when major disputes arise between the owners' committee and the property management company or owners. This collaboration and division of responsibilities reflect the complementarity between administrative power and primary autonomy in community governance. The primary-level CPC organization serves as the political core of community governance, coordinating resources through organizational embedding (e.g. community Party branches guiding neighborhood committee and owners' committee) and ideological leadership. The CPC organization encourages Party members to play an exemplary role in owners' committee, establishes a negotiation platform among the subdistrict office, property management company and residents to mediate conflicts and implement policies. Through "Party-building guidance," it ensures the correct direction of primary self-governance while strengthening the Party's political leadership and mass mobilization capabilities, serving as a bridge between the government and social autonomy.

9.3.1 Governance Organization with the Government as the Main Entity

Subdistrict office is an essential component of China's urban primary government, serving as an outpost agency of the district (county) government responsible for implementing urban community management and social services. The subdistrict office generally oversees multiple community neighborhood committees (neighborhood committees), which specifically manage community affairs, implement government functions, mediate residents' interests, and provide public services. They serve as a bridge between the government and residents in social governance.

The neighborhood committee is a mass self-governing organization that assists the government in community governance and service provision. The most basic level of social management in our country is the urban and rural "neighborhood committee" or "villagers' committee". Historically, the scale of a neighborhood committee resembled that of a modern residential compound. However, during the urban community reform from the late 20th century to the early 21st century, to meet the hardware requirements set by the Ministry of Civil Affairs (such as having elderly facilities, meeting rooms of a certain size, entertainment service facilities, etc., and having a specified number of resident households), many neighborhood committees were merged. If the service facilities or household scale of a neighborhood committee failed to meet the standards, residential compound-level committees would be consolidated. As a result, a single neighborhood committee now often governs multiple residential compounds. From then on, the urban neighborhood committee system gradually transitioned to a model combining "resident autonomy" with "government guidance." After the reform, the merged neighborhood committees oversee excessively large residential compounds, resulting in a lack of detailed understanding of specific conditions in individual compounds and diminishing the autonomous characteristics of many neighborhood committees. First, the majority of members in neighborhood committees are not residents of the community or neighborhood they serve. A significant portion of these committees do not include a single resident from the local community, with most members being external workers. Second, the salaries of neighborhood committee members are issued by the higher-level subdistrict (government). Furthermore, their daily work primarily involves executing directives from superiors, and the routine tasks assigned by higher-level governments are relatively cumbersome. In such circumstances, many neighborhood committees have lost the legally mandated function of resident autonomy, effectively becoming primary organizations tasked with fulfilling assignments from higher authorities, while lacking the actual resources and capacity to engage in genuine resident autonomy. This reality stands in stark contrast to the elections of villagers' committees in the rural areas, where all 533,000 administrative villages' villagers' committees are elected by villagers through voting under the supervision of the civil affairs departments at various levels responsible for managing primary political power.

9.3.2 *Governance Organization with the Owners as the Main Entity*

Owners' committee (owners' committee system) is an autonomous organization elected by all owners within a residential compound or commercial property, representing them to exercise collective management rights and safeguard their legitimate interests. The core objective is to achieve owner autonomy, oversee property service enterprises, manage communal affairs of the compound, and coordinate the relationship between owners and property management company. Comparing the neighborhood committee with the owners' committee reveals that the neighborhood committee is a hybrid of "administration + autonomy" in primary governance. Its service objects encompass all residents (without property ownership restrictions), including registered residents, non-registered populations, tenants, etc. Its core objective is to maintain community harmony and public interests. Funding sources include both community service revenues and government allocations, necessitating adherence to government guidance and the undertaking of certain administrative support functions. The owners' committee, on the other hand, is an autonomous organization representing the rights and interests of owners. Its service objects are limited to property owners and their shared interests, with the core objective of safeguarding owners' private and common property rights. Its functions are solely related to property management affairs, and it operates as a purely autonomous entity without direct subordination to the government. Although the legal framework for owners' committees has been continuously improved, with clear regulations at both national and local levels, in reality, there are not many residential compounds that have established such committees. Due to insufficient participation and governance capabilities among owners, the effectiveness of these committees remains quite limited in practical operation. According to data released by the Beijing Municipal Commission of Housing and Urban-Rural Development in April 2021, the establishment rate of owners' committees and property management committees in the city reached 90.1%, but owners' committees accounted for only 25% of the total.

A successful case of owners' committee governance is the Xinqing Yayuan residential compound located in Beilin District, Xi'an, Shaanxi. The scale of this residential compound is relatively small, with just over 200 households, which objectively reduces the difficulty of owner's autonomy. Through the active organization of a retired resident and the effective operation of the owners' committee, the living environment of the community has been significantly improved, safeguarding the rights and interests of the owners. The owners' committee of the residential compound has been in operation since 2011, supplemented by a board of supervisors comprising over a dozen members. New owners or those who are suitable and willing to engage in residential compound affairs may first join the board of supervisors. The owners' committee holds monthly regular meetings, and the board of supervisors also participates in the discussions to provide oversight. In addition to the owners' committee and the board of supervisors, an advisory committee has been established to include former, older members of the owners' committee and the board of supervisors. To ensure legal compliance, lawyers, community governance advisors, and full-time secretaries were also appointed. Under the active organization and arrangement of the owners' committee, the residential compound abandoned the traditional "all-inclusive" property management model in 2017 and adopted a remuneration-based system to hire the property company. This means all income (including utility fees and property management fees) belongs to the collective owners. The property company's annual budget must be approved by the owners meeting, and its revenues and expenditures must be reported to the owners. At the end of the year, they receive a certain percentage of remuneration as remuneration. Under the effective organization and governance of the owners' committee, the residential compound has not only significantly improved its environment but also managed to accumulate a considerable amount of public revenue annually. As of 2024, the owners' committee has completed hundreds of renovation projects without utilizing the principal of the compound's major repair fund. Over the span of 13 years, it has generated over RMB 6 million in shared benefits for all owners of the compound while ensuring that property management fees have remained unchanged for many years.

With the increasing demand for demolition and renewal of old residential compounds, some regions in China have begun experimenting with the owner renewal committee system, which has been successfully implemented in pilot projects. The owner renewal committee is essentially a manifestation of the owners' committee system in the specific context of renewal and renovation within a residential compound. It serves as the main organizing and implementing body for the autonomous renewal project of the residential compound, playing a significant role in advancing the renewal and renovation of older compounds. Since 2024, regions such as Zhejiang and Hubei in China have initiated the autonomous renewal model for aging compounds, where the autonomous renewal committee established by the owners plays a pivotal role.

The ZJUT New Village Project is Hangzhou's first pilot project led by the owner renewal committee for autonomous renewal, which successfully achieved in-situ reconstruction of the residential compound and significantly improved living standards. The residential compound was originally built as a faculty dormitories compound for Zhejiang University of Technology in the 1980s, with outdated supporting facilities. Four of its buildings have been identified as dangerous structures, posing significant safety hazards. In 2023, under government guidance, Gongshu District of Hangzhou demolished and reconstructed 13 dilapidated buildings in ZJUT New Village in line with the principles of "resident-centered, government-led, and financially balanced." The renewal model is a resident-centered project, organized by the Zhaohui Subdistrict Office of the Gongshu District People's Government in Hangzhou, and implemented by the Hangzhou Gongshu District Urban Development Group Co., Ltd. Each building elects a representative to form a resident-led renewal committee. With nearly 100% resident approval, the committee submits a renovation proposal to the government. The total reconstruction cost is approximately RMB 530 million, with residents contributing RMB 470 million (accounting for 80%), and the remaining portion covered by government policy subsidies, achieving an overall balance of revenue and expenditure. The successful promotion of this model demonstrates that the owner renewal committee, as the main entity carrying out compound-level in-situ reconstruction, not only effectively safeguards the rights and interests of owners in terms of governance but also accelerates the urban renewal process, tangibly improving the living standards and housing quality of the residents in the compound.

The Dangerous and Dilapidated Housing Renovation Project in Gezhouba Area at No. 10 Wangzhougang Road, Xiling District in Hubei was also implemented by using a similar resident renovation cooperative model, achieving market-oriented and collaborative transformation of dangerous and dilapidated buildings. The area originally contained the residential compounds for the families of the Gezhouba Water Control Project, with old compounds accounting for 47% of the total. The buildings were severely aged, and the supporting facilities were outdated. The renovation faces three major challenges: first, residents relied heavily on expropriation policies and lacked motivation for independent renovation; second, economic balance was difficult to achieve due to constraints from urban planning controls, resulting in low development indicators; third, there was insufficient policy and regulatory support, including a lack of supporting policies such as property rights cancellation, value assessment, and tax exemptions, making it difficult for the government to form a cohesive effort. To this end, a resident renovation cooperative composed of resident representatives was first established as the main entity of the renovation project. The cooperative then hired design units and consultants to draft the financial balance plan and to determine housing layouts, construction scale, and contribution ratios. The government was responsible for approving plans, providing policy incentives and subsidies, and coordinating relevant approval processes; market entities were responsible for offering professional services such as engineering construction, financing loans, and asset operation. After the cooperative signed the contract with the development enterprise, the Ministry of Housing and Urban-Rural Development would establish a dedicated account. Following the fund supervision model for guaranteed housing delivery, it ensured that project funds were used exclusively for their designated purposes. Each expenditure must be reviewed and approved by the relevant industry authorities and the cooperative, thereby enhancing the trust of the owners. At the same time, the government established standards and a tiered reward-subsidy system for each area, redirecting profits from higher-revenue areas to support plots with restricted floor area ratios or financial difficulties, and ensuring overall economic balance within the areas and reducing investment risk. Through this renovation model, the residential compound successfully achieved in-situ reconstruction, significantly improving living standards.

The aforesaid cases further emphasize the importance of an organizational entity capable of representing the owners' interests and possessing governance capabilities. Such an entity can act on behalf of owners in daily communal affairs management, investment decision-making, interfacing with regulatory authorities for approval processes, and procuring and coordinating professional services with market entities. This approach not only comprehensively safeguards the rights and interests of owners but also maximizes policy support and mobilizes market forces to efficiently advance daily management and improvement initiatives of the residential compounds.

9.3.3 Primary-level CPC Organization

The primary-level CPC organization serves as the cornerstone of community governance. By strengthening its political functions and organizational capacity, the CPC organization integrates Party leadership into all aspects of community governance, providing both directional guidance and institutional support for the process. In urban communities, a CPC organization (such as a Party committee or general Party branch) is typically established to lead all tasks within the community. Through CPC-led initiatives, the primary-level CPC organization unites various organizations and residents, transforming organizational strengths into governance effectiveness, which is a key to achieving community co-development, co-governance, and shared benefits. During his inspection in Ningxia, General Secretary Xi Jinping pointed out: "The CPC organization at the community level serves as the nerve endings connecting the Party with the grassroots masses, and must play a leading role within communities." According to the CPC's statistical communiqué on internal affairs, the coverage rate of grassroots CPC organizations has exceeded 99.9%. On one hand, the community CPC organization shoulders responsibilities such as promoting the Party's policies, implementing decisions from higher authorities, leading grassroots governance, and mobilizing and serving the masses. It integrates entities such as district-based units, "two new" organizations (new economic and new social organizations), as well as property management company and owners' committee into the Party-building network, facilitating resource sharing and collective decision-making. On the other hand, the CPC organization plays a role of political guidance and coordination support in residents' self-governance: Party members participate in community affairs by serving as building supervisors or volunteers, playing an exemplary role in conflict mediation, public security patrols, and public welfare services, and thereby enhancing the organizational level of residents' engagement in community governance. For example, during the pandemic control period, the community CPC organization coordinated volunteer teams to organize nucleic acid tests and material distribution, fully demonstrating its role as a stronghold in the battle.

The successful practices of level-level CPC organizations in community governance demonstrate that Party building and leadership can effectively enhance the organizational capacity, self-governance, and collaborative efficiency of communities. For instance, in the Baocui Huadu Community of Ruijing Subdistrict, Beichen District, Tianjin, a four-tier framework of "Community Party Committee - Grid Party Branch - Building Party Group - Party Member Center Household" has been established to deepen the integration of Party building with grid governance. Through micro-grid management, services are extended to the smallest units. By establishing a community "Greater Party Committee" and a Party building joint meeting system, the neighborhood committee, property management company, enterprises, and other stakeholders are coordinated to achieve resource integration, enabling efficient and routine resolution of community issues. Additionally, by introducing social organizations and volunteer teams, an upgraded version of the "Five Constants and Five Deliveries" service has been created, covering various public services such as child care and elderly support.

Another example is the Tinghu District, Yancheng, Jiangsu, which addressed the issue of "leaderless groups" through integrated Party-building efforts, transforming governance from a "one-man show" into a "grand chorus." The community has established a "red grid" of Party branches at all levels, with the subdistrict CPC organization at its core and the community CPC organization as the main artery. It has established a "red model" for urban community Party building demonstration sites and activated the "red cells" driven by Party members. This approach has effectively extended the community CPC organization downward, streamlined the organizational relationships of Party members, and constructed a four-tier organizational system: "Grassroots Party Committee - Community Party Committee - Grid (Area, Residential) Party Branch - Building Party Group." This ensures the comprehensive coverage of the Party's organization and work forces in residential areas, extending to buildings and reaching the people, and thereby smoothing the "nerve endings" that connect and serve the masses and achieving dual improvements in management and service efficiency. The second is to establish a community consultation and deliberation system, which has completely resolved the previous issues of "leaderless groups" and "fragmented efforts" in community governance. This system continuously motivates the enthusiasm, initiative, and creativity of the general public and local institutions in participating in primary-level community affairs, transforming neighborhood governance from a "one-man show" into a "grand chorus." As a result, a positive situation of co-development, co-governance, and shared benefits has been formed. Third is that they have expanded the service capabilities of primary-level CPC organizations from multiple angles, levels, and dimensions. Pioneering the "Tinghu Model of Integrated Childcare and Early Education" nationwide, they explored a new "elderly and child care" system led by the government, integrating enterprises, communities, families, and social organizations, and achieving remarkable results.

The aforesaid cases demonstrate that primary-level CPC organizations serve as the leading core and coordinating hub in community governance. First, the primary-level CPC organizations have played a leading role and strengthened the governance framework, such as by establishing and improving community CPC organizations, extending the CPC organization network to residential buildings and compounds, and forming "CPC organization grid" governance units. At the same time, institutional design is employed to "organize" various governance forces, ensuring the efficient operation of collaborative governance through rules and procedures. The second key measure is for primary-level CPC organizations to leverage their political strengths, taking the lead in coordinating various community stakeholders to form a cohesive governance force. For instance, the Party committee of Liangwu residential compound in Shanghai applied the "Three-Committee Collaboration" model, which integrates the neighborhood committee (community self-governance body), the owners' committee (owner self-governance entity), and the property management company into a joint operation. The CPC organization serves as the central coordinator for deliberation and decision-making, fostering a collaborative environment for shared governance and collective decision-making. CPC organizations can also engage diverse entities such as social organizations and local enterprises to jointly participate, meeting residents' varied needs through coordinated efforts of government, market, and society. Moreover, by leveraging the exemplary role of CPC organization members, primary-level CPC organizations can further cultivate self-governance leaders and establish resident self-governance teams, thereby enhancing residents' sense of ownership and revitalizing their capacity for self-governance.

In summary, primary-level CPC organizations serve as the leading and coordinating center in community governance. By integrating resources and mobilizing the masses, they foster a governance ecosystem characterized by multi-party participation and positive interaction under the guidance of Party building. These valuable experiences demonstrate that as long as the guiding role of Party building is fully leveraged, the people-centered approach is upheld, and continuous innovation is pursued, China's community governance capacity can be consistently enhanced, and the modernization goals of primary-level social governance will surely be gradually achieved.

9.4 Outlook on Future Residential Compound Governance Model

Community governance is the foundational component of social governance. A scientific and efficient community governance model is the key to maintaining a favorable living environment in residential compounds, achieving sustainable development, and continuously improving the community environment. The enhancement of community governance systems and capabilities is also a microcosmic manifestation of the modernization and deepening of the national governance system and governance capacity. China urgently needs to strengthen the modern community governance capabilities to better undertake the daily operation, maintenance, and renovation of residential areas, thereby improving the quality of living. A review of the issues in the operation and management of urban communities in China indicates that the primary contradictions lie in the following. First, the lack of robust organizational management systems and governance entities to coordinate the interests of residents and owners, leverage the strengths of various social forces, form unified opinions and decisions, and facilitate the implementation and advancement of community governance affairs; second, the ambiguous property rights, management rights, and revenue rights of community public facilities, which severely hinder the proactive engagement of various stakeholders in governance and the realization of a virtuous cycle.

Summarizing the existing governance models and institutional arrangements, along with successful case studies, we believe the future direction for the operation and maintenance of residential compounds and the community governance model should evolve as follows:

I. **It is necessary to establish a governance entity at the compound level that can harmonize the interests of owners and residents. Give full play to the leading role of Party building, as well as the communication and coordination capabilities of Party members and activists in the neighborhood committee and owners' committee.** With the residential compound as the basic organizational unit, establish a primary-level governance organization that fully represents the property rights and residential interests of the owners. This organization will serve as the main governance entity, uniformly undertaking the operation and management of various public facilities and equipment within the community. The organization should leverage the advantages of grid-based management to reach every owner and resident in the community, thereby safeguarding the owners' property rights and income rights over housing facilities, while also meeting the residents' right to housing and daily living needs.

II. **The entity shall act as the operating body to conduct unified management of public space and facilities within the residential compound.** On one hand, this entity represents the owners in income distribution and maintains a favorable living environment for residents. This entity should play a central coordinating role, capable of liaising with governments at all levels to promptly communicate and implement policies, staying abreast of the latest national policies and management requirements. It is also responsible for coordinating various types of energy infrastructure (water, heating, electricity), environmental infrastructure (greenery, waste), public service infrastructure (elderly care and child care), and new energy infrastructure (PV, charging points). Acting as the main interface for government, social organizations, and enterprises, it oversees public consultations, decision-making on plans, fund management, and income distribution.

III. **The entity shall establish a sustainable funding operation model, utilizing revenue from public space and facilities to reinvest in routine maintenance and renovation of the residential compound, as well as investments in PV systems or charging points, thereby enhancing the living environment and preserving/increasing property value.** In the process of infrastructure renovation, PV system, and charging point investment and construction: this entity represents the interests of the owners, coordinating residents and the owners to reach unified decisions. The government can play a role in providing policy support, standardizing qualifications and procedures, and offering institutional safeguards, while also addressing regional imbalances and providing financial guarantees for older communities. Enterprises provide the services as entrusted, such as carrying out renovation and construction projects, offering facility inspection and maintenance services, and installing and operating PV systems. In this mode, coupled with a well-designed fund operation and management plan to incentivize rational energy usage among users, it not only facilitates easier unification of the residential compound owners in terms of operation, maintenance, renovation, and investment but also further motivates residents' enthusiasm. This promotes the implementation of various renewal measures and the installation and operation of PV facilities.

This ensures the unified and orderly operation of various communal facilities in management, fully leveraging the strengths of diverse forces. Through the effective operation of various facilities, it can not only enhance the residential environment, optimize community services, and fulfill the owners' needs for property value preservation and appreciation, but also tangibly improve the living experience of residents. This modern community governance system characterized by "Party-building guidance, resident autonomy, government support, and market collaboration" not only aligns with China's national conditions and the realities of urban community evolution but also meets the aspirations of the people for a better life in the new era.

9.5 Distributed PV Application Modes in Residential Compounds

In recent years, the BIPV in China's urban residential buildings has experienced rapid development, ushering in a new era of BIPV technology. In the new development era, distributed PV systems in urban residential buildings are also facing new challenges. The primary issues include the inability to fully utilize distributed PV power and the problem where the "basket" is more expensive than the "vegetables."

First, the consumption of distributed PV power across various regions has become a critical issue. The grid-connected capacity of PV systems in multiple regions has reached the stability limit, leading to prominent issues such as insufficient regulation capability, restricted reverse power flow, and excessive voltage deviation. These conditions currently make residential distributed PV systems unsuitable for grid connection, resulting in the designation of "red zones" for distributed PV grid bearing capacity. Consequently, various regions have imposed requirements on distributed PV grid connection and consumption, encouraging distributed PV systems to prioritize self-consumption and local energy utilization.

Second, the second "bottleneck" of distributed PV is the "vegetable basket is more expensive than the vegetables" problem. The "vegetable basket" of distributed PV refers to the initial investment during the installation phase. Although the price of PV modules is below RMB 1/W, the cost of mounting materials and construction remains at least RMB 2/W, making the initial investment still relatively high. The "vegetable" of distributed PV refers to their price, with the actual cost being less than half of the initial investment. In addition, there is a significant disparity in the urban-rural development of distributed PV systems. Rural distributed PV faces challenges such as insufficient local consumption capacity and difficulties in power transmission, while urban building-integrated distributed PV grapples with the issue of "prioritizing installation over utilization." Although the installation scale of building-integrated PV systems is rapidly expanding, research on their usage patterns remains relatively scarce, with insufficient attention paid to the critical issue of how to achieve effective PV power consumption.

From the perspective of the management model of distributed PV systems in urban residential buildings, it primarily involves three major stakeholders: the community developer, property management company, and residents. The interest relationships among the three parties are illustrated in the figure. The construction of distributed PV systems is initially invested by the community developer, and this cost will be transferred into housing prices, effectively making it an investment by the residents. At the same time, the building external envelope required for distributed BIPV installation, such as roofs and exterior walls, belongs to the community residents. The residents lease their roofs to the property management company, which then provides BIPV services to users. These services can take various forms, including domestic hot water (DHW), electric vehicle charging points, and electricity for communal areas within the community. Therefore, a reasonable community management model must take into account the relationship between community property management company and residents, where determining ownership and benefits is the key to the future management model of community PV (Fig. 9.1).

In summary, it is necessary to optimize both technology and management models to achieve local generation and local consumption of distributed PV power, thereby enhancing their value. The new technologies for building-integrated distributed PV systems in communities have been elaborated earlier in this book. Next, we will take the management model of building-integrated PV systems as an example to focus on analyzing and discussing community management approaches, and finally propose recommendations for the future management model of community PV. Depending on the application of PV, there are currently four community PV management models: (1) BIPV + decentralized DHW; (2) BIPV + charging points; (3) BIPV + public area consumption; (4) BIPV + public building consumption. This section will provide a detailed introduction and discussion of the four community-based building distributed PV management models. It includes stakeholder analysis for each management model, calculates the distributed PV power consumption and value under different models, and finally offers recommendations for future community-based building distributed PV system management models.

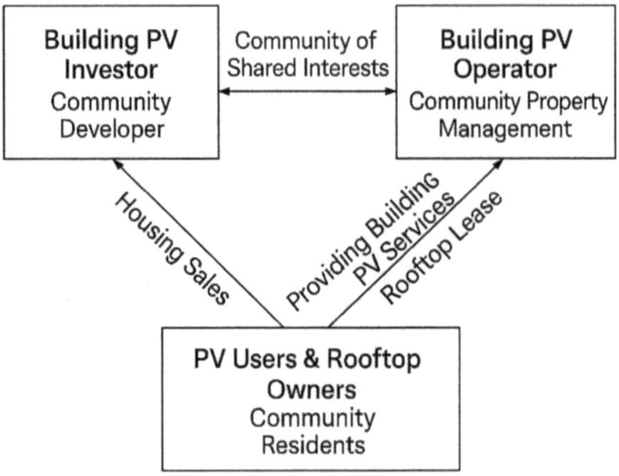

Fig. 9.1 Diagram of stakeholder relationship in community PV management model

9.5.1 BIPV + Decentralized DHW Model

The BIPV + decentralized DHW model refers to the use of BIPV for decentralized domestic hot water heating in each household of the community. Figure 9.2 shows the schematic diagram of a PV direct-drive AC/DC electric water heater connection. This type of product is already commercially available and can be applied to the domestic hot water system in residential communities. This model is suitable for newly developed communities, where the developer invests in installing BIPV and decentralized domestic hot water systems for each household during the community development process. The developer may also use this to increase property prices, thus ultimately making the investment effectively borne by residents, similar to traditional decentralized solar thermal hot water systems. In this operational model of the system, no PV fee is charged usually. Residents only need to pay the electricity bill generated when there is no solar heating. From the perspective of economic benefits, the investment and operation entities of this model are essentially the residents themselves. Therefore, the key to the sustainable development of this model lies in achieving 100% consumption of distributed domestic hot water by BIPV.

Specifically, field research on actual engineering cases indicates that the average daily domestic hot water consumption for bathing per capita ranges between 20–40 L. For a three-person household, the DHW demand would be

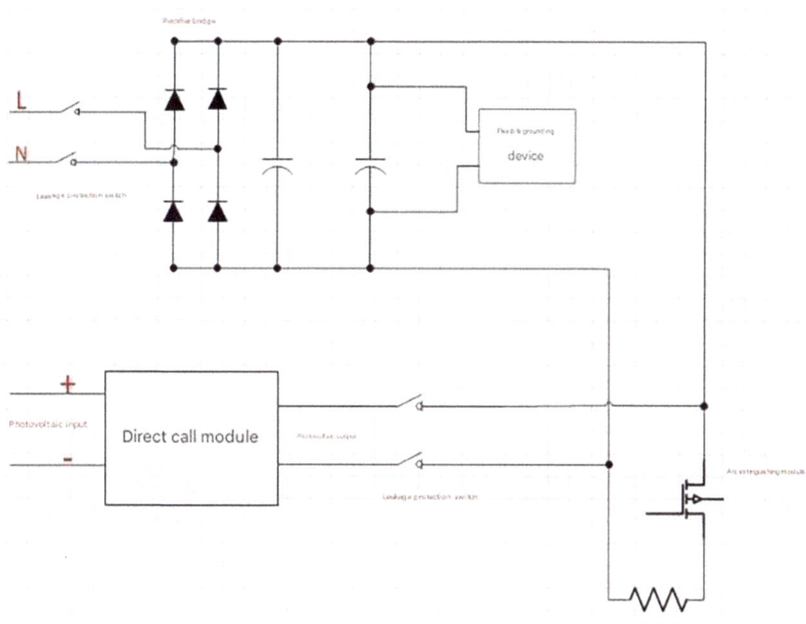

Fig. 9.2 Schematic diagram of PV direct-drive AC/DC electric water heater connection

60–120 L. Therefore, the capacity of the decentralized DHW storage tank can be set at 100 L (the tank is designed as a flat ceiling-mounted unit to maximize space utilization in residential areas, such as being installed in the toilet ceiling), with a corresponding temperature rise of 40 °C. Under ideal conditions without considering heat dissipation, the required rooftop PV electricity generation is calculated to be 4 kWhe. The PV area needed per household is 5–7 m^2. If a building's rooftop PV utilization area reaches 200 m^2 (corresponding to 3–4 households per floor with 6–13 storeys), 100% of the generated PV power can be consumed by the decentralized DHW system, achieving complete local generation and local consumption.

In the future, as living standards improve, the uses of DHW will continue to diversify, and DHW consumption will correspondingly increase. To achieve local generation and consumption and sustainable development under the BIPV + decentralized DHW model, this approach is more suitable for newly built communities with buildings under 13 storeys. During implementation, it is necessary to monitor the water tank temperature of each household when DHW is used. If higher DHW consumption leads to faster temperature drops in the tank, any additional heating energy costs incurred due to insufficient PV power generation should be borne by the respective users. In this model, the BIPV replaces the electricity typically needed for DHW heating. Therefore, the PV value in this mode is equivalent to the residential electricity price of RMB 0.5/kWhe, slightly higher than the grid feed-in value of PV power. This can establish a sustainable PV power utilization model characterized by local generation and consumption.

9.5.2 BIPV + Charging Point Model

The BIPV + charging point model involves directly connecting the BIPV system to electric vehicle charging points in the community parking lot, meeting the daily charging needs of residents for commuting. In this model, the initial investment and operation of the BIPV system and charging points are both managed by the charging point company, with the primary goal of achieving local generation and consumption of BIPV power. On one hand, electric vehicles, as distributed mobile energy storage units, have their battery capacity significantly increased in recent years. Taking private electric vehicles as an example, statistics indicate that in 2023, the average battery capacity of private passenger electric vehicles in China reached 45–50 kWh, with some models even exceeding 100 kWh [1], while the average daily electricity consumption for urban private cars was only about 7 kWh [2]. If each household is equipped with an average of 10 m^2 of PV panels, its daily electricity generation would be 6–8 kWhe, meeting the average daily travel electricity consumption of urban private cars, which is approximately 7 kWhe. However, not every household in the community owns an electric vehicle, so the average PV design scheme of 10 m^2 per household can be further optimized based on the electric vehicle penetration rate in the corresponding area to achieve full PV consumption.

In terms of total volume, the daily PV electricity generation may roughly match the average daily charging demand of electric vehicles in the community. If the system is equipped with an energy storage device, the operational stability can be further enhanced. In this case, the additional initial investment and operating costs are attributed to the installation, control, and maintenance of the energy storage system. From the perspective of controlling initial investment, to achieve full consumption of PV power without installing an energy storage device, it is necessary to consider the proportion of residents' electric vehicles parked in the community during the day. Extensive statistical data indicates [3] that private electric vehicles in typical cities spend approximately 39%–67% of their time parked in residential building parking lots. Among these, the estimated proportion of daytime parking in communities is around 50%. Further consideration is given to whether the peak PV power generation during the day aligns with the capacity of community charging points and whether the capacity of the electric vehicles parked in communities can consume the real-time peak PV power output. This issue requires further research and design.

The BIPV in this model replaces the user's usual demand for charging electric vehicles. This model is suitable for newly built communities, as well as existing communities with sufficient BIPV and parking spaces. The PV value in this model is equivalent to the charge price of RMB 1.15/kWhe (electricity price of RMB 0.5/kWhe + service fee of RMB 0.65/kWhe) at a charging station. The community's charging point company can reasonably reduce prices based on the utilization of BIPV, maximizing the PV value and fostering sustainable development of the PV charging point services for the residents (Fig. 9.3).

9.5.3 BIPV + Public Area Consumption Model

The BIPV + public area consumption model refers to the application of community-based distributed PV systems to public area electricity usage, such as the operation of public area elevators, underground parking lot lighting, security systems, and public area illumination. The investment for this model comes from the community developer, while the operation is managed by the community property management company. Both parties often form a collaborative entity. The advantage of this model lies in avoiding complex interactions with the power grid while helping the property management company achieve electricity cost savings. However, its primary limitation lies in the temporal mismatch between electrical load and PV power generation. For instance, during peak daylight hours when PV generation is high, many residents are away from home, resulting in low electrical load. The demand for the lighting system in public areas is minimal, making it challenging to fully utilize PV output solely through systems such as elevators in public spaces. Therefore, it is crucial to thoroughly consider the peak generation patterns of existing PV resources within the community. Poor design could lead to significant waste of these power generation resources. The focus of this model is the alignment

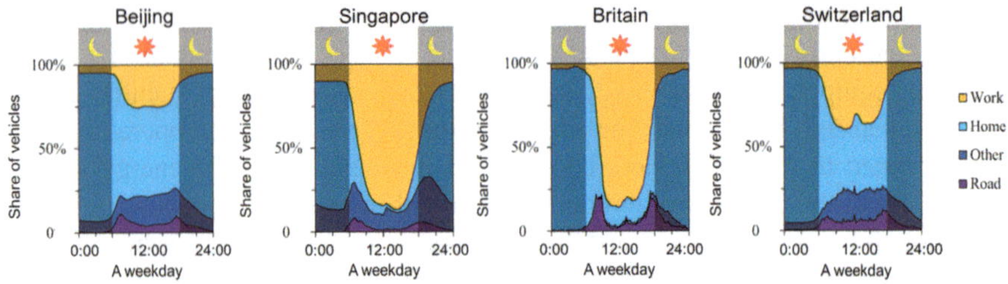

Fig. 9.3 Vehicle parking locations in different regions on weekdays

between the load curve of public area electricity consumption and the distributed PV power consumption curve of buildings, requiring specific design and optimization from the early stages of community development.

In this model, BIPV replaces the public area electricity demand of the community. Since it requires connecting BIPV to the public area power consumption system, it is more suitable for newly built communities. Certainly, it is also applicable to existing communities where the difficulty of retrofitting public area electricity system is low. The PV value in this model is equivalent to the community public electricity price of RMB 0.5/kWh, slightly higher than the feed-in tariff. This model should prioritize real-time PV power consumption, as adding an energy storage system might reduce the PV value. Regarding the community management model, the electricity consumption in public areas is covered by the community and reflected in the residents' property fees. Once public area electricity can be supplied by PV power generation, transparent billing information can be used to correspondingly reduce residents' property fees, achieving a benign operational model.

9.5.4 BIPV + Public Building Consumption Model

The BIPV + public building consumption model refers to the application of community-based BIPV systems to small-scale public buildings with commercial electricity usage in mixed-use residential-commercial communities, as well as nearby large-scale public buildings. Relevant incentive policies have already been introduced for this model. On December 5, 2024, the National Energy Administration issued the Guiding Opinions on Supporting the Innovative Development of New Business Entities in the Power Sector. The Opinions point out that new types of business entities refer to various resources in the distribution segment with power and electricity regulation capabilities, characterized by new technologies and operational models. They are categorized into single-technology new business entities and resource-aggregated new business entities. Among them, single-technology new business entities primarily include distributed energy resources such as distributed PV, decentralized wind power, and energy storage, as well as adjustable loads; resource aggregation-based new business entities mainly include virtual power plants (load aggregators) and smart microgrids. A virtual power plant is an advanced power operation and organization model that utilizes digital and intelligent technologies to aggregate distributed energy resources and adjustable loads, enabling coordinated participation in system operation and market transactions. A smart microgrid is a small-scale generation, distribution, and consumption system that primarily utilizes renewable energy sources, possesses certain intelligent regulation and self-balancing capabilities, and can operate independently or in conjunction with the main power grid. A source-grid-load-storage integration project in which the power distribution process has the aforesaid characteristics can be regarded as a smart microgrid.

In this model, BIPV replaces the electricity demand of public buildings in the commercial district formed around the community. This approach is applied because the BIPV system needs to be connected to the power grid of the adjacent commercial district. It is particularly suitable for newly constructed buildings where residential communities and neighboring commercial district are planned as an integrated unit during the design phase. The equivalent value of PV energy in this model is calculated at the commercial electricity rate of RMB 1/kWhe. This model still needs further research and practical application. Technically, there are still problems in regulating and operating the consumption of community PV by public buildings, which needs to be studied. In terms of the management mode, it is necessary to study how to enhance the value of community PV and achieve a win-win model for public buildings and the community.

9.5.5 Summary of Application Models for Distributed PV in Residential Compounds

At present, the new challenges faced by distributed PV in urban residential building communities in the new development period are the difficult consumption of distributed PV and the problem that the "vegetable basket more expensive than the vegetables." To address these problems, it is necessary to achieve the local generation and consumption of distributed PV power through the optimization of technology and management models, and to enhance the value of distributed PV. The discussion and exploration of four BIPV management models are summarized in the table below. Communities need to select an appropriate distributed BIPV management model based on the specific needs of residents during the development and operation process, thereby achieving full utilization of distributed BIPV and enhancing its value (Table 9.1).

This article takes the management mode of distributed BIPV in community buildings as an example to conduct a preliminary discussion and analysis on the future management mode of the community. The key element in the management mode is to achieve the self-consumption and full matching of renewable resources in the community, and to enhance the service value of renewable resources to community residents as much as possible. Therefore, this model encourages community property management to provide better comprehensive services and management. The property management company can also effectively return the value created by

Table 9.1 Four types and three grid connection modes of distributed PV power

Management mode	Applicable type	Equivalent PV value (yuan/kWhe)	Operation mode
BIPV + decentralized DHW	Newly built community	0.5	Developer investment, residents' use, no separate charges, only electricity fees for non-solar heating
BIPV + charging point	Newly built community and existing community	1.15	Investment and operation by charging point company, charged price adjusted according PV power consumption
BIPV + public area consumption	Newly built community and existing community	0.5	Investment and operation by community developer and property management company, no need for complex interaction with the power grid, lower electricity price
BIPV + public building consumption	Newly built community	1.0	Investment and operation by community developer, connection to public building power system, unified planning and optimized consumption efficiency

BIPV to community residents, so as to improve residents' satisfaction with the services of renewable resources, form a good cooperative relationship between community property management company and residents, and ultimately achieve the sustainable development of renewable resources such as BIPV in community buildings.

In the future, a comprehensive management approach needs to be established: First, clarify the property rights of renewable resources, and through forms similar to community conventions, reasonably allocate the utilization and benefits of renewable resources; second, establish a negotiation and management mechanism between the community property management company and community residents to realize reasonable joint decision-making; finally, during the operation and maintenance process, make the benefits transparent and public to improve the satisfaction of community residents and achieve the sustainable development of community renewable resources.

Chapter 10
Policy Standards for Urban Residential Buildings

10.1 Urban Residential Building Energy Efficiency Standards and Policies

10.1.1 Overview

Energy efficiency standards and labeling, as a crucial energy-saving management system in the field of end-use energy consumption, serve as an important means for all countries to address energy supply shortages and achieve sustainable development. International experience has proven that when the energy efficiency standards and labeling are applied to end-use energy products such as household appliances, which have a high penetration rate and consume a large amount of energy, the benefits obtained by the country will far exceed the costs increased for implementing the energy efficiency standard label project and producing high-energy-efficiency products. China's energy efficiency standards and labeling programme originated in the 1980s and the beginning of this century respectively. The former is the technical basis and criterion for the industry, while the latter is the information transfer and product promotion carrier for consumers. The two are closely coordinated and dovetailed with each other, jointly promoting the advancement of energy conservation efforts in the field of end-use energy consumption.

As of now, through standards integration, China has issued 60 energy efficiency standards (see Table 10.1), covering five major categories of products, including 17 for household appliances, 9 for commercial equipment, 7 for office equipment, 8 for lighting fixtures, and 19 for industrial equipment. In addition, energy efficiency standards have been established for key energy-consuming systems such as motors, pumps, and lighting, promoting energy-saving efforts in an orderly manner from individual equipment to energy-consuming systems. This facilitates the coordinated improvement of overall energy efficiency in key industrial sectors. The issuance and implementation of energy efficiency standards play a crucial role in guiding orderly

Building Energy Research Center of Tsinghua University, *Decarbonizing Urban Residential Buildings*, https://doi.org/10.1007/978-981-95-8200-6_10

Table 10.1 List of published energy efficiency standards

No.	Product field	Product name
1	Household appliances	Household refrigerators, room air conditioners, electric washing machines, electric fans, electric water heaters, gas water heaters, flat-panel TVs and set-top boxes, solar water heating systems, range hoods, heat pump water heaters, water dispensers, household gas cooking appliances, AC ventilation fans, dishwashers, smart toilets, air cleaners, kitchen appliances
2	Commercial equipment	Unitary air conditioners, multi-split air conditioners, commercial gas cooking appliances, motor-compressors for air conditioners, air duct air conditioning units, commercial induction cookers, heat pumps and water chillers, commercial refrigerators, cold storage (boxes), and compressor-condensing units
3	Office equipment	Computer displays, copiers, printers and fax machines, external power supplies, microcomputers, projectors, data centers, tower and rack servers
4	Lighting fixtures	Ballasts for gas discharge lamps for general lighting, fluorescent lamps for general lighting, high-pressure sodium lamps, metal halide lamps, LED products for interior lighting, tungsten halogen lamps, LED flat panel lamps for general lighting, LED luminaires for road and tunnel lighting
5	Industrial equipment	Motors, air compressors, blowers, centrifugal pumps, power transformers, AC contactors, industrial boilers, centrifugal blowers, electric welding machines, heating furnaces for petroleum industry, permanent magnet synchronous motors, high-voltage three-phase squirrel-cage asynchronous motors, submersible pumps, water electrolysis hydrogen production systems, petrochemical centrifugal pumps, regenerative rolling reheating furnaces, wastewater mixers, dust collectors, and aerators

market competition, promoting advancements in energy-saving technologies, and balancing the "green barriers" in international trade.

10.1.2 Progress in Energy Efficiency Improvement of Major End-Use Products

(1) **Household refrigerator**

The first edition of the energy efficiency standard for household refrigerators *Minimum Allowable Values of Energy Efficiency and Testing Methods for Household Refrigerators* (GB 12021.2-1989) was implemented on December 1, 1990, with the energy efficiency index limit for refrigerators and freezers set at 120%. The second edition of the energy efficiency standard took effect on April 1, 2000, followed by the third edition on November 1, 2003, and the fourth edition on April 1, 2008. The fifth edition of the energy efficiency standard was implemented on November 1, 2016, with the energy efficiency index limit for refrigerators and freezers reduced to 70%. A comparison of the minimum allowable value of energy efficiency for refrigerators and freezers across various editions of the standards is shown in Fig. 10.1. Compared to the first edition of the energy efficiency standard in 1989, the absolute value of the mini-

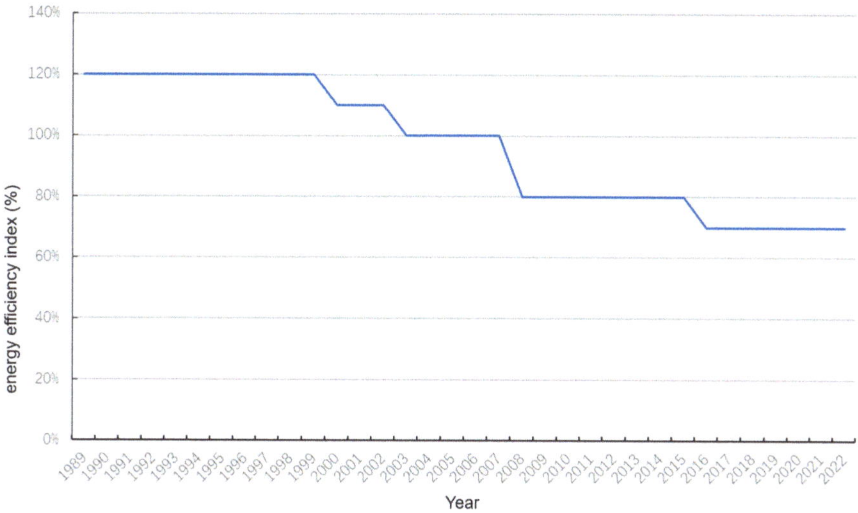

Fig. 10.1 Refrigerator-freezer energy efficiency standards across editions: energy efficiency index threshold values

mum allowable value of energy efficiency for refrigerators and freezers has been significantly reduced.

The household refrigerator industry has been using energy labels for eighteen years, with the models of Grades 1 and 2 energy efficiency now accounting for as high as 83%. Taking the mainstream refrigerators and freezers as an example, the proportion of Grade 1 energy-saving products has undergone nearly three tidal-wave-like surges, as shown in Fig. 10.2. From March 2005 to mid-2009, the refrigerator market completed its first round of high-efficiency transformation, with the proportion of Grade 1 product models reaching a peak of 77%. After the upgrade and implementation of the energy efficiency label in 2009, with the coordinated promotion of the People Benefit Project in 2012, the proportion of Class 1 product models rose rapidly again, reaching 90% during the People Benefit Project period. After the conclusion of the People Benefit Project, the policy effect slightly receded, and the average energy efficiency grade of products experienced a minor decline. On October 1, 2016, after the implementation of the new standard for household refrigerators, the energy efficiency requirements for Level 1 were significantly raised to meet EU standards, while the requirements for Level 2 were also elevated. The threshold for energy-saving products increased, marking the beginning of the third round of high-efficiency transformation in the refrigerator market. Currently, Grade 1 models account for 56.8% of the market. Overall, the energy efficiency grade of refrigerator products is relatively high, and the market optimization and upgrading effect is significant.

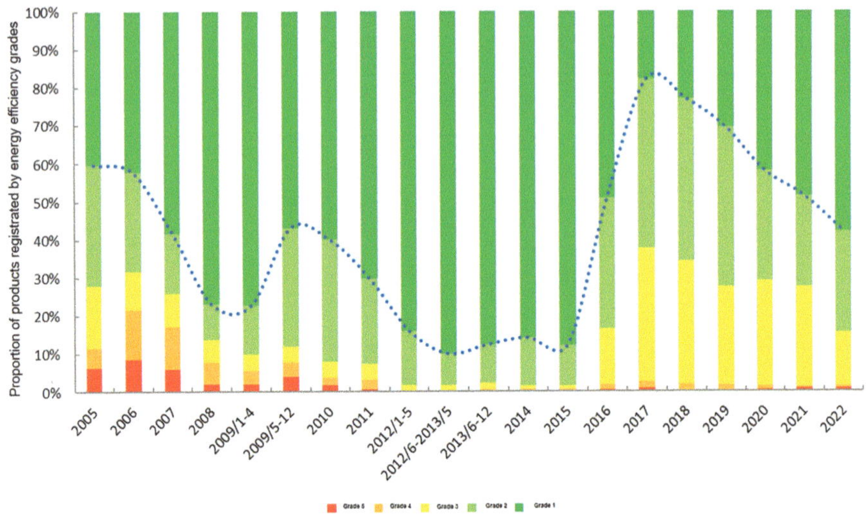

Fig. 10.2 Energy efficiency structure changes of household refrigerators over the years

(2) **Room air conditioner**

The first edition of the energy efficiency standard for fixed-frequency room air conditioners *Minimum Allowable Value of the Energy Efficiency and Energy Efficiency Grades for Room Air Conditioners* (GB 12021.3-1989) was implemented on April 1, 1990. The second edition of the energy efficiency standard came into effect on April 1, 2001, followed by the third edition on March 1, 2005. The fourth edition was implemented on June 1, 2010, and the fifth edition took effect on July 1, 2020. Taking a fixed-frequency room air conditioner with a heat pump and refrigerating capacity of 3500–4500 W as an example, considering the differences in test methods and evaluation indicators across various versions of the standards, the comparison of the minimum allowable value of energy efficiency for room air conditioners in each edition after fitting is shown in Fig. 10.3. Compared to the first edition of the energy efficiency standards in 1989, the minimum allowable value of energy efficiency has seen an improvement of nearly 80%.

The fixed-frequency room air conditioner is not only the earliest product to bear the energy efficiency label but also the product with the longest duration under the Energy-Saving Products People Benefit Project. It underwent two rounds of the project: from June 2009 to May 2011 and from June 2012 to May 2013. The coordinated implementation of energy efficiency standards and labeling and the People Benefit Project has played a significant and powerful role in driving the transformation of the fixed-frequency air conditioning product industry and market structure. This synergy has led to four tidal-wave-like shifts in the market share of Grade 1 and Grade 2 energy-efficient products, as illustrated in Fig. 10.4. After the implementation of the energy efficiency label on March 1, 2005, the market share of Grades 1 and 2 energy-saving products continued to

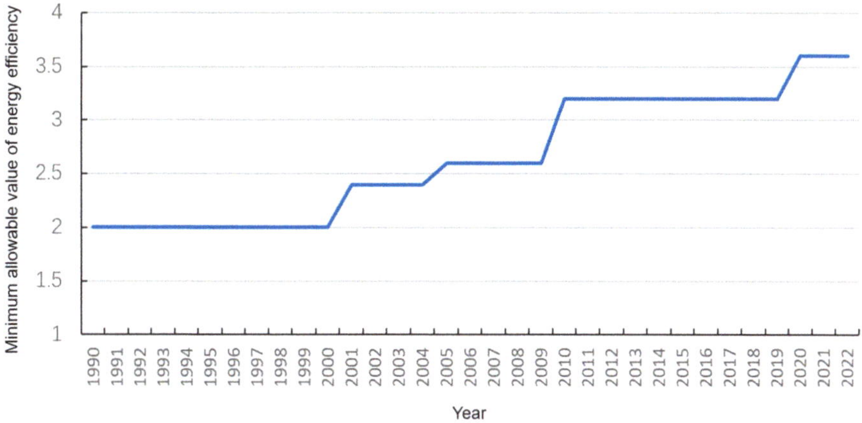

Fig. 10.3 Minimum allowable value of energy efficiency of fixed-frequency room air conditioner typical products (3500–4500 W refrigerating capacity, heat pump type) in each edition of energy efficiency standard

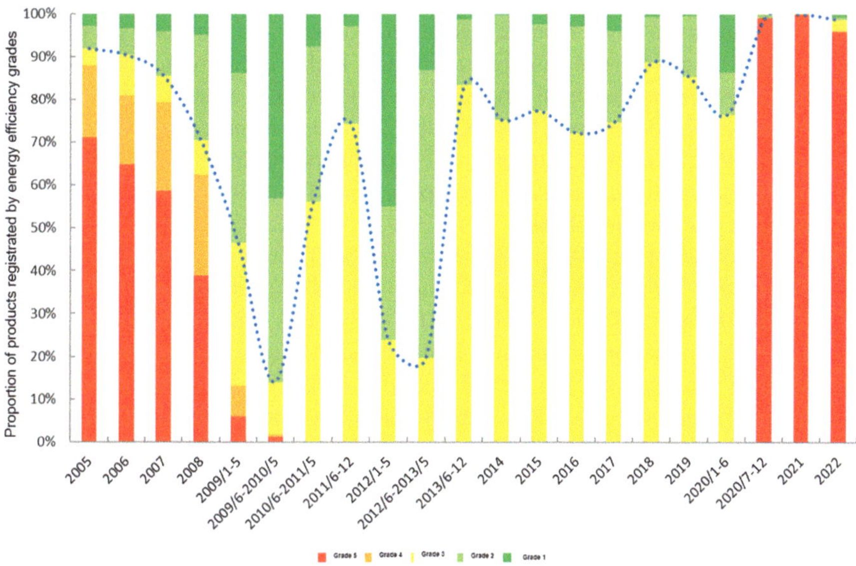

Fig. 10.4 Energy efficiency structure changes of fixed-frequency room air conditioners over the years

rise, reaching 53% before the implementation of the People Benefit Project in 2009. The overall energy efficiency ratio of products increased from 2.85 to 3.3, a growth of 14%. Under the combined effect of the first round of the People Benefit Project, the market share of Grades 1 and 2 energy-saving products reached a peak of 85% during 2009–2010, with the overall energy efficiency ratio of products reaching 3.47, a 22% increase. On June 1, 2010, the upgraded energy efficiency standard and labeling were implemented, with increased energy efficiency requirements. The market share of energy-saving products dropped back to a new reference value before restarting the shift toward high efficiency. In the first half of 2012, the proportion of energy-saving products rose again to 76%. With the synergistic effect of the second round of the People Benefit Project, it reached the peak of 80% during 2012–2013, and the overall energy efficiency of products reached around 3.5. Around 2013, the market share of variable-frequency air conditioners began to increase rapidly, while the market share of fixed-frequency air conditioners gradually shrank. The new energy efficiency standard for room air conditioners, *Minimum Allowable Values of the Energy Efficiency and Energy Efficiency Grades for Room Air Conditioners* (GB 21455-2019), consolidates *The Minimum Allowable Value of the Energy Efficiency and Energy Efficiency Grades for Room Air Conditioners* (GB 12021.3-2010) and *The Minimum Allowable Values of the Energy Efficiency and Energy Efficiency Grades for Variable Speed Room Air Conditioners* (GB 21455-2013). It officially came into effect on July 1, 2020. The new standard presents a more comprehensive and accurate evaluation method for the energy efficiency of air conditioning products, expands the scope of energy efficiency requirements to cover more refrigeration products, and promotes the development of air conditioning equipment towards higher energy efficiency. With the upgrading of the industrial structure, the proportion of energy-saving variable-frequency air conditioners (Levels 1 and 2) has significantly increased, reaching 68.2%; 98.2% of fixed-frequency air conditioners are Grade 5 energy efficiency products, which are gradually being phased out from the market.

(3) **Electric washing machine**

The first edition of the energy efficiency standard for electric washing machines, *The Limited Value and Testing Method of Energy Consumption for Household Electric Washing Machines* (GB 12021.4-1989), was implemented on December 1, 1990. The energy consumption limit for pulsator-type washing machines was set at 38 Wh/cycle·kg. The second edition of the energy efficiency standard was implemented on May 1, 2005, and the third edition on October 1, 2013, with the power consumption limit for pulsator washing machines set at 22 Wh/cycle·kg. Given the differences in testing methods across various editions of the standards, the comparison of the minimum allowable value of energy efficiency for pulsator washing machines in each edition after fitting is shown in Fig. 10.5. Compared to the energy efficiency grade corresponding to the minimum allowable value of energy efficiency in the first edition of the energy efficiency standard in 1989, there has been an improvement of nearly 56%.

The energy efficiency labeling for electric washing machines was implemented on March 1, 2007, and upgraded on October 1, 2013. As shown in Fig. 10.6, the proportion of Grades 1 and 2 energy-saving products steadily increased from 2007 to 2013, reaching its peak at approximately 85% before the implementation of the People Benefit Project in 2012. In June 2013, the People Benefit Project ended. In October, the upgraded energy efficiency standard and labeling were implemented, and the energy efficiency requirements were increased. The

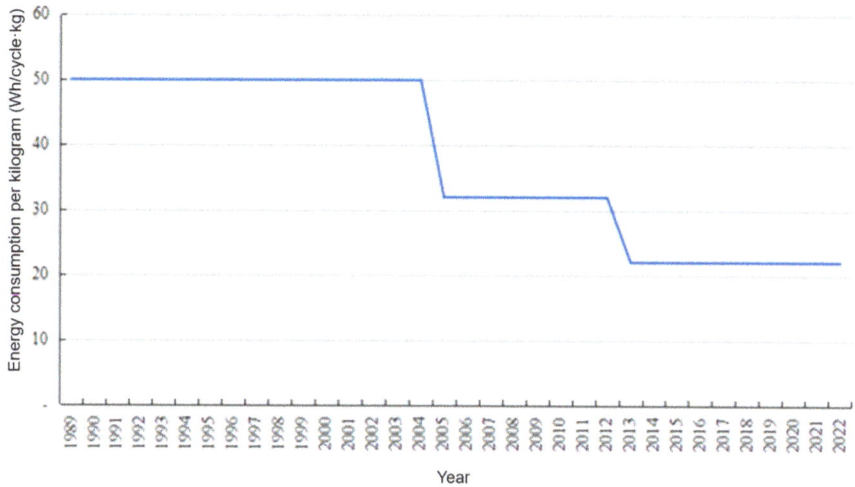

Fig. 10.5 Minimum allowable value of energy efficiency for pulsator washing machines in each edition of energy efficiency standard for electric washing machine

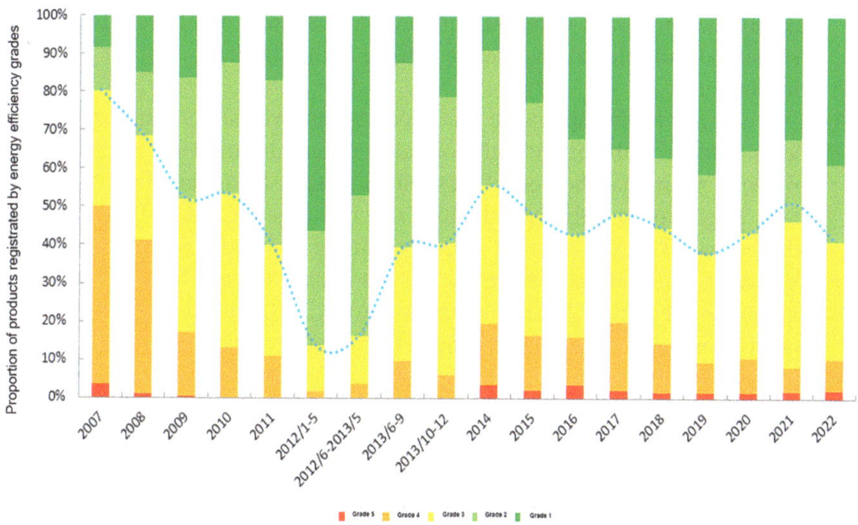

Fig. 10.6 Energy efficiency structure changes of electric washing machines over the years

proportion of registration for Grades 1 and 2 energy-saving products dropped back to a new reference value, initiating a new gradual improvement process in energy efficiency grades from low to high. Therefore, the upgrade of the technical requirements for energy efficiency standard and labeling has played the most crucial role in improving the energy efficiency level of washing machine products. Overall, since the implementation of the energy efficiency label for washing machines, the proportion of energy-saving products at Grades 1 and 2 has increased from 20% under lower energy efficiency requirements in 2007 to 59% under higher energy efficiency requirements in 2022. The effect of market structure optimization is evident.

(4) **Smart toilet**

A smart toilet is an innovative product that combines a traditional toilet with an intelligent control module. It not only features basic flushing functions but also integrates advanced user experience enhancements such as warm water cleaning (including rear wash and front wash), seat heating, and warm air drying. On December 31, 2019, China released its first national standard for the energy efficiency and water efficiency of smart toilets, titled *Minimum Allowable Values and Grades of the Energy Efficiency and Water Efficiency for Smart Water Closets* (GB 38448-2019), which officially came into effect on July 1, 2020. This standard stipulates three water efficiency grades of smart toilets, with Grade 1 indicating the highest water efficiency. The energy efficiency indicator is evaluated based on the energy consumption per unit cycle, which involves measuring and calculating the power consumption of the smart toilet over a 1.5-hour test period using the test method and calculation formula specified in the standard. The water efficiency indicator is assessed based on the water consumption for flushing and cleaning, which entails measuring and calculating the single-use water consumption of the smart toilet according to the test method and calculation formula defined in the standard.

(5) **Air cleaner**

As a specialized health appliance designed to improve indoor ambient air and address indoor ambient air pollution, air cleaners are used in various indoor environments such as homes, offices, public places, industrial plants, and hospitals. They play an active and effective role in mitigating indoor $PM_{2.5}$, formaldehyde, and VOC air pollution. Rapidly gaining popularity as special-purpose products, air cleaners have entered countless households and become one of the commonly used household appliances. From 2010 to the present, China's annual sales volume and sales revenue of air cleaners have maintained a growth rate of over 20%, while annual production has grown by more than 25%. The industry's development pace is remarkably rapid, and the growth rate along with the resulting total energy consumption cannot be overlooked. The energy efficiency standard *Minimum Allowable Values of the Energy Efficiency and Energy Efficiency Grades for Air Cleaners* (GB 36893-2018) was first released on November 19, 2018, and was revised in 2024. The updated standard will come into effect on October 1, 2025. The technical indicators specified in the minimum allowable value of energy efficiency and energy efficiency grade of air cleaners in this standard include the energy efficiency ratio and standby power. The energy efficiency grade is divided into 3 levels, with level 1 being the highest. Since the release and implementation of the standard along with the energy efficiency labeling, the proportion changes of air cleaners' energy efficiency grades in the market are shown in Fig. 10.7.

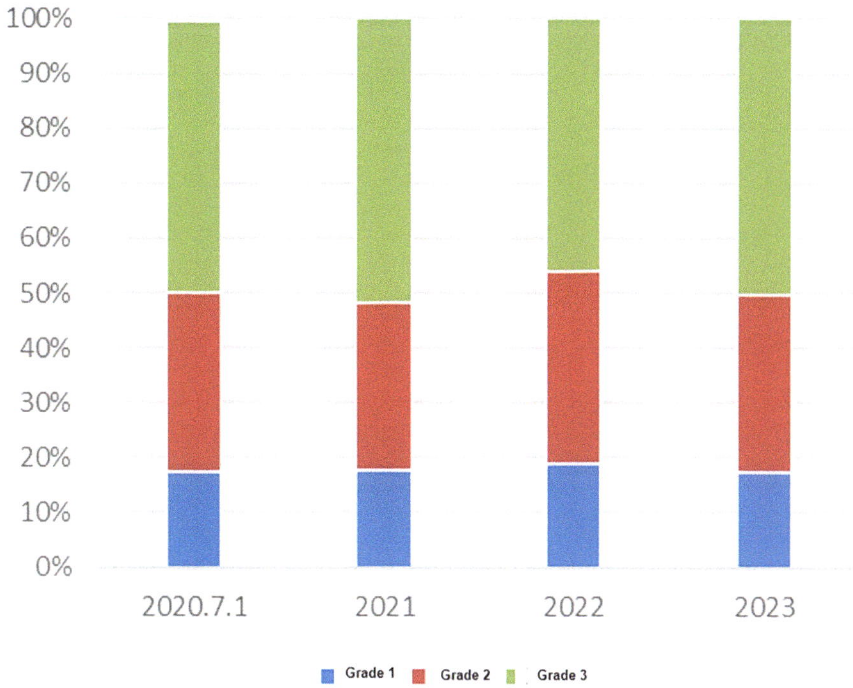

Fig. 10.7 Proportion changes of different energy efficiency grades of air cleaners

(6) **Indoor LED products for general lighting**

Starting from October 1, 2016, China began implementing energy efficiency labeling for non-directional self-ballasted LED lamps. On November 1, 2020, based on the newly released energy efficiency standards, energy efficiency labeling was extended to LED products for interior lighting (including LED downlights, directional integrated LED lamps, and non-directional self-ballasted LED lamps) as well as LED luminaires for road and tunnel lighting. Currently, non-directional self-ballasted LED lamps account for approximately 4% of models at energy efficiency level 1, 42% at level 2, and 54% at level 3. As shown in Fig. 10.8, the majority of registered models are dominated by energy efficiency Grade 3. From the perspective of technological R&D and product registration by leading enterprises, the initial luminous efficiency of top-tier non-directional self-ballasted LED lamps in the market is on par with international standards. However, judging by the overall energy efficiency registration status of the industry, the sector remains relatively conservative. Considering the factors such as product quality stability and market supervision, most manufacturers tend to register their products with efficiency grades lower than their actual performance. Consequently, the proportion of energy-saving products has shown a year-on-year decline in recent years. On November 1, 2020, the upgraded version of the energy efficiency standards and labeling were imple-

Fig. 10.8 Energy efficiency structure changes of non-directional self-ballasted LED lamp in the market over the years

mented, with higher energy efficiency requirements and an expanded range of product categories. The proportion of energy-saving products slightly decreased to a new reference value. Over the following two years, the share of energy-saving products increased year by year, reaching 49% by 2022.

10.1.3 Outlook

During the "15th Five-Year Plan" period, China's energy efficiency standards and labeling programme faces new development opportunities and challenges. The optimization and upgrading of standards will always be required. The requirements in the following aspects will have to be fulfilled in the future.

(1) Strengthen basic research and optimize implementation mechanisms. First, continuously optimize the organizational management and implementation system mechanisms, further clarify the responsibilities of all relevant parties, and enhance implementation effectiveness. Conduct research on incorporating online commodity transactions into the regulatory framework, and clarifying regulatory entities, models, and departmental responsibilities to effectively safeguard consumer rights. Second, improve the energy efficiency labeling implementation system. Implement the requirements of the national reform of "streamlining administration, delegating powers, and improving regulation and

services," continuously improve and optimize the functionality and process design of the energy efficiency label registration information system, and enhance the quality and efficiency of registration services. Enhance information disclosure, improve the readability and credibility of the information on energy efficiency label, serve consumers in selecting efficient and energy-saving products, and assist market supervision law enforcement officers in improving enforcement efficiency. Third, continuously conduct evaluations on the labeling programme. Summarize and analyze the implementation scope, efficiency, and energy-saving effects of energy efficiency standards and labels to provide data support for related policy implementation and project research. Support research and introduce supporting policies to expand the scope of energy efficiency label recognition and strengthen the implementation effectiveness of the labeling programme.

(2) Timely carry out the formulation and revision of standards, and expand the scope of labeling implementation. Implement the requirements of documents such as the *Green and High-Efficiency Cooling Action Plan, Implementation Plan for Establishing a Sound Standard Measurement System for Carbon Peaking and Carbon Neutrality, Guidelines for the Development of a Carbon Peaking and Carbon Neutrality Standard System, Guiding Opinions on Coordinating Energy Conservation, Carbon Reduction, and Recycling to Accelerate the Upgrading and Transformation of Key Product and Equipment*, and the *Notice on Further Strengthening the Update, Upgrade, and Application of Energy Efficiency Standards*. Aligning with the trends in consumer markets and industry development, accelerate the formulation of new energy efficiency standards for cold storage and servers, and promptly update the energy efficiency standards for products such as microcomputers, transformers, high-voltage motors, permanent magnet synchronous motors, water chillers, and small kitchen appliances. This will promote the application of energy-saving technologies and drive the high-quality development of the industry. Summarize the implementation experience of existing energy efficiency labeling for target products, continue to investigate the feasibility of applying energy labels to new products, and steadily expand the coverage of energy efficiency labeling.

(3) Improve the closed-loop management mechanism for energy efficiency standards throughout the entire process to better leverage the foundational and leading role of standards. At the current stage, there are still weaknesses in the process management of standard formulation, implementation, supervision, and evaluation. The updating, upgrading and application cannot fully meet the demands of the carbon peaking and nutrality strategy, and the comprehensive benefits of the green and low-carbon transition have yet to be fully realized. We will further improve the closed-loop management mechanism for the formulation, implementation, supervision, and evaluation of mandatory energy efficiency standards. Strengthen and complement industrial chains, accelerate the establishment of a statistical survey and monitoring system for standard implementation, build an evaluation framework for energy-saving technologies and standards, and create a rapid-response mechanism for standard formulation and

revision triggered by information feedback. Based on the PDCA cycle concept, achieve smooth operation, tight connection, positive feedback loop, and systematic iterative improvement in the four stages of standard formulation, implementation, supervision, and evaluation. All parties involved must ensure timely problem control through sensitive and accurate information feedback and rapid response improvements, preventing issues such as lagging standard development updates behind industry progress, inadequate supervision of standard implementation, and insufficient support and leadership. This guarantees that the standard content remains advanced, promptly upgraded, and effectively applied, and achieves the desired outcomes.

(4) Strengthen the development of experiment and testing capabilities, and promote the improvement of energy label information compliance. Strengthen the management, capability, and integrity awareness of testing laboratories, encourage consistency comparison of data, promote transparent and open testing, and enhance the effectiveness of energy label information. Accelerate the development of the technical foundation supporting the intelligent supervision of energy labels, promote the comprehensive application of new-generation information technologies such as the Internet of Things, blockchain, and industrial internet, as well as intelligent detection technologies in testing institutions, and speed up the establishment of smart networked testing laboratories. This will support the visualization and transparency of the testing process, ensure traceability and verifiability of test data, enable efficient aggregation of testing information, and provide credible evidence storage, thereby ensuring the reliability of test results and supporting intelligent product supervision.

(5) Enrich the display of energy label information and improve the promotion mechanism for green and low-carbon products. Accelerate the development of a series of standards for energy conservation and carbon reduction in high-efficiency end-use energy-consumption products, and explore the models for substitution of environmentally friendly refrigerants for refrigeration products. By leveraging the QR code on energy labels to convey comprehensive quality information, enhance the interpretation of enegy label information and improve its usability. Also promote the "Top Runner" initiative for product energy efficiency benchmarking. Conduct research to explore methods for displaying additional information on the energy label information platform, such as energy efficiency benchmarking data, green and low-carbon performance, annual energy consumption, electricity savings and carbon reduction metrics, testing agency ratings, and market supervision results. This will provide consumers with more reference information for comparing green and low-carbon products, supporting the implementation of market-driven mechanisms such as green credits and the Carbon Generalized System of Preferences.

(6) Strengthen the supervision of energy efficiency standards and labels, and improve the regulatory system. First, comprehensively implement the "double random, one public" supervision mechanism, and continuously carry out market supervision and inspection of energy efficiency standards and labels. Strengthen the supervision of energy labels, enhance local oversight, expand

the scope of regulation, and intensify enforcement efforts. Second, investigate and improve the market supervision system and mechanisms, strengthen communication, coordination, and resource integration among various departments, promote the implementation of social supervision activities, build a multidimensional supervision system, and fully leverage the role of social oversight. Third, explore and improve the relevant models and legal regulations for e-commerce participation in energy efficiency labeling implementation, providing crucial support for the enforcement of the system. Fourth, strengthen the disclosure of supervision and inspection information, the application of results, and the coordination of penalties. Investigate and establish a mechanism for sharing regulatory results between supervisory authorities and energy efficiency labeling authorization agencies, and facilitating authorized agencies to implement coordinated penalties such as revoking registration announcements, thereby effectively deterring illegal activities. Fifth, by integrating new technologies such as the Internet of Things, blockchain, and big data, strengthen research and application of smart market supervision technologies, build information platforms, and promote the modernization and precision of energy label market regulation.

(7) Conduct promotional activities and strengthen international cooperation. First, prioritize publicity and promotion as a key task, and adequately engage industry institutions and resources, to carry out such publicity and promotion systematically, persistently, and consistently. Enhancing publicity and raising social awareness are crucial measures to foster a favorable environment for the promotion of high-efficiency and energy-saving products and technological advancement, as well as to amplify the impact of energy efficiency standards and labeling initiatives. Second, further strengthen international cooperation and steadily expand the institutional opening up of standards. Drawing on the latest international best practices in energy efficiency standards and labeling programmes, actively promote international and regional coordination and mutual recognition of standards and labels. This will enhance the global influence of China's energy efficiency standards and labeling programme, establish an international cooperation and exchange platform for relevant enterprises and institutions, and facilitate the integration of China's high-efficiency, high-quality products into the global market.

Bibliography

Beijing Municipal Development and Reform Commission. (2022) *Beijing's 14th Five-Year Plan for Climate Change Response and Energy Conservation* [Z]. 2022-08-14. Retrieved from https://m.thepaper.cn/baijiahao_19453480

Beijing Municipal Development and Reform Commission, Beijing Municipal Commission of Urban Management. *Action plan for the development of high-quality supercharging stations for new energy vehicles in Beijing* [M]. Beijing: Beijing Municipal Government, 2023.

Che, Y., Yang, J., Zhou, Y., Zhao, Y., He, W., & Wu, J. (2019). Demand response from the control of aggregated inverter air conditioners. IEEE Access, 7, 88163–88173.

Chen, W.,. Huang, X., Cui, K., et al. (2016). Application of microchannel heat exchangers in R290 heat pump dryer systems [C]. In Proceedings of the 2016 China Household Appliances Technology Conference. Hangzhou Sanhua Home Appliances Heat Management Systems Co., Ltd., 2016: 4.

China Automotive Power Battery Industry Innovation Alliance. (2025) *Monthly report on power battery information, December 2024* [R/OL]. 2025-01-13 [Accessed 2025-04-30]. Available at: https://mp.weixin.qq.com/s/W7181pLemWq1eThd4o2vew

China Building Energy Efficiency Association, Building Energy Consumption and Carbon Emission Data Committee. (2024). Research Report on Carbon Emissions in China's Urban and Rural Construction Sector (2024 Edition) [R]. Chongqing, China, 2024.

China Energy Conservation Association, Energy Foundation. (n.d.). China Heat Pump Development Roadmap—Building Sector [R], Beijing, China.

Dai Y, Liu XC, Li H, et al. Building-related electric vehicle charging behaviors and energy consumption patterns: An urban-scale analysis[J]. Transportation Research Part D: Transport and Environment, 2025, 141: 104663.

Francisco, F. H., José, M., Alberto, F., et al. (2016). A new terminal unit combining a radiant floor with an underfloor air system: Experimentation and numerical model. Energy and Buildings, 133, 70-78.

Fu Z, Liu XC, et al. Orderly solar charging of electric vehicles and its impact on charging behavior: A year-round field experiment.[J] Applied Energy 2025, 381: 125211.

GB/T 18883-2022 Indoor Air Quality Standards (n.d.).

A. Geslin, L. Xu, et al. Dynamic cycling enhances battery lifetime.[J] Nat Energy 2025 10:172–180.

Hong, Z. (2023). Study on the refrigerant leakage safety of R290 cycle heat pump water heaters [D]. Huazhong University of Science and Technology.

Howard L. The Climate of London: Deduced from Meteorological Observations Made in the Metropolis and at Various Places Around it[M]. Cambridge University Press, 1833.

Jiang, Y. (2021). Photovoltaic storage and direct current flexibility—A new building power distribution system to achieve zero-carbon electricity. HVAC, 51(10), 1–12.

Building Energy Research Center of Tsinghua University, *Decarbonizing Urban Residential Buildings*, https://doi.org/10.1007/978-981-95-8200-6

Jiang, A., & Wang, B. (2013). Study on the adjustment characteristics of the scroll compressor with refrigerant leakage [D]. Beijing: Tsinghua University.

Jiang, Y., Zhang, J., Zhang, T., et al. (2024). Dynamic carbon emission responsibility factors for electricity. China Electrical Engineering Journal, 44(17), 7024–7039.

Jinhua Municipal Development and Reform Commission. *Action plan for promoting charging infrastructure and new energy vehicle adoption in rural areas of Jinhua (2023–2025)* [Z]. Jinhua: Jinhua Municipal Development and Reform Commission, 2023.

Kalnay E, Cai M. Impact of urbanization and land-use change on climate[J]. Nature, 2003, 423(6939): 528–531. https://doi.org/10.1038/nature01675.

Kim, D., Joe, G., Park, S., et al. (2017). Experimental evaluation of the thermal performance of raised floor integrated radiant heating panels. Energies, 10, 10101632.

Krol, S., Zabiegala, B., & Namiesnik, J. (2011). Monitoring and analytics of semivolatile organic compounds (SVOCs) in indoor air. Analytical and Bioanalytical Chemistry, 400(6), 1751-1769.

Lam V N, Cui X, Stroebl F, et al. A decade of insights: Delving into calendar aging trends and implications[J]. Joule, 2025, 9(1): 101796.

Lecomte, J.-F., et al. (2014). ICRP Publication 126: Radiological protection against radon exposure. Annals of the ICRP, 43, 5-73.

Li, G., Yin, S., Bing, J., et al. (2021). Application of R290 in heat pump tumble dryers [C]. In Proceedings of the 2021 China Household Appliances Technology Conference. Qingdao Haier Washing Machine Co., Ltd.; Digital National Key Laboratory, 2021: 5.

Li, X., Ma, L., Shi, L., et al. (2023a). CO2 cascade air source heat pump heating operation test. Journal of Refrigeration, 44(02), 39-46.

Li, X., He, G., Ning, Q., et al. (2023b). Refrigerant leakage safety study of R290 heat pump water heaters with external coil [C]. In Proceedings of the 2022 China Household Appliances Technology Conference. Huazhong University of Science and Technology, School of Energy and Power Engineering, 2023: 10.

Li, X., Zheng, N., Zhang, W., An, Q., Ji, Y., Chen, C., Wang, S., & Peng, L. (2024). Comprehensive assessment of phthalates in indoor dust across China between 2007 and 2019: Benefits from regulatory restrictions. Environmental Pollution, 342, 123147.

Li, Y. et al. The potentials of vehicle-grid integration on peak shaving of a community considering random behavior of aggregated vehicles.[J] Next Energy 2025, 7: 100233.

Liu, Y., Xu, H., Li, M., et al. (2022a). Advantages and application challenges of CO2 heat pump water heaters and a defrosting solution. Hydraulics Pneumatics and Sealing, 42(03), 79-82.

Liu, W., Sun, Y., Liu, N., Hou, J., Huo, X., Zhao, Y., Zhang, Y., Deng, F., Kan, H., Zhao, Z., Huang, C., Zhao, B., Zeng, X., Qian, H., Zheng, X., Liu, W., Mo, J., Sun, C., Su, C., et al. (2022b). Indoor exposure to phthalates and its burden of disease in China. Indoor Air, 32(4).

Liu, X. C., Liu, X. H., Jiang, Y., et al. (2023a). PEDF (photovoltaics, energy storage, direct current, flexibility): A power distribution system for buildings in grid decarbonization—Definition, technology review, and applications. CSEE Journal of Power and Energy Systems, 9(3), 829–845.

Liu, X. C., Liu, X. H., Zhang, T., et al. (2023b). Characterization and design methods of generalized energy storage resources in building areas. China Electrical Engineering Journal, 1–13. [2023-11-20]. Retrieved from https://doi.org/10.13334/j.0258-8013.pcsee.222949.

Liu XC, Fu Z, Qiu SY, et al. Building-centric investigation into electric vehicle behavior: A survey-based simulation method for charging system design[J]. Energy, 2023c, 271: 127010.

Luo, Z., Peng, J., Cao, J., Yin, R., Zou, B., Tan, Y., & Yan, J. (2022). Demand flexibility of residential buildings: Definitions, flexible loads, and quantification methods. Engineering, 16, 123–140.

Ma, R., Mao, C., Shan, M., et al. (2018). Heating application of low-temperature air source heat pump hot air units in rural Beijing. Regional Heating, (01), 24-31.

Martinez, S., Vellei, M., & Le Dréau, J. (2022). Demand-side flexibility in a residential district: What are the main sources of uncertainty? Energy and Buildings, 255, 111595.

Mehrtash, M., Hobbs, B., & Ela, E. (2023). Reserve and energy scarcity pricing in United States power markets: A comparative review of principles and practices. Renewable and Sustainable Energy Reviews, 183, 113465.

Mhoff ML, Zhang P, Wolfe RE, et al. Remote sensing of the urban heat island effect across biomes in the continental USA[J]. Remote Sensing of Environment, 2010, 114(3): 504–513. https://doi.org/10.1016/j.rse.2009.10.008.
Ministry of Ecology and Environment. (2023). Notice on issuing the "Recommended List of Substitutes for Ozone Depleting Substances in China" [EB/OL].
National Development and Reform Commission. (2023). Power demand-side management measures [EB/OL]. (2023-09-15). [2023-11-06]. Retrieved from https://www.gov.cn/xinwen/2017-09/26/5227721/files/264228a3ce924f70afd268ce373fc989.pdf?eqid=aebb25470004162e00000006645a0d84.
Oke T R. City size and the urban heat island[J]. Atmospheric Environment (1967), 1973, 7(8): 769–779. https://doi.org/10.1016/0004-6981(73)90140-6.
Pope III, C. A., & Dockery, D. W. (2006). Health effects of fine particulate air pollution: Lines that connect. Journal of the American Medical Association, 56, 709-742.
Rosenzweig C, Solecki W D, Parshall L, et al. Characterizing the urban heat island in current and future climates in New Jersey[J]. Environmental Hazards, 2005, 6(1):51–62. https://doi.org/10.1016/j.hazards.2004.12.001.
Santamouris M. Analyzing the heat island magnitude and characteristics in one hundred Asian and Australian cities and regions[J]. Science of The Total Environment, 2015, 512–513: 582–598. https://doi.org/10.1016/j.scitotenv.2015.01.060.
Shanghai Municipal People's Government. *Measures for the construction and management of EV charging facilities in residential communities of Shanghai* [S]. Shanghai: Shanghai Municipal Government, 2023.
Sheng L, Tang X, You H, et al. Comparison of the urban heat island intensity quantified by using air temperature and Landsat land surface temperature in Hangzhou, China[J]. Ecological Indicators, 2017, 72: 738–746. https://doi.org/10.1016/j.ecolind.2016.09.009.
Shenzhen Municipal Development and Reform Commission. *Special plan for supercharging facilities for new energy vehicles in Shenzhen (2023–2025)* [M]. Shenzhen: Shenzhen Municipal Government, 2023.
Sun, H., Duan, M., Wu, Y., et al. (2021). Thermal performance investigation of a novel heating terminal integrated with flat heat pipe and heat transfer enhancement. Energy, 236, 121411.
The General Office of the State Council of China (2023). *Guiding opinions on further building a high-quality charging infrastructure system* [Z]. 2023-06-08. Retrieved from https://www.gov.cn/zhengce/content/202306/content_6887167.htm
Tsinghua University. (2025) *Strategic development path for China's PV-storage-DC-flexible buildings (Phase II): Subtopic 5—Research on the potential and models for electric vehicles and buildings to participate in grid regulation* [R/OL]. 2023-12-15 [Accessed 2025-04-30]. Available at: https://www.efchina.org/Reports-zh/report-lccp-20231215-zh.
Wang M. Study on the influence of cold island effect and ambient wind field on the performance of air source heat pump array. [D]. Shandong University, 2019.
Wang, X. (2024). Research and application of air source heat pump hot air system defrosting technology with direct hot air [D]. Tsinghua University.
Wang, C., Zhang, H., Jiang, R., Jiang, Q., & Song, W. (2005). Current status of formaldehyde and total volatile organic compound pollution in indoor air after decoration. Journal of Environmental and Health, (05), 356-358.
Wang, L., Zhao, B., Liu, C., Lin, H., Yang, X., & Zhang, Y. (2010). Review of indoor SVOCs pollution in China. Science Bulletin, 55(11), 967-977.
Wang, L., Ding, N., Yang, G., et al. (2015). Analysis of formaldehyde purification effects of various air purifiers. Chinese Journal of Health Inspection, 25(04), 567-568+571.
Wang, Z., Liu, C., Su, X., et al. (2021). Study on heating temperature based on the least comfortable body parts. HVAC, 51(11), 95–98+47.
Wang, T., Miao, S., Yao, F., et al. (2023). Day-ahead and intra-day joint dispatch strategy of high-penetration wind power systems considering dynamic frequency response constraints. IEEE

Transactions on Power Engineering, 1–19. [2023-11-06]. Retrieved from http://kns.cnki.net/kcms/detail/11.2107.TM.20230307.1802.002.html

World Health Organization (WHO). (2009). WHO Handbook on Indoor Radon: A Public Health Perspective. World Health Organization.

World Health Organization (WHO). (2014). WHO Guidelines for Indoor Air Quality: Household Fuel Combustion [R]. Geneva: World Health Organization.

Wu, Y., Sun, H., Yang, Z., et al. (2023). Dynamic process simulation of indoor temperature distribution in radiant-convective heating terminals. Building and Environment, 244, 110843.

Xiong, J., Yu, K., Zhao, B., et al. (2023). Zhuhai Gree Electric Appliances Inc., Air Conditioning Equipment and System Operation Energy Efficiency National Key Laboratory; Zhuhai Gree Energy-saving Environmental Refrigeration Technology Research Center Co., Ltd. Key technologies and applications of efficient heat pumps based on floor thermal storage [Z]. Identification Date: 2023-02-04.

Xu, D., Shang, B., & Cao, Z. (2007). Survey on major pollutants in residential indoor air in some Chinese cities. Health Research, (04), 473-476.

Xu X, Tang S, Han X, et al. Flexible bidirectional pulse charging regulation achieving long-life lithium-ion batteries[J]. Journal of Energy Chemistry, 2024, 96: 59–71.

Yang G. PM2.5 Influence on Urban Heat Island (UHI) Effect in major Chinese cities and the Mechanisms. China University of Geosciences, 2023.

Yang, X., Hu, Z., Wan, Z., Li, X., Zhang, L., & Li, Y. (2007). Investigation of benzene series in indoor air in Jiangxi Province. Jiangxi Science, No. 104(06), 737-740.

Yang P, Ren G, Liu W. Spatial and Temporal Characteristics of Beijing Urban Heat Island Intensity[J]. Journal of Applied Meteorology and Climatology, 2013, 52(8): 1803–1816. https://doi.org/10.1175/JAMC-D-12-0125.1.

Yang P, Ren G, Yan P. Evidence for a Strong Association of Short-Duration Intense Rainfall with Urbanization in the Beijing Urban Area[J]. Journal of Climate, 2017,30(15):5851–5870. https://doi.org/10.1175/JCLI-D-16-0671.1.

Yang, Z., Chi, J., Luo, B., et al. (2024a). Analyzing excess heat factors of convective/radiant terminals: Balancing beginning and steady stage. Journal of Building Engineering, 84, 108576.

Yang, Z., Hu, Q., Li, Y., et al. (2024b). Thermal performance of novel convective-adjustable flat panel radiant unit. Applied Thermal Engineering, 244, 122689.

Yu, S., Fu, S., Behrendt, J., et al. (2022). China Carbon Neutrality Comprehensive Report 2022: Deep Electrification to Support Carbon Neutrality [R]. Energy Foundation, Beijing, China. Retrieved from https://www.efchina.org/14FYP-zh/Reports-zh/report-lceg-20201210-zh.

Zhang, A. (2024). Air Purifiers in Chinese Residences: Operating Behavior, Effect, and Optimization [D]. Tsinghua University.

Zhang, W., Hao, S., Huang, P., Yin, Z., Lu, S., & Jiang, L. (2007). Changes of total volatile organic compounds in newly decorated indoor air over time. Journal of Environmental and Health, (10), 782-783.

Zhang X, Zou Y, Fan J, et al. Usage pattern analysis of Beijing private electric vehicles based on real-world data[J]. Energy, 2019, 167: 1074–1085.

Zhao, S., Zhu, Y., Zhang, X., et al. (2022). Research on key technologies for the revision of the national standard GB/T 18801-2015 "Air Purifiers". Home Appliance Science and Technology (01), 32-36. https://doi.org/10.19784/j.cnki.issn1672-0172.2022.01.004.2.

Zheng, C., Liu, J., Cui, Y., et al. (2024). Research status and prospects of phase change thermal storage floor. China Plastics, 38(10), 70–74.

Zhou, X., Zhao, Q., & Zhang, Y. (2021). Development prospects and key technologies of China's energy and power system under the "dual carbon" target. China Electric Power Enterprise Management, (31), 14–17.

Zhu, Y., Song, X., Wu, R., et al. (2021). A review on reducing indoor particulate matter concentrations from personal-level air filtration intervention under real-world exposure situations. Indoor Air, 31(6), 1707-1721.

Zhu, L., Zhang, C., Tao, X., et al. (2024). A high-power single-unit dual-stage scroll ultra-low temperature air source heat pump [P]. Chinese Patent: CN201910655874.3, 2024-06-11.

MIX
Papier aus verantwortungsvollen Quellen
Paper from responsible sources
FSC® C105338

If you have any concerns about our products,
you can contact us on
ProductSafety@springernature.com

In case Publisher is established outside the EU,
the EU authorized representative is:
Springer Nature Customer Service Center GmbH
Europaplatz 3, 69115 Heidelberg, Germany

Printed by Libri Plureos GmbH
in Hamburg, Germany